MEMOIRS
of the
American Mathematical Society

Number 953

Mixed-Norm Inequalities and
Operator Space L_p Embedding Theory

Marius Junge
Javier Parcet

January 2010 • Volume 203 • Number 953 (second of 5 numbers) • ISSN 0065-9266

American Mathematical Society
Providence, Rhode Island

Library of Congress Cataloging-in-Publication Data

Junge, Marius.
  Mixed-norm inequalities and operator space L_p embedding theory / Marius Junge, Javier Parcet.
       p. cm. — (Memoirs of the American Mathematical Society, ISSN 0065-9266 ; no. 953)
  "Volume 203, number 953 (second of 5 numbers)."
  Includes bibliographical references.
  ISBN 978-0-8218-4655-1 (alk. paper)
    1. Harmonic analysis. 2. Banach spaces. 3. L_p spaces. 4. Noncommutative function spaces. 5. Martingales (Mathematics). 6. Quantum statistics. I. Parcet, Javier, 1975- II. Title.

QA403.J86   2010
515′.2433—dc22   2009041393

Memoirs of the American Mathematical Society

This journal is devoted entirely to research in pure and applied mathematics.

Publisher Item Identifier. The Publisher Item Identifier (PII) appears as a footnote on the Abstract page of each article. This alphanumeric string of characters uniquely identifies each article and can be used for future cataloguing, searching, and electronic retrieval.

Subscription information. Beginning with the January 2010 issue, *Memoirs* is accessible from www.ams.org/journals. The 2010 subscription begins with volume 203 and consists of six mailings, each containing one or more numbers. Subscription prices are as follows: for paper delivery, US$709 list, US$567 institutional member; for electronic delivery, US$638 list, US$510 institutional member. Upon request, subscribers to paper delivery of this journal are also entitled to receive electronic delivery. If ordering the paper version, subscribers outside the United States and India must pay a postage surcharge of US$65; subscribers in India must pay a postage surcharge of US$95. Expedited delivery to destinations in North America US$57; elsewhere US$160. Subscription renewals are subject to late fees. See www.ams.org/customers/macs-faq.html#journal for more information. Each number may be ordered separately; *please specify number* when ordering an individual number.

Back number information. For back issues see www.ams.org/bookstore.

Subscriptions and orders should be addressed to the American Mathematical Society, P. O. Box 845904, Boston, MA 02284-5904 USA. *All orders must be accompanied by payment.* Other correspondence should be addressed to 201 Charles Street, Providence, RI 02904-2294 USA.

Copying and reprinting. Individual readers of this publication, and nonprofit libraries acting for them, are permitted to make fair use of the material, such as to copy a chapter for use in teaching or research. Permission is granted to quote brief passages from this publication in reviews, provided the customary acknowledgment of the source is given.

Republication, systematic copying, or multiple reproduction of any material in this publication is permitted only under license from the American Mathematical Society. Requests for such permission should be addressed to the Acquisitions Department, American Mathematical Society, 201 Charles Street, Providence, Rhode Island 02904-2294 USA. Requests can also be made by e-mail to reprint-permission@ams.org.

Memoirs of the American Mathematical Society (ISSN 0065-9266) is published bimonthly (each volume consisting usually of more than one number) by the American Mathematical Society at 201 Charles Street, Providence, RI 02904-2294 USA. Periodicals postage paid at Providence, RI. Postmaster: Send address changes to Memoirs, American Mathematical Society, 201 Charles Street, Providence, RI 02904-2294 USA.

© 2009 by the American Mathematical Society. All rights reserved.
Copyright of individual articles may revert to the public domain 28 years after publication. Contact the AMS for copyright status of individual articles.
This publication is indexed in *Science Citation Index*®, *SciSearch*®, *Research Alert*®, *CompuMath Citation Index*®, *Current Contents*®/*Physical, Chemical & Earth Sciences*.
Printed in the United States of America.

∞ The paper used in this book is acid-free and falls within the guidelines
established to ensure permanence and durability.
Visit the AMS home page at http://www.ams.org/

10 9 8 7 6 5 4 3 2 1   14 13 12 11 10 09

Contents

Introduction	1
0.1. Noncommutative function spaces	2
0.2. Amalgamated L_p spaces	3
0.3. Conditional L_p spaces	5
0.4. Intersection spaces	7
0.5. Mixed-norm inequalities	8
0.6. Operator space L_p embeddings	9
Chapter 1. Noncommutative integration	13
1.1. Noncommutative L_p spaces	13
1.2. Pisier's vector-valued L_p spaces	17
1.3. The spaces $L_p^r(\mathcal{M}, \mathsf{E})$ and $L_p^c(\mathcal{M}, \mathsf{E})$	20
Chapter 2. Amalgamated L_p spaces	27
2.1. Haagerup's construction	29
2.2. Triangle inequality on $\partial_\infty \mathsf{K}$	31
2.3. A metric structure on the solid K	38
Chapter 3. An interpolation theorem	43
3.1. Finite von Neumann algebras	44
3.2. Conditional expectations on $\partial_\infty \mathsf{K}$	48
3.3. General von Neumann algebras I	55
3.4. General von Neumann algebras II	61
3.5. Proof of the main interpolation theorem	66
Chapter 4. Conditional L_p spaces	71
4.1. Duality	72
4.2. Conditional L_∞ spaces	73
4.3. Interpolation results and applications	74
Chapter 5. Intersections of L_p spaces	79
5.1. Free Rosenthal inequalities	79
5.2. Estimates for BMO type norms	83
5.3. Interpolation of 2-term intersections	99
5.4. Interpolation of 4-term intersections	103
Chapter 6. Factorization of $\mathcal{J}_{p,q}^n(\mathcal{M}, \mathsf{E})$	107
6.1. Amalgamated tensors	108
6.2. Conditional expectations and ultraproducts	112
6.3. Factorization of the space $\mathcal{J}_{\infty,1}^n(\mathcal{M}, \mathsf{E})$	115

Chapter 7. Mixed-norm inequalities — 119
  7.1. Embedding of $\mathcal{J}_{p,q}^n(\mathcal{M}, \mathsf{E})$ into $L_p(\mathcal{A}; \ell_q^n)$ — 119
  7.2. Asymmetric L_p spaces and noncommutative (Σ_{pq}) — 126

Chapter 8. Operator space L_p embeddings — 129
  8.1. Embedding Schatten classes — 129
  8.2. Embedding into the hyperfinite factor — 132
  8.3. Embedding for general von Neumann algebras — 144

Bibliography — 153

Abstract

Let $f_1, f_2, \ldots, f_n$ be a family of independent copies of a given random variable f in a probability space $(\Omega, \mathcal{F}, \mu)$. Then, the following equivalence of norms holds whenever $1 \le q \le p < \infty$

$$(\Sigma_{pq}) \qquad \Big(\int_\Omega \Big[\sum_{k=1}^n |f_k|^q\Big]^{\frac{p}{q}} d\mu\Big)^{\frac{1}{p}} \sim \max_{r \in \{p,q\}} \Big\{ n^{\frac{1}{r}} \Big(\int_\Omega |f|^r d\mu\Big)^{\frac{1}{r}} \Big\}.$$

We prove a noncommutative analogue of this inequality for sums of free random variables over a given von Neumann subalgebra. This formulation leads to new classes of noncommutative function spaces which appear in quantum probability as square functions, conditioned square functions and maximal functions. Our main tools are Rosenthal type inequalities for free random variables, noncommutative martingale theory and factorization of operator-valued analytic functions. This allows us to generalize (Σ_{pq}) as a result for noncommutative L_p in the category of operator spaces. Moreover, the use of free random variables produces the right formulation of $(\Sigma_{\infty q})$, which has not a commutative counterpart. In the last part of the paper, we use our mixed-norm inequalities to construct a completely isomorphic embedding of L_q –equipped with its natural operator space structure– into some sufficiently large L_p space for $1 \le p < q \le 2$. The construction of such embedding has been open for quite some time. We also show that hyperfiniteness and the QWEP are preserved in our construction.

Received by the editor November 4, 2005 and, in revised form, October 10, 2007.

Article electronically published on August 26, 2009; S 0065-9266(09)00570-5.

2000 *Mathematics Subject Classification.* Primary 46L07, 46L09, 46L51, 46L52, 46L53, 46L54.

Key words and phrases. Noncommutative L_p, free random variables, complete embedding.

Marius Junge is partially supported by the National Science Foundation DMS-0556120. Affiliation at the time of publication: Department of Mathematics, University of Illinois at Urbana-Champaign, 273 Altgeld Hall, 1409 W. Green Street, Urbana, IL 61801. USA; junge@math.uiuc.edu.

Javier Parcet is partially supported by 'Programa Ramón y Cajal, 2005' and also by the Grants MTM2007-60952, CCG07-UAM/ESP-1664 and CCG08-CSIC/ESP-3485, Spain. Affiliation at time of publication: Instituto de Ciencias Matemáticas CSIC-UAM-UC3M-UCM Consejo Superior de Investigaciones Científicas C/ Serrano 121, 28006, Madrid, Spain; javier.parcet@uam.es.

©2009 American Mathematical Society

Introduction

Probabilistic methods play an important role in harmonic analysis and Banach space theory. Let us just mention the relevance of sums of independent random variables, p-stable processes or martingale inequalities in both fields. The analysis of subspaces of the classical L_p spaces is specially benefited from such probabilistic notions. Viceversa, Burkholder's martingale inequality for the conditional square function has been discovered in view of Rosenthal's inequality for the norm in L_p of sums of independent random variables. This is only one example of the fruitful interplay between harmonic analysis, probability theory and Banach space geometry carried out mostly in the 70's by Burkholder, Gundy, Kwapień, Maurey, Pisier, Rosenthal and many others.

More recently it became clear that a similar endeavor for noncommutative L_p spaces requires an additional insight from quantum probability and operator space theory [15, 17, 50]. A noncommutative theory of martingale inequalities finds its beginnings in the work of Lust-Piquard [36] and Lust-Piquard/Pisier [37] on the noncommutative Khintchine inequality. The seminal paper of Pisier and Xu on the noncommutative analogue of Burkholder-Gundy inequality [51] started a new trend in quantum probability. Nowadays, most classical martingale inequalities have a satisfactory noncommutative analogue, see [16, 28, 41, 53]. In the proof of these results the classical stopping time arguments are no longer available, essentially because point sets disappear after quantization. These arguments are replaced by functional analytic or combinatorial arguments. In the functional analytic approach we often encounter new spaces. Indeed, maximal functions in the noncommutative context can only be understood and defined through analogy with vector-valued L_p spaces. A careful analysis of these spaces is crucial in establishing basic results such as Doob's inequality [16] for noncommutative martingales and the noncommutative maximal theorem behind Birkhoff's ergodic theorem [29]. The proof of maximal theorems and noncommutative versions of Rosenthal's inequality often uses square function and conditioned square function estimates, see [26] and the references therein. These are examples of more general classes of noncommutative function spaces to be defined below. However, all of them illustrate our main motto in this paper. Namely, *certain problems can be solved by finding and analyzing the appropriate class of Banach spaces*. We shall develop in this paper a new theory of *generalized noncommutative L_p spaces* with three problems in mind for a given von Neumann algebra $\mathcal{A}$.

PROBLEM 1. Calculate the $L_p(\mathcal{A}; \ell_q)$ norm for sums of free random variables.

PROBLEM 2. If $1 \leq p < q \leq 2$, find a cb-embedding of $L_q(\mathcal{A})$ into some L_p space.

PROBLEM 3. Any reflexive subspace of $\mathcal{A}_*$ embeds into some L_p for certain $p > 1$.

Our main contribution in this paper is the calculation of mixed norms for sums of free random variables and its application to construct a complete embedding of L_q into L_p. Unfortunately, the generalization of Rosenthal's theorem [57] to noncommutative L_p spaces –whose simplest version is the content of Problem 3 above– is beyond the scope of this paper and we analyze it in [23]. We should nevertheless note that its solution is also deeply related to the main results in this paper. Namely, the interplay of interpolation and intersection is at the heart of both results. On the other hand, operator space L_p embedding theory is motivated by the classical notion of q-stable variables and norm estimates for sums of independent random variables. Let us briefly explain this. The construction of the cb-embedding for $(p,q) = (1,2)$ was obtained in [17]. The simplest model of 2-stable variables is provided by normalized gaussians (g_k). In this particular case and after taking operator coefficients (a_k) in some noncommutative L_1 space, the noncommutative Khintchine inequality [37] tells us that

$$(1) \quad \mathbb{E}\Big\|\sum_k a_k g_k\Big\|_1 \sim \inf_{a_k = r_k + c_k} \Big\|\Big(\sum_k r_k r_k^*\Big)^{\frac{1}{2}}\Big\|_1 + \Big\|\Big(\sum_k c_k^* c_k\Big)^{\frac{1}{2}}\Big\|_1.$$

Let us point out that operator spaces are a very appropriate framework for analyzing noncommutative L_p spaces and linear maps between them. Indeed, inequality (1) describes the operator space structure of the subspace spanned by the g_k's in L_1 as the sum $R+C$ of row and column subspaces of $\mathcal{B}(\ell_2)$. We refer to [11] and [47] for background information on operator spaces. In the language of noncommutative probability many operator space inequalities translate into module valued versions of scalar inequalities, this will be further explained below. The only drawback of inequality (1) is that it does not coincide with Pisier's definition of the operator space ℓ_2

$$(2) \quad \Big\|\sum_k a_k \otimes \delta_k\Big\|_{L_1(\mathcal{M};\ell_2)} = \inf_{a_k = \alpha \gamma_k \beta} \|\alpha\|_{L_4(\mathcal{M})} \Big(\sum_k \|\gamma_k\|_{L_2(\mathcal{M})}^2\Big)^{\frac{1}{2}} \|\beta\|_{L_4(\mathcal{M})}.$$

However, it was proved by Pisier that the right side in (2) is obtained by complex interpolation between the row and the column square functions appearing on the right of (1). One of the main results in this article is a far reaching generalization of this observation. In fact, the solution of Problem 2 in full generality is closely related to this analysis.

Following our guideline we will now introduce and discuss the new class of spaces relevant for these problems and martingale theory. These generalize Pisier's theory of $L_p(L_q)$ spaces over hyperfinite von Neumann algebras. We begin with a brief review of some noncommutative function spaces which have lately appeared in the literature, mainly in noncommutative martingale theory. We refer to Chapter 1 below for a more detailed exposition.

0.1. Noncommutative function spaces

Inspired by Pisier's theory [46], several noncommutative function spaces have been recently introduced in quantum probability. The first motivation comes from some of Pisier's fundamental equalities, which we briefly review. Let $\mathcal{N}_1$ and $\mathcal{N}_2$ be two hyperfinite von Neumann algebras. Then, given $1 \le p, q \le \infty$ and defining $1/r = |1/p - 1/q|$, we have

i) If $p \leq q$, the norm of $x \in L_p(\mathcal{N}_1; L_q(\mathcal{N}_2))$ is given by
$$\inf\Big\{\|\alpha\|_{L_{2r}(\mathcal{N}_1)}\|y\|_{L_q(\mathcal{N}_1\bar{\otimes}\mathcal{N}_2)}\|\beta\|_{L_{2r}(\mathcal{N}_1)} \,\Big|\, x = \alpha y\beta\Big\}.$$

ii) If $p \geq q$, the norm of $x \in L_p(\mathcal{N}_1; L_q(\mathcal{N}_2))$ is given by
$$\sup\Big\{\|\alpha x\beta\|_{L_q(\mathcal{N}_1\bar{\otimes}\mathcal{N}_2)} \,\Big|\, \alpha, \beta \in \mathsf{B}_{L_{2r}(\mathcal{N}_1)}\Big\}.$$

On the other hand, the row and column subspaces of L_p are defined as follows

$$L_p(\mathcal{M}; R_p^n) = \Big\{\sum_{k=1}^n x_k \otimes e_{1k} \,\Big|\, x_k \in L_p(\mathcal{M})\Big\} \subset L_p(\mathcal{M}\bar{\otimes}\mathcal{B}(\ell_2)),$$

$$L_p(\mathcal{M}; C_p^n) = \Big\{\sum_{k=1}^n x_k \otimes e_{k1} \,\Big|\, x_k \in L_p(\mathcal{M})\Big\} \subset L_p(\mathcal{M}\bar{\otimes}\mathcal{B}(\ell_2)),$$

where (e_{ij}) denotes the unit vector basis of $\mathcal{B}(\ell_2)$. These spaces are crucial in the noncommutative Khintchine/Rosenthal type inequalities [**26, 37, 40**] and in noncommutative martingale inequalities [**28, 51, 53**], where the row and column spaces are traditionally denoted by $L_p(\mathcal{M}; \ell_2^r)$ and $L_p(\mathcal{M}; \ell_2^c)$. Now, considering a von Neumann subalgebra $\mathcal{N}$ of $\mathcal{M}$ with a normal faithful conditional expectation $\mathsf{E}: \mathcal{M} \to \mathcal{N}$, we may define L_p norms of the conditional square functions

$$\Big(\sum_{k=1}^n \mathsf{E}(x_k x_k^*)\Big)^{\frac{1}{2}} \quad \text{and} \quad \Big(\sum_{k=1}^n \mathsf{E}(x_k^* x_k)\Big)^{\frac{1}{2}}.$$

The expressions $\mathsf{E}(x_k x_k^*)$ and $\mathsf{E}(x_k^* x_k)$ have to be defined properly for $1 \leq p \leq 2$, see [**16**] or Chapter 1 below. Note that the resulting spaces coincide with the row and column spaces defined above when $\mathcal{N}$ is $\mathcal{M}$ itself. When $n = 1$ we recover the spaces $L_p^r(\mathcal{M}, \mathsf{E})$ and $L_p^c(\mathcal{M}, \mathsf{E})$, which have been instrumental in proving Doob's inequality [**16**], see also [**21, 29**] for more applications.

0.2. Amalgamated L_p spaces

The definition of amalgamated L_p spaces is algebraic. We recall that by Hölder's inequality $L_u(\mathcal{M})L_q(\mathcal{M})L_v(\mathcal{M})$ is contractively included in $L_p(\mathcal{M})$ when $1/p = 1/u + 1/q + 1/v$. Let us now assume that $\mathcal{N}$ is a von Neumann subalgebra of $\mathcal{M}$ with a normal faithful conditional expectation $\mathsf{E}: \mathcal{M} \to \mathcal{N}$. Then we have natural isometric inclusions $L_s(\mathcal{N}) \subset L_s(\mathcal{M})$ for $0 < s \leq \infty$ and we may consider the *amalgamated L_p space*

$$L_u(\mathcal{N})L_q(\mathcal{M})L_v(\mathcal{N})$$

as the subset of elements x in $L_p(\mathcal{M})$ which factorize as $x = \alpha y\beta$ with $\alpha \in L_u(\mathcal{N})$, $y \in L_q(\mathcal{M})$ and $\beta \in L_v(\mathcal{N})$. The natural "norm" is then given by the following expression

$$\|x\|_{u \cdot q \cdot v} = \inf\Big\{\|\alpha\|_{L_u(\mathcal{N})}\|y\|_{L_q(\mathcal{M})}\|\beta\|_{L_v(\mathcal{N})} \,\Big|\, x = \alpha y\beta\Big\}.$$

However, the triangle inequality for the homogeneous expression $\|\ \|_{u \cdot q \cdot v}$ is by no means trivial. Moreover, it is not clear a priori that this subset of $L_p(\mathcal{M})$ is indeed a linear space. Before explaining these difficulties in some detail, let us consider some examples. We fix an integer $n \geq 1$ and the subalgebra $\mathcal{N}$ embedded in the

diagonal of the direct sum $\mathcal{M} = \mathcal{N} \oplus_\infty \mathcal{N} \oplus_\infty \cdots \oplus_\infty \mathcal{N}$ with n terms. The natural conditional expectation is
$$\mathsf{E}_n(x_1, \ldots, x_n) = \frac{1}{n} \sum_{k=1}^n x_k.$$
Then it is easy to see that for $1 \leq p \leq 2$ and $1/p = 1/2 + 1/w$ we have
$$L_p(\mathcal{N}; R_p^n) = \sqrt{n}\, L_w(\mathcal{N}) L_2(\mathcal{M}) L_\infty(\mathcal{N}),$$
$$L_p(\mathcal{N}; C_p^n) = \sqrt{n}\, L_\infty(\mathcal{N}) L_2(\mathcal{M}) L_w(\mathcal{N}),$$
isometrically. Here we use the notation $\gamma \mathsf{X}$ to denote the space X equipped with the norm $\gamma \|\cdot\|_\mathsf{X}$. At the time of this writing and with independence of this paper, a result of Pisier [44] on interpolation of these spaces for $p = \infty$ was generalized by Xu [69] for arbitrary p's
$$\Big\|\sum_{k=1}^n x_k \otimes \delta_k\Big\|_{[L_p(\mathcal{N}; R_p^n), L_p(\mathcal{N}; C_p^n)]_\theta} = \inf_{x_k = \alpha y_k \beta} \|\alpha\|_{u_\theta} \Big(\sum_{k=1}^n \|y_k\|_2^2\Big)^{\frac{1}{2}} \|\beta\|_{v_\theta}.$$
Here $(1/u_\theta, 1/v_\theta) = ((1-\theta)/w, \theta/w)$ and for $\theta = 1/2$ we find Pisier's definition of $L_p(\mathcal{N}; \ell_2)$. That is, we obtain the space $\sqrt{n}\, L_{u_\theta}(\mathcal{N}) L_2(\mathcal{M}) L_{v_\theta}(\mathcal{N})$. Our definition is flexible enough to accommodate the conditional square function. Indeed, given a von Neumann subalgebra $\mathcal{N}_0$ of $\mathcal{N}$ with a normal faithful conditional expectation $\mathcal{E}_0 : \mathcal{N} \to \mathcal{N}_0$, we find
$$\Big\|\sum_{k=1}^n x_k \otimes \delta_k\Big\|_{L_w(\mathcal{N}_0) L_2(\mathcal{M}) L_\infty(\mathcal{N}_0)} = n^{-\frac{1}{2}} \Big\|\Big(\sum_{k=1}^n \mathcal{E}_0(x_k x_k^*)\Big)^{\frac{1}{2}}\Big\|_{L_p(\mathcal{N}_0)},$$
$$\Big\|\sum_{k=1}^n x_k \otimes \delta_k\Big\|_{L_\infty(\mathcal{N}_0) L_2(\mathcal{M}) L_w(\mathcal{N}_0)} = n^{-\frac{1}{2}} \Big\|\Big(\sum_{k=1}^n \mathcal{E}_0(x_k^* x_k)\Big)^{\frac{1}{2}}\Big\|_{L_p(\mathcal{N}_0)}.$$
Xu's interpolation does not apply in this more general setting, which appears in the context of the noncommutative Rosenthal inequality. This illustrates how certain amalgamated L_p spaces occur naturally in quantum probability. Now we want to understand for which range of indices (u, q, v) we have the triangle inequality. In fact, our proof intertwines with the proof of our main interpolation result which can be stated as follows. Let us consider the solid K in $\mathbb{R}^3$ defined by
$$\mathsf{K} = \Big\{(1/u, 1/v, 1/q) \,\Big|\, 2 \leq u, v \leq \infty,\ 1 \leq q \leq \infty,\ 1/u + 1/q + 1/v \leq 1\Big\}.$$

THEOREM A. *The amalgamated space $L_u(\mathcal{N}) L_q(\mathcal{M}) L_v(\mathcal{N})$ is a Banach space for any $(1/u, 1/v, 1/q) \in \mathsf{K}$. Moreover, if $(1/u_j, 1/v_j, 1/q_j) \in \mathsf{K}$ for $j = 0, 1$, the space $L_{u_\theta}(\mathcal{N}) L_{q_\theta}(\mathcal{M}) L_{v_\theta}(\mathcal{N})$ is isometrically isomorphic to*
$$\Big[L_{u_0}(\mathcal{N}) L_{q_0}(\mathcal{M}) L_{v_0}(\mathcal{N}),\, L_{u_1}(\mathcal{N}) L_{q_1}(\mathcal{M}) L_{v_1}(\mathcal{N})\Big]_\theta.$$

The triangle inequality follows from Theorem A. On the other hand we will need the triangle inequality in order to apply interpolation and factorization techniques in proving Theorem A. This intriguing interplay makes our proof quite involved. Our first step is showing that the triangle inequality holds in the boundary region
$$\partial_\infty \mathsf{K} = \Big\{(1/u, 1/v, 1/q) \in \mathsf{K} \,\Big|\, \min\{1/u, 1/q, 1/v\} = 0\Big\}.$$
Our argument uses the operator-valued analogue of Szegö's factorization theorem, a technique which will be used repeatedly throughout this paper. The triangle

inequality for other indices follows by convexity since K is the convex hull of $\partial_\infty K$, so that any other point in K is associated to an interpolation space between two spaces living in $\partial_\infty K$. Another technical difficulty is the fact that the intersection of two amalgamated L_p spaces is in general quite difficult to describe. Thus, any attempt to use a density argument meets this obstacle. The second step is to prove Theorem A for finite von Neumann algebras, where the intersections are easier to handle. Moreover, most of the factorization arguments (as Szegö's theorem) a priori only apply in the finite setting. In the third step we consider general von Neumann algebras using Haagerup's crossed product construction [**12**] to approximate σ-finite von Neumann algebras by direct limits of finite von Neumann algebras. Finally, we need a different argument for the case $\min(q_0, q_1) = \infty$, which is out of the scope of Haagerup's construction. The main technique here is a Grothendieck-Pietsch version of the Hahn-Banach theorem.

Let us observe that in the hyperfinite case Pisier was able to establish many of his results using the Haagerup tensor product. Though similar in nature, we can not directly use tensor product formulas for our interpolation results due to the complicated structure of general von Neumann subalgebras. Theorem A will also be useful in understanding certain interpolation spaces in martingale theory. Let us mention some open problems, for partial results see Chapter 5 below.

PROBLEM 4. Let $\mathcal{M}$ be a von Neumann algebra and denote by $\mathcal{H}_p^r(\mathcal{M})$ and $\mathcal{H}_p^c(\mathcal{M})$ the row and column Hardy spaces of noncommutative martingales over $\mathcal{M}$. Let us consider an interpolation parameter $0 < \theta < 1$.

(a) Calculate the interpolation norms $[\mathcal{H}_p^r(\mathcal{M}), \mathcal{H}_p^c(\mathcal{M})]_\theta$.
(b) If $x \in [\mathcal{H}_1^r(\mathcal{M}), \mathcal{H}_1^c(\mathcal{M})]_\theta$, the maximal function is in L_1.

0.3. Conditional L_p spaces

Once we know which amalgamated L_p spaces are Banach spaces it is natural to investigate their dual spaces. We assume as above that $\mathcal{N}$ is a von Neumann subalgebra of $\mathcal{M}$ and $\mathsf{E} : \mathcal{M} \to \mathcal{N}$ is a normal faithful conditional expectation. Let

$$1/s = 1/u + 1/p + 1/v \leq 1.$$

The *conditional* L_p space

$$L^p_{(u,v)}(\mathcal{M}, \mathsf{E})$$

is defined as the completion of $L_p(\mathcal{M})$ with respect to the norm

$$\|x\|_{L^p_{(u,v)}(\mathcal{M},\mathsf{E})} = \sup \Big\{ \|axb\|_{L_s(\mathcal{M})} \,\Big|\, \|a\|_{L_u(\mathcal{N})}, \|b\|_{L_v(\mathcal{N})} \leq 1 \Big\}.$$

In our next result we show that amalgamated and conditional L_p are related by anti-linear duality. This will allow us to translate the interpolation identities in Theorem A in terms of conditional L_p spaces. In this context the correct set of parameters is given by

$$\mathsf{K}_0 = \Big\{ (1/u, 1/v, 1/q) \in \mathsf{K} \,\Big|\, 2 < u, v \leq \infty,\ 1 < q < \infty,\ 1/u + 1/q + 1/v < 1 \Big\}.$$

THEOREM B. *Let $1 < p < \infty$ given by $1/q' = 1/u + 1/p + 1/v$, where the indices (u, q, v) belong to the solid K_0 and q' is conjugate to q. Then, the following isometric isomorphisms hold via the anti-linear duality bracket $\langle x, y \rangle = \mathrm{tr}(x^*y)$*

$$\big(L_u(\mathcal{N}) L_q(\mathcal{M}) L_v(\mathcal{N})\big)^* = L^p_{(u,v)}(\mathcal{M}, \mathsf{E}),$$

$$\left(L^p_{(u,v)}(\mathcal{M}, \mathsf{E})\right)^* = L_u(\mathcal{N}) L_q(\mathcal{M}) L_v(\mathcal{N}).$$

In particular, we obtain isometric isomorphisms

$$\left[L^{p_0}_{(u_0,v_0)}(\mathcal{M}, \mathsf{E}), L^{p_1}_{(u_1,v_1)}(\mathcal{M}, \mathsf{E})\right]_\theta = L^{p_\theta}_{(u_\theta,v_\theta)}(\mathcal{M}, \mathsf{E}).$$

As we shall see, Theorem B generalizes the interpolation results obtained by Pisier [**44**] and Xu [**69**] mentioned above. Pisier and Xu's results provide an explicit expression for the operator space structure of $[C_p^n, R_p^n]_\theta$ with $0 < \theta < 1$. Theorem B also provides explicit formulas for $[\ell_p^n, C_p^n]_\theta$ and $[\ell_p^n, R_p^n]_\theta$. In fact, a large variety of interesting formulas of this kind arise from Theorem B. A detailed analysis of these applications is out of the scope of this paper. On the other hand, the analogue of Theorem B for $p = \infty$ (which we will investigate separately) has been already applied to study the noncommutative John-Nirenberg theorem [**21**].

Exactly as it happens with amalgamated L_p spaces, several noncommutative function spaces arise as particular forms of conditional L_p spaces. Let us review the basic examples in both cases.

(a) The spaces $L_p(\mathcal{M})$ satisfy

$$L_p(\mathcal{M}) = L_\infty(\mathcal{N}) L_p(\mathcal{M}) L_\infty(\mathcal{N}) \quad \text{and} \quad L_p(\mathcal{M}) = L^p_{(\infty,\infty)}(\mathcal{M}, \mathsf{E}).$$

(b) The spaces $L_p(\mathcal{N}_1; L_q(\mathcal{N}_2))$:
- Let $p \le q$ and $1/r = 1/p - 1/q$. Then

$$L_p(\mathcal{N}_1; L_q(\mathcal{N}_2)) = L_{2r}(\mathcal{N}_1) L_q(\mathcal{N}_1 \bar\otimes \mathcal{N}_2) L_{2r}(\mathcal{N}_1).$$

- Let $p \ge q$ and $1/r = 1/q - 1/p$. Then

$$L_p(\mathcal{N}_1; L_q(\mathcal{N}_2)) = L^p_{(2r,2r)}(\mathcal{N}_1 \bar\otimes \mathcal{N}_2, \mathsf{E}),$$

where $\mathsf{E}: \mathcal{N}_1 \bar\otimes \mathcal{N}_2 \to \mathcal{N}_1$ is given by $\mathsf{E} = 1_{\mathcal{N}_1} \otimes \varphi_{\mathcal{N}_2}$.

(c) The spaces $L_p^r(\mathcal{M}, \mathsf{E})$ and $L_p^c(\mathcal{M}, \mathsf{E})$:
- Let $1 \le p \le 2$ and $1/p = 1/2 + 1/s$. Then

$$L_p^r(\mathcal{M}, \mathsf{E}) = L_s(\mathcal{N}) L_2(\mathcal{M}) L_\infty(\mathcal{N}),$$
$$L_p^c(\mathcal{M}, \mathsf{E}) = L_\infty(\mathcal{N}) L_2(\mathcal{M}) L_s(\mathcal{N}).$$

- Let $2 \le p \le \infty$ and $1/p + 1/s = 1/2$. Then

$$L_p^r(\mathcal{M}, \mathsf{E}) = L^p_{(s,\infty)}(\mathcal{M}, \mathsf{E})$$
$$L_p^c(\mathcal{M}, \mathsf{E}) = L^p_{(\infty,s)}(\mathcal{M}, \mathsf{E}).$$

In particular, taking $\mathsf{E}_n(x_1, ..., x_n) = \frac{1}{n} \sum_k x_k$ we find

$$L_p(\mathcal{M}; R_p^n) = \sqrt{n} \, L_p^r(\ell_\infty^n(\mathcal{M}), \mathsf{E}_n),$$
$$L_p(\mathcal{M}; C_p^n) = \sqrt{n} \, L_p^c(\ell_\infty^n(\mathcal{M}), \mathsf{E}_n).$$

(d) As we shall see through the text, asymmetric L_p spaces (a non-standard operator space structure on L_p which will be crucial in this paper) also have representations in terms of amalgamated or conditional L_p spaces.

0.4. Intersection spaces

Intersection of L_p spaces appear naturally in the theory of noncommutative Hardy spaces. These spaces are also natural byproducts of Rosenthal's inequality for sums of independent random variables. Let us first illustrate this point in the commutative setting and then provide the link to the spaces defined above. Let us consider a finite collection $f_1, f_2, \ldots, f_n$ of independent random variables on a probability space $(\Omega, \mathcal{F}, \mu)$. The Khintchine inequality implies for $0 < p < \infty$

$$\Big(\int_\Omega \Big[\sum_{k=1}^n |f_k|^2\Big]^{\frac{p}{2}} d\mu\Big)^{\frac{1}{p}} \sim_{c_p} \mathbb{E}\Big\|\sum_{k=1}^n \varepsilon_k f_k\Big\|_p.$$

Therefore, Rosenthal's inequality [**56**] gives for $2 \leq p < \infty$

$$(\Sigma_{p2}) \qquad \Big(\int_\Omega \Big[\sum_{k=1}^n |f_k|^2\Big]^{\frac{p}{2}} d\mu\Big)^{\frac{1}{p}} \sim_{c_p} \Big(\sum_{k=1}^n \|f_k\|_p^p\Big)^{\frac{1}{p}} + \Big(\sum_{k=1}^n \|f_k\|_2^2\Big)^{\frac{1}{2}}.$$

Here (ε_k) is an independent sequence of Bernoulli random variables equidistributed on ± 1. We can easily generalize this result for calculating ℓ_q sums of independent random variables. Indeed, consider $1 \leq q \leq p < \infty$ and define $g_1, g_2, \ldots, g_n$ by the relation $g_k = |f_k|^{q/2}$ for $1 \leq k \leq n$. Then we have the following identity for the index $s = 2p/q$

$$\Big(\int_\Omega \Big[\sum_{k=1}^n |f_k|^q\Big]^{\frac{p}{q}} d\mu\Big)^{\frac{1}{p}} = \Big(\int_\Omega \Big[\sum_{k=1}^n |g_k|^2\Big]^{\frac{s}{2}} d\mu\Big)^{\frac{2}{qs}}.$$

Therefore (Σ_{p2}) implies

$$(\Sigma_{pq}) \qquad \Big(\int_\Omega \Big[\sum_{k=1}^n |f_k|^q\Big]^{\frac{p}{q}} d\mu\Big)^{\frac{1}{p}} \sim_{c_p} \max_{r \in \{p,q\}} \Big\{\Big(\sum_{k=1}^n \int_\Omega |f_k|^r d\mu\Big)^{\frac{1}{r}}\Big\}.$$

In particular, Rosenthal's inequality provides a natural realization of

$$\mathcal{J}_{p,q}^n(\Omega) = n^{\frac{1}{p}} L_p(\Omega) \cap n^{\frac{1}{q}} L_q(\Omega)$$

into $L_p(\Omega; \ell_q^n)$. More precisely, if $f_1, f_2, \ldots, f_n$ are taken to be independent copies of a given random variable f, the right hand side of (Σ_{pq}) is the norm of f in the intersection space $\mathcal{J}_{p,q}^n(\Omega)$ and inequality (Σ_{pq}) provides an isomorphic embedding of $\mathcal{J}_{p,q}^n(\Omega)$ into the space $L_p(\Omega; \ell_q^n)$.

Quite surprisingly, replacing independent variables by matrices of independent variables in (Σ_{pq}) requires to intersect *four* spaces using the so-called *asymmetric* L_p spaces. In other words, the *natural* operator space structure of $\mathcal{J}_{p,q}^n$ comes from a 4-term intersection space. We have already encountered such a phenomenon in [**22**] for the case $q = 1$. To justify this point, instead of giving precise definitions we note that Hölder inequality gives $L_p = L_{2p} L_{2p}$, meaning that the p-norm of f is the infimum of $\|g\|_{2p} \|h\|_{2p}$ over all factorizations $f = gh$. If L_p^r and L_p^c denote the row and column quantizations of L_p (see Chapter 1 for the definition), the operator space analogue of the isometry above is given by the complete isometry $L_p = L_{2p}^r L_{2p}^c$, see Chapter 7 for more details. In particular, according to the algebraic definition of $L_p(\ell_q)$, the space $\mathcal{J}_{p,q}^n$ has to be redefined as the product

$$\mathcal{J}_{p,q}^n = \Big(n^{\frac{1}{2p}} L_{2p}^r \cap n^{\frac{1}{2q}} L_{2q}^r\Big)\Big(n^{\frac{1}{2p}} L_{2p}^c \cap n^{\frac{1}{2q}} L_{2q}^c\Big).$$

We shall see in this paper that
$$\mathcal{J}_{p,q}^n = n^{\frac{1}{p}} L_{2p}^r L_{2p}^c \cap n^{\frac{1}{2q}+\frac{1}{2p}} L_{2q}^r L_{2p}^c \cap n^{\frac{1}{2p}+\frac{1}{2q}} L_{2p}^r L_{2q}^c \cap n^{\frac{1}{q}} L_{2q}^r L_{2q}^c.$$
On the Banach space level we have the isometries
$$L_{2p}^r L_{2q}^c = L_s = L_{2q}^r L_{2p}^c \quad \text{with} \quad 1/s = 1/2p + 1/2q.$$
Moreover, again by Hölder inequality it is clear that
$$n^{\frac{1}{s}} \|f\|_s \le \max\left\{ n^{\frac{1}{p}} \|f\|_p, n^{\frac{1}{q}} \|f\|_q \right\}.$$

Therefore, the two cross terms in the middle disappear in the Banach space level. However, replacing scalars by operators in the context of independence/freness over a given von Neumann subalgebra, the Banach space estimates from above are no longer valid and all four terms may have a significant contribution.

It is the operator space structure of $\mathcal{J}_{p,q}^n$ what originally led us to introduce amalgamated and conditional L_p spaces. To be more precise, we consider a von Neumann algebra $\mathcal{M}$ equipped with a normal faithful state φ and a von Neumann subalgebra $\mathcal{N}$ with associated normal faithful conditional expectation E. Then, if we fix $1 \le q \le p \le \infty$, we define
$$\mathcal{J}_{p,q}^n(\mathcal{M}, \mathsf{E}) = \bigcap_{u,v \in \{2r, \infty\}} n^{\frac{1}{u}+\frac{1}{p}+\frac{1}{v}} L_{(u,v)}^p(\mathcal{M}, \mathsf{E}) \quad \text{with} \quad 1/r = 1/q - 1/p.$$
This definition is motivated by the fundamental isometry
$$(3) \qquad S_p^m\big(\mathcal{J}_{p,q}^n(\mathcal{M})\big) = \mathcal{J}_{p,q}^n\big(\mathrm{M}_m \otimes \mathcal{M}, 1_{\mathrm{M}_m} \otimes \varphi\big),$$
which will be proved in Chapter 7. Some preliminary results on $\mathcal{J}_{p,q}^n(\mathcal{M})$ (and vector-valued generalizations) are already contained in the recent paper [**22**]. We extend many results from [**22**] to the realm of free random variables including the limit case $p = \infty$. Our main result for the spaces $\mathcal{J}_{p,q}^n(\mathcal{M}, \mathsf{E})$ shows that we have an interpolation scale with respect to the index $1 \le q \le p$.

THEOREM C. *If $1 \le p \le \infty$, then*
$$\big[\mathcal{J}_{p,1}^n(\mathcal{M}, \mathsf{E}), \mathcal{J}_{p,p}^n(\mathcal{M}, \mathsf{E})\big]_\theta \simeq \mathcal{J}_{p,q}^n(\mathcal{M}, \mathsf{E})$$
with $1/q = 1 - \theta + \theta/p$ and with relevant constants independent of n.

There seems to be no general argument to make intersections commute with complex interpolation. For commutative L_p spaces or rearrangement invariant spaces one can often find concrete formulas of the resulting interpolation norms, see [**35**]. However, in the noncommutative context these arguments are no longer valid and we need genuinely new tools.

0.5. Mixed-norm inequalities

The central role played by Rosenthal inequality partially justifies why the index p must be finite in the commutative form of (Σ_{pq}). This also happens in [**22**], where we used the noncommutative Rosenthal inequality from [**28**]. However, mainly motivated by Problems 2 and 3, one of our main goals in this paper is to obtain a *right* formulation of $(\Sigma_{\infty q})$. As in several other inequalities involving independent random variables, such as the noncommutative Khintchine inequalities, the limit case as $p \to \infty$ holds when replacing classical independence by Voiculescu's concept of freeness [**67**]. Therefore, it is not surprising that we shall use in our proof the free analogue of Rosenthal inequality [**26**]. Following (Σ_{pq}) we have a natural candidate for a complemented embedding of $\mathcal{J}_{p,q}^n$ in $L_p(\ell_q^n)$ using free probability.

We define A_k to be the direct sum $\mathcal{M} \oplus \mathcal{M}$ for $1 \leq k \leq n$. Then we consider the reduced amalgamated free product $\mathcal{A} = *_{\mathcal{N}} \mathsf{A}_k$ where the conditional expectation $\mathsf{E}_{\mathcal{N}} : \mathcal{A} \to \mathcal{N}$ has the form $\mathsf{E}_{\mathcal{N}}(x_1, x_2) = \frac{1}{2}\big(\mathsf{E}(x_1) + \mathsf{E}(x_2)\big)$ when restricted to the algebra A_k. Let $\pi_k : \mathsf{A}_k \to \mathcal{A}$ denote the natural embedding of A_k into $\mathcal{A}$. Moreover, given $x \in \mathcal{M}$ we shall write x_k as an abbreviation of $\pi_k(x, -x)$. Note that x_k is a mean-zero element for $1 \leq k \leq n$. Our main embedding result reads as follows.

THEOREM D. *Let $1 \leq q \leq p \leq \infty$. The map*

$$u : x \in \mathcal{J}_{p,q}^n(\mathcal{M}, \mathsf{E}) \mapsto \sum_{k=1}^n x_k \otimes \delta_k \in L_p(\mathcal{A}; \ell_q^n)$$

is an isomorphism with complemented image. The constants are independent of n.

The map u is of course reminiscent of the fundamental mappings employed in [**15, 17, 22**] constructing certain embeddings of L_p spaces. Theorem C follows as a consequence of Theorem D using the fact that

$$L_p(\mathcal{A}, \ell_q^n) = \big[L_p(\mathcal{A}, \ell_1^n), L_p(\mathcal{A}, \ell_p^n)\big]_\theta.$$

We have tried in vain to prove Theorem D directly using uniquely tools from free probability. The methods around Voiculescu's inequality seem to work perfectly fine in the limit case $p = \infty$, for which there is no commutative version. However, basic tools from free harmonic analysis are still missing for a direct proof of Theorem D. Instead, apart from the free analogue [**26**] of Rosenthal inequality, we also use factorization techniques and interpolation results for noncommutative Hardy and BMO spaces, see Chapter 5 for further details.

The interested reader might be surprised that we have not formulated our results in the category of operator spaces. However, as so often in martingale theory these results are automatic, provided the spaces carry the correct operator space structure, given in this case by (3) and

$$S_p^m\Big(L_p\big(*_k \mathsf{A}_k; \ell_q^n\big)\Big) = L_p\Big(\mathrm{M}_m \otimes (*_k \mathsf{A}_k); \ell_q^n\Big) = L_p(\mathcal{A}; \ell_q^n),$$

where the von Neumann algebra $\mathcal{A}$ is now given by

$$\mathcal{A} = *_{\mathrm{M}_m} \widehat{\mathsf{A}}_k \quad \text{with} \quad \widehat{\mathsf{A}}_k = \mathrm{M}_m \otimes \mathsf{A}_k = \big(\mathrm{M}_m \otimes \mathcal{M}\big) \oplus \big(\mathrm{M}_m \otimes \mathcal{M}\big).$$

0.6. Operator space L_p embeddings

In the last part of the paper, we shall use the mixed-norm inequalities obtained in Theorem D to construct a completely isomorphic embedding of L_q into L_p for $1 \leq p < q \leq 2$. Our construction solves positively a problem formulated by Pisier in 1996, whose precise statement is the following.

THEOREM E. *Let $1 \leq p < q \leq 2$ and let $\mathcal{M}$ be a von Neumann algebra. Then, there exists a sufficiently large von Neumann algebra $\mathcal{A}$ and a completely isomorphic embedding of $L_q(\mathcal{M})$ into $L_p(\mathcal{A})$, where both spaces are equipped with their respective natural operator space structures. Moreover, we have*

    i) *If $\mathcal{M}$ is QWEP, we can choose $\mathcal{A}$ to be QWEP.*
    ii) *If $\mathcal{M}$ is hyperfinite, we can choose $\mathcal{A}$ to be hyperfinite.*

This is the second part of a series of three papers in which we investigate the validity of some well-known results, from the embedding theory of L_p spaces, in the category of operator spaces. We refer to the Introduction of the first part [24] for the motivations which led to our result and also to place it in the right context. As explained there, the presence of noncommutative L_p spaces is not only necessary but natural. The major result in [24] is a complete embedding of discrete L_q spaces into the predual of a von Neumann algebra. This is the simplest case of our complete embedding and it allowed us to give an essentially self-contained approach for a wider audience. In the third paper [25], we shall prove that for $\mathcal{M}$ infinite dimensional the von Neumann algebra $\mathcal{A}$ must be of type III. This generalizes a previous result of Pisier [48] and requires different techniques from [23].

Our argument can be sketched as follows. We first construct a cb-embedding of the Schatten class S_q into $L_p(\mathcal{A})$ with $\mathcal{A}$ a QWEP von Neumann algebra. This will give rise to an L_p generalization of [24, Theorem D]. The drawback of this first approach is that this construction does not preserve hyperfiniteness. The argument to fix this is quite involved and requires recent techniques from [18] and [20]. We first apply a transference argument, via a noncommutative Rosenthal type inequality in L_1 for identically distributed random variables, to replace freeness in our construction by some sort of noncommutative independence. This allows to avoid free product von Neumann algebras and use tensor products instead. Then we combine the algebraic central limit theorem with the notion of noncommutative Poisson random measure to eliminate the use of ultraproducts in the process. After these modifications in our original argument, it is easily seen that hyperfiniteness is preserved. This more involved construction of the cb-embedding is the right one to analyze the finite dimensional case. In other words, we estimate the dimension of $\mathcal{A}$ in terms of the dimension of $\mathcal{M}$, see Remark 8.15 below for details. The proof for general von Neumann algebras requires to consider a 'generalized' Haagerup tensor product since we are not in the discrete case anymore. Our approach here follows Pisier's method in [49].

After a quick look at the main results in [3, 15], the problem of constructing an *isometric* cb-embedding of L_q into L_p arises in a natural way. This remains an open problem.

Structure of the paper. This paper has three natural parts. Part I contains Chapters 1–4, where we prove Theorems A and B. In Part II, which is given by Chapters 5–7, we discuss our mixed-norm inequalities. This includes Theorems C and D. Finally, Part III is the content of Chapter 8 where we construct our operator space L_p embeddings stated in Theorem E. Those readers interested on getting rapidly to the core of the paper might skip Part I –including Chapter 1 if the reader is familiar with recent advances on noncommutative integration– in a first reading. Indeed, the arguments there are quite technical and essentially independent from the rest of the paper. Thus, the definitions and statements that we have included in this Introduction should be enough to follow Parts II and III without too much trouble. Of course, although very rarely, from time to time the reader will have to go back and read the statement of some result from Part I in order to keep track of the argument.

Background and notation. Although we shall review some basic concepts along the paper, we assume some familiarity with some branches of operator algebra such as von Neumann algebras, noncommutative integration, operator space theory and free probability. Moreover, we shall also use specific techniques like Calderón's complex interpolation method [2], Haagerup's approximation theorem [12], the Grothendieck-Pietsch separation argument [43], Raynaud's results on ultraproducts of noncommutative L_p spaces [55], a noncommutative form of a Poisson process [20]... Some of these techniques will be briefly introduced within the text. In any case, our exposition intends to be as self-contained as possible. To conclude, the non-expert reader must think of [24] as a prerequisite to read Chapter 8, since it serves as a good motivation for most of our present approach.

We shall use standard notation from the literature such as e.g. [47, 62]. Apart from this and the terminology introduced along the paper, our notation will be as coherent with [24] as we can.

Acknowledgements. This work was mostly carried out in a one-year visit of the second-named author to the University of Illinois at Urbana-Champaign. The second-named author would like to thank the Math Department of the University of Illinois for its support and hospitality. The first-named author wants to thank the équipe d'analyse fonctionelle at Paris VI for the warm hospitality in the summer 2005, where some part of the arguments in this paper were found.

CHAPTER 1

Noncommutative integration

In this chapter we review some basic notions on noncommutative integration that will be frequently used through out this paper. We begin by recalling Haagerup and Kosaki's constructions of noncommutative L_p spaces. Then we briefly introduce Pisier's theory of vector-valued noncommutative L_p spaces, giving some emphasis to those aspects which are relevant in this work. Finally, we analyze some basic properties of certain L_p spaces associated to a conditional expectation, which were recently introduced in the literature and are basic for our further purposes. We shall assume some familiarity with von Neumann algebras. Basic concepts such as trace, state, commutant, affiliated operator, crossed product... can be found in [**31**] or [**62**] and will be freely used along the text.

1.1. Noncommutative L_p spaces

Noncommutative L_p spaces over non-semifinite von Neumann algebras will be used quite frequently in this paper. In the literature there exist two compatible constructions of L_p in such a general setting: Haagerup L_p spaces and Kosaki's interpolation spaces. These constructions and the associated notion of conditional expectation will be considered in this section.

1.1.1. Haagerup L_p spaces. A full-detailed exposition of this theory is given in Haagerup [**13**] and Terp [**63**] papers. We just present the main notions according to our purposes with an exposition similar to [**28**]. A preliminary restriction is that, in view of our aims, we can work in what follows with normal faithful (*n.f.* in short) states instead of normal semifinite faithful (*n.s.f.* in short) weights.

Let $\mathcal{M}$ be a von Neumann algebra equipped with a distinguished *n.f.* state φ. The GNS construction applied to φ yields a faithful representation ρ of $\mathcal{M}$ into $\mathcal{B}(\mathcal{H})$ so that $\rho(\mathcal{M})$ is a von Neumann algebra acting on $\mathcal{H}$ with a separating and generating unit vector u satisfying $\varphi(x) = \langle u, \rho(x)u \rangle$ for all $x \in \mathcal{M}$. Let us agree to identify $\mathcal{M}$ with $\rho(\mathcal{M})$ in the following. Then, the modular operator Δ is the (generally unbounded) operator obtained from the polar decomposition $S = J\Delta^{1/2}$ of the anti-linear map $S : \mathcal{M}u \to \mathcal{M}u$ given by $S(xu) = x^*u$, see Section 9.2 in [**31**] or [**60**]. We denote by $\sigma_t : \mathcal{M} \to \mathcal{M}$ the one-parameter modular automorphism group associated to the separating and generating unit vector u. That is, for any $t \in \mathbb{R}$ we have an automorphism of $\mathcal{M}$ given by

$$\sigma_t(x) = \Delta^{it} x \Delta^{-it}.$$

Then we consider the crossed product $\mathcal{R} = \mathcal{M} \rtimes_\sigma \mathbb{R}$, which is defined as the von Neumann algebra acting on $L_2(\mathbb{R}; \mathcal{H})$ and generated by the representations $\pi : \mathcal{M} \to \mathcal{B}(L_2(\mathbb{R}; \mathcal{H}))$ and $\lambda : \mathbb{R} \to \mathcal{B}(L_2(\mathbb{R}; \mathcal{H}))$ with

$$\big(\pi(x)\xi\big)(t) = \sigma_{-t}(x)\xi(t) \qquad \text{and} \qquad \big(\lambda(s)\xi\big)(t) = \xi(t-s)$$

for $t \in \mathbb{R}$ and $\xi \in L_2(\mathbb{R}; \mathcal{H})$. Note that the representation π is faithful so that we can identify $\mathcal{M}$ with $\pi(\mathcal{M})$. The dual action of $\mathbb{R}$ on $\mathcal{R}$ is defined as follows. Let $W : \mathbb{R} \to \mathcal{B}(L_2(\mathbb{R}; \mathcal{H}))$ be the unitary representation $(W(t)\xi)(s) = e^{-its}\xi(s)$. Then we define the one-parameter automorphism group $\hat{\sigma}_t : \mathcal{R} \to \mathcal{R}$ by

$$\hat{\sigma}_t(x) = W(t) x W(t)^*.$$

It turns out that $\mathcal{M}$ is the space of fixed points of the dual action

$$\mathcal{M} = \Big\{ x \in \mathcal{R} \,\Big|\, \hat{\sigma}_t(x) = x \text{ for all } t \in \mathbb{R} \Big\}.$$

Following [42], the crossed product $\mathcal{R}$ is a semifinite von Neumann algebra and admits a unique n.s.f. trace τ satisfying $\tau \circ \hat{\sigma}_t = e^{-t}\tau$ for all $t \in \mathbb{R}$. Let $L_0(\mathcal{R}, \tau)$ denote the topological $*$-algebra of τ-measurable operators affiliated with $\mathcal{R}$ and let $0 < p \le \infty$. The *Haagerup noncommutative L_p space* over $\mathcal{M}$ is defined as

$$L_p(\mathcal{M}, \varphi) = \Big\{ x \in L_0(\mathcal{R}, \tau) \,\Big|\, \hat{\sigma}_t(x) = e^{-t/p} x \text{ for all } t \in \mathbb{R} \Big\}.$$

It is clear from the definition that $L_\infty(\mathcal{M}, \varphi)$ coincides with $\mathcal{M}$. Moreover, as it is to be expected, $L_1(\mathcal{M}, \varphi)$ can be canonically identified with the predual $\mathcal{M}_*$ of the von Neumann algebra $\mathcal{M}$. This requires a short explanation. Given a normal state $\omega \in \mathcal{M}_*^+$, the dual state $\tilde{\omega} : \mathcal{R}_+ \to [0, \infty]$ is defined by

$$\tilde{\omega}(x) = \omega\Big(\int_\mathbb{R} \hat{\sigma}_s(x) \, ds \Big).$$

Note that, by the translation invariance of the Lebesgue measure, the operator valued integral above is invariant under the dual action. In particular, it can be regarded as an element of $\mathcal{M}$. As a n.s.f. weight on $\mathcal{R}$ and according to [42], the dual weight $\tilde{\omega}$ has a Radon-Nikodym derivative h_ω with respect to τ so that

$$\tilde{\omega}(x) = \tau(h_\omega x)$$

for any $x \in \mathcal{R}_+$. The operator h_ω so defined belongs to $L_1(\mathcal{M}, \varphi)_+$. Indeed,

$$\tau(h_\omega \hat{\sigma}_t(x)) = \omega\Big(\int_\mathbb{R} \hat{\sigma}_s(\hat{\sigma}_t(x)) \, ds \Big) = \omega\Big(\int_\mathbb{R} \hat{\sigma}_s(x) \, ds \Big) = \tau(h_\omega x).$$

In particular,

$$\tau(\hat{\sigma}_t(h_\omega)\hat{\sigma}_t(x)) = e^{-t}\tau(h_\omega x) = e^{-t}\tau(h_\omega \hat{\sigma}_t(x)) \quad \text{for all} \quad x \in \mathcal{R},$$

which implies that $\hat{\sigma}_t(h_\omega) = e^{-t} h_\omega$. Therefore, there exists a bijection between $\mathcal{M}_*^+$ and $L_1(\mathcal{M}, \varphi)_+$ which extends to a bijection between the predual $\mathcal{M}_*$ and $L_1(\mathcal{M}, \varphi)$ by polar decomposition

$$\omega = u|\omega| \in \mathcal{M}_* \mapsto u h_{|\omega|} = h_\omega \in L_1(\mathcal{M}, \varphi).$$

In fact, after imposing on $L_1(\mathcal{M}, \varphi)$ the norm

$$\|h_\omega\|_1 = |\omega|(1) = \|\omega\|_{\mathcal{M}_*},$$

we obtain an isometry between $\mathcal{M}_*$ and $L_1(\mathcal{M}, \varphi)$. There is however a nicer way to describe this norm. As we have seen, for any $x \in L_1(\mathcal{M}, \varphi)$ there exists a unique $\omega_x \in \mathcal{M}_*$ such that $h_{\omega_x} = x$. This gives rise to the functional $\mathrm{tr} : L_1(\mathcal{M}, \varphi) \to \mathbb{C}$ called *trace* and defined by

$$\mathrm{tr}(x) = \omega_x(1).$$

The functional tr is continuous since $|\text{tr}(x)| \leq \text{tr}(|x|) = \|x\|_1$ and satisfies the tracial property
$$\text{tr}(xy) = \text{tr}(yx).$$
Our distinguished state φ can be recovered from tr as follows. First we note as above that its dual weight $\tilde{\varphi}$ admits a Radon-Nikodym derivative D_φ with respect to τ, so that $\tilde{\varphi}(x) = \tau(D_\varphi x)$ for $x \in \mathcal{R}_+$. Then, it turns out that
$$\varphi(x) = \text{tr}(D_\varphi x) \quad \text{for} \quad x \in \mathcal{M}.$$
According to this, we will refer in what follows to D_φ as the *density* of φ. Moreover, we shall write D instead of D_φ whenever the dependence on φ is clear from the context. Given $0 < p < \infty$ and $x \in L_p(\mathcal{M}, \varphi)$, we define
$$\|x\|_p = \left(\text{tr}|x|^p\right)^{1/p} \quad \text{and} \quad \|x\|_\infty = \|x\|_\mathcal{M}.$$
$\|\ \|_p$ is a norm (resp. p-norm) on $L_p(\mathcal{M}, \varphi)$ for $1 \leq p \leq \infty$ (resp. $0 < p < 1$).

LEMMA 1.1. *The Haagerup L_p spaces satisfy the following properties:*
  i) *If $0 < p, q, r \leq \infty$ with $1/r = 1/p + 1/q$, we have*
  $$\|xy\|_r \leq \|x\|_p \|y\|_q \quad \text{for all} \quad x \in L_p(\mathcal{M}, \varphi), \ y \in L_q(\mathcal{M}, \varphi).$$
  ii) *If $1 \leq p < \infty$, $L_p(\mathcal{M}, \varphi)^*$ is anti-linear isometrically isomorphic to $L_{p'}(\mathcal{M}, \varphi)$ via*
  $$x \in L_{p'}(\mathcal{M}, \varphi) \mapsto \text{tr}(x^* \cdot) \in L_p(\mathcal{M}, \varphi)^*.$$

An element $x \in \mathcal{M}$ is called analytic if the function $t \in \mathbb{R} \mapsto \sigma_t(x) \in \mathcal{M}$ extends to an analytic function $z \in \mathbb{C} \to \sigma_z(x) \in \mathcal{M}$. By [42] we know that the subspace $\mathcal{M}_a$ of analytic elements in $\mathcal{M}$ is a weak* dense *-subalgebra of $\mathcal{M}$. The proof of the following result can be found in [28]. It will be useful in the sequel.

LEMMA 1.2. *If $0 < p < \infty$, we have*
  i) $\mathcal{M}_a D^{1/p}$ *is dense in* $L_p(\mathcal{M}, \varphi)$.
  ii) $D^{(1-\eta)/p} \mathcal{M}_a D^{\eta/p} = \mathcal{M}_a D^{1/p}$ *for all $0 \leq \eta \leq 1$.*

1.1.2. Kosaki's interpolation. The given definition of Haagerup L_p space has the disadvantage that the intersection of $L_p(\mathcal{M}, \varphi)$ and $L_q(\mathcal{M}, \varphi)$ is trivial for $p \neq q$. In particular, these spaces do not form an interpolation scale. All these difficulties disappear with Kosaki's construction. As above, we only consider von Neumann algebras equipped with n.f. states. The general construction for any von Neumann algebra can be found in [32] and [64]. Let us consider a von Neumann algebra $\mathcal{M}$ equipped with a n.f. state φ. First we define
$$L_1(\mathcal{M}) = \mathcal{M}_*^{\text{op}}.$$
Note that the natural operator space structure for $L_1(\mathcal{M})$ requires to consider $\mathcal{M}_*^{\text{op}}$ instead of $\mathcal{M}_*$, we refer the reader to [47] for a detailed explanation. Then, given any real number t, we consider the map
$$j_t : x \in \mathcal{M} \mapsto \sigma_t(x)\varphi \in L_1(\mathcal{M}) \quad \text{with} \quad (\sigma_t(x)\varphi)(y) = \varphi(\sigma_t(x)y).$$
According to [32] there exists a unique extension $j_z : \mathcal{M} \to L_1(\mathcal{M})$ such that, for any $0 \leq \eta \leq 1$, the map $j_{-i\eta}$ is injective. In particular, $(j_{-i\eta}(\mathcal{M}), L_1(\mathcal{M}))$ is compatible for complex interpolation and we define the *Kosaki noncommutative L_p spaces* as follows
$$\mathcal{L}_p(\mathcal{M}, \varphi, \eta) = \left[j_{-i\eta}(\mathcal{M}), L_1(\mathcal{M})\right]_{\frac{1}{p}}$$

by specifying
$$\|x\|_0 = \|j_{-i\eta}^{-1}(x)\|_{\mathcal{M}} \quad \text{and} \quad \|x\|_1 = \|x\|_{L_1(\mathcal{M})}.$$
The following result was proved in [13] except for the last isometry [32].

THEOREM 1.3. *Let $1 \leq p \leq \infty$ and let $\mathcal{M}$ be any von Neumann algebra:*

i) *If φ_1 and φ_2 are two n.f. states on $\mathcal{M}$, we have an isometric isomorphism*
$$L_p(\mathcal{M}, \varphi_1) = L_p(\mathcal{M}, \varphi_2).$$

ii) *If φ is a n.f. state and $0 \leq \eta \leq 1$, we have an isometric isomorphism*
$$L_p(\mathcal{M}, \varphi) = \mathcal{L}_p(\mathcal{M}, \varphi, \eta).$$

More concretely, given $x \in \mathcal{M}$
$$\|j_{-i\eta}(x)\|_{\mathcal{L}_p(\mathcal{M},\varphi,\eta)} = \|\mathrm{D}^{\eta/p} x \mathrm{D}^{(1-\eta)/p}\|_{L_p(\mathcal{M},\varphi)}.$$

According to Theorem 1.3, Haagerup and Kosaki noncommutative L_p spaces can be identified. We shall write $L_p(\mathcal{M})$ to denote in what follows *any* of the spaces defined above. In particular, after the corresponding identifications, we may use the complex interpolation method for Haagerup L_p spaces. We should also note that Kosaki's definition of L_p presents some other disadvantages. The main lacks in this paper will be the absence of positive cones and the fact that the case $0 < p < 1$ is excluded from the definition. In particular, we will use Haagerup L_p spaces and we will apply Theorem 1.3 when needed.

1.1.3. Conditional expectations. Let us consider a von Neumann algebra $\mathcal{M}$ equipped with a *n.f.* state φ and a von Neumann subalgebra $\mathcal{N}$ of $\mathcal{M}$. A *conditional expectation* $\mathsf{E} : \mathcal{M} \to \mathcal{N}$ is a positive contractive projection. E is called faithful if $\mathsf{E}(x^*x) \neq 0$ for any $x \in \mathcal{M}$ and normal when it has a predual $\mathsf{E}_* : \mathcal{M}_* \to \mathcal{N}_*$. According to Takesaki [61], if $\mathcal{N}$ is invariant under the action of the modular automorphism group (i.e. $\sigma_t(\mathcal{N}) \subset \mathcal{N}$ for all $t \in \mathbb{R}$), there exists a unique faithful normal conditional expectation $\mathsf{E} : \mathcal{M} \to \mathcal{N}$ satisfying $\varphi \circ \mathsf{E} = \varphi$. Moreover, by Connes [6] it commutes with the modular automorphism group
$$\mathsf{E} \circ \sigma_t = \sigma_t \circ \mathsf{E}.$$

The required invariance of $\mathcal{N}$ under the action of σ_t implies that the modular automorphism group associated to $\mathcal{N}$ coincides with the restriction of σ to $\mathcal{N}$. It follows that $\mathcal{N} \rtimes_\sigma \mathbb{R}$ is a von Neumann subalgebra of $\mathcal{M} \rtimes_\sigma \mathbb{R}$. In particular, the space $L_p(\mathcal{N})$ can be identified isometrically with a subspace of $L_p(\mathcal{M})$, see [28] for details. In this paper we shall permanently assume the existence of a normal faithful conditional expectation $\mathsf{E} : \mathcal{M} \to \mathcal{N}$.

It is well-known that in the tracial case, the conditional expectation E extends to a contractive projection from $L_p(\mathcal{M})$ onto $L_p(\mathcal{N})$ for any $1 \leq p \leq \infty$, which is still positive and has the modular property $\mathsf{E}(axb) = a\mathsf{E}(x)b$ for all $a, b \in \mathcal{N}$ and $x \in L_p(\mathcal{M})$. These properties remain valid in this context. We summarize in the following lemma the main properties of E that will be used in the sequel and refer the reader to [28] for a proof of these facts.

LEMMA 1.4. *Let $\mathcal{M}$ and $\mathcal{N}$ be as above:*

i) $\mathsf{E} : L_p(\mathcal{M}) \to L_p(\mathcal{N})$ *is a positive contractive projection.*
ii) *If $2 \leq p \leq \infty$ and $x \in L_p(\mathcal{M})$, we have $\mathsf{E}(x)^* \mathsf{E}(x) \leq \mathsf{E}(x^*x)$.*

iii) If $a \in L_p(\mathcal{N}), x \in L_q(\mathcal{M}), b \in L_r(\mathcal{N})$ and $\frac{1}{p} + \frac{1}{q} + \frac{1}{r} \leq 1$, $\mathsf{E}(axb) = a\mathsf{E}(x)b$.
In particular, if $1 \leq p \leq \infty$ and $x \in \mathcal{M}$

$$\mathsf{E}(\mathrm{D}^{1/p}x) = \mathrm{D}^{1/p}\mathsf{E}(x) \quad and \quad \mathsf{E}(x\mathrm{D}^{1/p}) = \mathsf{E}(x)\mathrm{D}^{1/p}.$$

1.2. Pisier's vector-valued L_p spaces

Noncommutative vector-valued L_p spaces $L_p(\mathcal{M}; \mathrm{X})$ appeared quite recently with Pisier's work [46]. The reason for the novelty of such a natural concept lies in the fact that X must be equipped with an operator space structure rather than a Banach space one. Moreover, many natural properties such as duality, complex interpolation, etc... must be formulated in the category of operator spaces. Let us begin by recalling the concept of operator space.

1.2.1. Operator spaces. Operator space theory plays a central role in this paper. It can be regarded as a noncommutative generalization of Banach space theory and it has proved to be an essential tool in operator algebra as well as in noncommutative harmonic analysis. An *operator space* X is defined as a closed subspace of the space $\mathcal{B}(\mathcal{H})$ of bounded operators on some Hilbert space $\mathcal{H}$. Let $\mathrm{M}_n(\mathrm{X})$ denote the vector space of $n \times n$ matrices with entries in X. If we have an isometric embedding $j: \mathrm{X} \to \mathcal{B}(\mathcal{H})$, let us consider the sequence of norms on $\mathrm{M}_n(\mathrm{X})$ for $n \geq 1$ given by

$$\|(x_{ij})\|_{\mathrm{M}_n(\mathrm{X})} = \|(j(x_{ij}))\|_{\mathcal{B}(\mathcal{H}^n)}.$$

Ruan's axioms [11, 47, 58] describe axiomatically those sequences of matrix norms which can occur from an isometric embedding into $\mathcal{B}(\mathcal{H})$. Any such sequence of norms provides X with a so-called operator space structure. Every Banach space can be equipped with several operator space structures. In particular, the most important information carried by an operator space is not the space itself but the way in which it embeds isometrically into $\mathcal{B}(\mathcal{H})$. For this reason, the main difference between Banach and operator space theory lies on the morphisms rather than on the spaces. The morphisms in the category of operator spaces are *completely bounded* linear maps $u: \mathrm{X} \to \mathrm{Y}$. That is, linear maps satisfying that the quantity

$$\|u\|_{cb(\mathrm{X},\mathrm{Y})} = \sup_{n \geq 1} \|id_{\mathrm{M}_n} \otimes u\|_{\mathcal{B}(\mathrm{M}_n(\mathrm{X}),\mathrm{M}_n(\mathrm{Y}))}$$

is finite. In this paper we shall assume some familiarity with basic notions such as duality, Haagerup tensor products, the OH spaces... that can be found in the recent books [11, 47].

1.2.2. The hyperfinite case. Before any other consideration, let us recall the natural operator space structure (*o.s.s.* in short) on $L_p(\mathcal{M})$. Proceeding as in Chapter 3 of [46], we regard $\mathcal{M}$ as a subspace of $\mathcal{B}(\mathcal{H})$ with $\mathcal{H}$ being the Hilbert space arising from the GNS construction. Similarly, by embedding the predual von Neumann algebra $\mathcal{M}_*$ on its bidual $\mathcal{M}^*$, we obtain an *o.s.s.* for $\mathcal{M}_*$. The *o.s.s.* on $L_1(\mathcal{M})$ is then given by that of $\mathcal{M}_*^{\mathrm{op}}$, see page 139 in [47] for a detailed justification of this definition. Then, the complex interpolation method for operator spaces developed in [45] provides a natural *o.s.s.* on $L_p(\mathcal{M})$

$$L_p(\mathcal{M}) = \big[L_\infty(\mathcal{M}), L_1(\mathcal{M})\big]_{1/p} = \big[\mathcal{M}, \mathcal{M}_*^{\mathrm{op}}\big]_{1/p}.$$

Let us now describe the natural operator space structure for vector-valued L_p spaces. Let $\mathcal{M}$ be an hyperfinite von Neumann algebra and let X be any operator space. Then, recalling the definitions of the projective and the minimal tensor product in the category of operator spaces, we define
$$L_1(\mathcal{M};X) = L_1(\mathcal{M}) \widehat{\otimes} X \quad \text{and} \quad L_\infty(\mathcal{M};X) = L_\infty(\mathcal{M}) \otimes_{\min} X.$$
Then, the *space* $L_p(\mathcal{M};X)$ is defined by complex interpolation
$$L_p(\mathcal{M};X) = \big[L_\infty(\mathcal{M};X), L_1(\mathcal{M};X)\big]_{1/p}.$$
As it was explained in [46], the hyperfiniteness of $\mathcal{M}$ leads to obtain some expected properties of these spaces. We shall discuss the non-hyperfinite case in the next paragraph. We are not reviewing here the basic results from Pisier's theory, for which we refer the reader to Chapters 1,2 and 3 in [46].

Let us fix some notation. R and C denote the row and column operator spaces (*c.f.* Chapter 1 of [47]) constructed over ℓ_2. Identifying $\mathcal{B}(\ell_2)$ (via the canonical basis of ℓ_2) with a space of matrices with infinitely many rows and columns, R and C are the first row and first column subspaces of $\mathcal{B}(\ell_2)$. Similarly, if $1 \le p \le \infty$ and S_p denotes the Schatten p-class over ℓ_2, the spaces R_p and C_p denote the row and column subspaces of S_p. The finite-dimensional versions over ℓ_2^n will be denoted by R_p^n and C_p^n respectively. Note that, as in the infinite-dimensional case, we have $R_n = R_\infty^n$ and $C_n = C_\infty^n$. Given an operator space X, the vector-valued Schatten classes with values in X will be denoted by $S_p(X)$ and $S_p^n(X)$ respectively. Among the several characterizations of these spaces given in [46], we have
$$S_p(X) = C_p \otimes_h X \otimes_h R_p \quad \text{and} \quad S_p^n(X) = C_p^n \otimes_h X \otimes_h R_p^n.$$

1.2.3. The non-hyperfinite case. One of the main restrictions in Pisier's theory [46] comes from the fact that the construction of $L_p(\mathcal{M};X)$ requires $\mathcal{M}$ to be hyperfinite. This excludes for instance free products of von Neumann algebras, a very relevant tool in this paper. There exists however a very recent construction in [19] of $L_p(\mathcal{M};X)$ which is valid for any QWEP von Neumann algebra. Nevertheless, since we only deal with very specific operator spaces, we briefly discuss them.

THE SPACES $L_p(\mathcal{M}; R_p^n)$ AND $L_p(\mathcal{M}; C_p^n)$. As we explained in the Introduction, these spaces are of capital importance. Let us recall that the spaces $R_p^n(X)$ and $C_p^n(X)$ are defined as subspaces of $S_p^n(X)$ for any operator space X. Therefore, motivated by Fubini's theorem for noncommutative L_p spaces [46], we obtain the following definition valid for arbitrary von Neumann algebras
$$L_p(\mathcal{M}; R_p^n) = R_p^n(L_p(\mathcal{M})) \quad \text{and} \quad L_p(\mathcal{M}; C_p^n) = C_p^n(L_p(\mathcal{M})).$$
These spaces appear quite frequently in the theory of noncommutative martingale inequalities. However, in that context they are respectively denoted by $L_p(\mathcal{M}; \ell_2^r)$ and $L_p(\mathcal{M}; \ell_2^c)$, see e.g. [51] or [53] for more details.

THE SPACES $L_p(\mathcal{M}; \ell_q^n)$. The spaces $L_p(\mathcal{M}; \ell_1^n)$ and $L_p(\mathcal{M}; \ell_\infty^n)$ were defined in [16] for general von Neumann algebras to study Doob's maximal inequality for noncommutative martingales. Indeed, since the notion of maximal function does not make any sense in the noncommutative setting, the norm in $L_p(\Omega)$ of Doob's maximal function
$$f^*(w) = \sup_{n \ge 1} |f_n(w)|$$

was reinterpreted as the norm in $L_p(\Omega; \ell_\infty)$ of the sequence $(f_1, f_2, \ldots)$. This was the original motivation to study the spaces $L_p(\mathcal{M}; \ell_1)$ and $L_p(\mathcal{M}; \ell_\infty)$. A detailed exposition of these spaces can also be found in [**29**]. The spaces $L_p(\mathcal{M}; \ell_q^n)$ will be of capital importance in the second part of this paper. These spaces were recently defined in [**30**] over non-hyperfinite von Neumann algebras as follows

$$L_p(\mathcal{M}; \ell_q^n) = \left[L_p(\mathcal{M}; \ell_\infty^n), L_p(\mathcal{M}; \ell_1^n)\right]_{\frac{1}{q}}.$$

This formula trivially generalizes by the reiteration theorem [**2**]. A relevant property of these spaces proved in [**30**] is Fubini's isometry $L_p(\mathcal{M}; \ell_p^n) = \ell_p^n(L_p(\mathcal{M}))$. It is also proved in [**30**] that Pisier's identities hold. In other words, defining the auxiliary index $1/r = |1/p - 1/q|$ we find

i) If $p \leq q$, we have

$$\|x\|_{pq} = \inf \left\{ \|\alpha\|_{L_{2r}(\mathcal{M})} \|y\|_{L_q(\mathcal{M}; \ell_q^n)} \|\beta\|_{L_{2r}(\mathcal{M})} \mid x = \alpha y \beta \right\}.$$

ii) If $p \geq q$, we have

$$\|x\|_{pq} = \sup \left\{ \|\alpha y \beta\|_{L_q(\mathcal{M}; \ell_q^n)} \mid \alpha, \beta \in \mathsf{B}_{L_{2r}(\mathcal{M})} \right\}.$$

REMARK 1.5. Let $\mathcal{M}$ be an hyperfinite von Neumann algebra and

$$\begin{aligned} \mathsf{A}_p(\mathcal{M}; n) &= \left[L_p(\mathcal{M}; \ell_\infty^n), L_p(\mathcal{M}; \ell_1^n)\right]_{\frac{1}{2}}, \\ \mathsf{B}_p(\mathcal{M}; n) &= \left[L_p(\mathcal{M}; C_p^n), L_p(\mathcal{M}; R_p^n)\right]_{\frac{1}{2}}. \end{aligned}$$

According to [**46**], we have

$$\mathsf{A}_p(\mathcal{M}; n) = L_p(\mathcal{M}; \mathrm{OH}_n) = \mathsf{B}_p(\mathcal{M}; n).$$

It is therefore natural to ask whether these identities remain valid with our definition of $L_p(\mathcal{M}; \mathrm{OH}_n)$ for non-hyperfinite $\mathcal{M}$. This is indeed the case since, following the same terminology already defined in the Introduction, we have

i) If $1 \leq p \leq 2$ and $1/p = 1/2 + 1/q$, we have

$$\mathsf{A}_p(\mathcal{M}; n) = L_{2q}(\mathcal{M}) \ell_2^n(L_2(\mathcal{M})) L_{2q}(\mathcal{M}) = \mathsf{B}_p(\mathcal{M}; n).$$

ii) If $2 \leq p \leq \infty$ and $1/2 = 1/p + 1/q$, we have

$$\mathsf{A}_p(\mathcal{M}; n) = \sqrt{n}\, L^p_{(2q, 2q)}(\ell_\infty^n(\mathcal{M}), \mathsf{E}_n) = \mathsf{B}_p(\mathcal{M}; n),$$

where E_n is given by $\mathsf{E}_n : \sum_{k=1}^n x_k \otimes \delta_k \in \ell_\infty^n(\mathcal{M}) \mapsto \frac{1}{n} \sum_{k=1}^n x_k \in \mathcal{M}$.

The isometric identities for the A_p's follow from Pisier's identities above while the identities for the B_p's follow from [**69**] or Theorem 3.2 below. Thus, we agree to define the *o.s.s.* on $L_p(\mathcal{M}; \mathrm{OH}_n)$ as any of the interpolation spaces above.

REMARK 1.6. As we shall see through this paper, all the spaces mentioned so far arise as particular cases of our definition below of amalgamated and conditional L_p spaces.

1.3. The spaces $L_p^r(\mathcal{M}, \mathsf{E})$ and $L_p^c(\mathcal{M}, \mathsf{E})$

The spaces $L_p^r(\mathcal{M}, \mathsf{E})$ and $L_p^c(\mathcal{M}, \mathsf{E})$ were introduced in [16], where they turned out to be quite useful in the context of noncommutative probability. Both spaces will play a central role in this paper since they are the most significant examples of the so-called amalgamated and conditional L_p spaces, to be defined below. Let $\mathcal{M}$ be a von Neumann algebra equipped with a n.f. state φ and let $\mathcal{N}$ be a von Neumann subalgebra of $\mathcal{M}$. Let $\mathsf{E}: \mathcal{M} \to \mathcal{N}$ denote the conditional expectation of $\mathcal{M}$ onto $\mathcal{N}$. Then, given $1 \le p \le \infty$ and $(\alpha, a) \in \mathcal{N} \times \mathcal{M}$, we define

$$
\begin{aligned}
\|\alpha \mathrm{D}^{\frac{1}{p}} a\|_{L_p^r(\mathcal{M}, \mathsf{E})} &= \|\alpha \mathrm{D}^{\frac{1}{p}} \mathsf{E}(aa^*) \mathrm{D}^{\frac{1}{p}} \alpha^*\|_{L_{p/2}(\mathcal{N})}^{1/2}, \\
\|a \mathrm{D}^{\frac{1}{p}} \alpha\|_{L_p^c(\mathcal{M}, \mathsf{E})} &= \|\alpha^* \mathrm{D}^{\frac{1}{p}} \mathsf{E}(a^*a) \mathrm{D}^{\frac{1}{p}} \alpha\|_{L_{p/2}(\mathcal{N})}^{1/2}.
\end{aligned}
\tag{1.1}
$$

$L_p^r(\mathcal{M}, \mathsf{E})$ and $L_p^c(\mathcal{M}, \mathsf{E})$ will stand for the completions with respect to these norms.

REMARK 1.7. Note that for $1 \le p < 2$ we have $0 < p/2 < 1$ so that we are forced to use Haagerup L_p spaces in the definition (1.1). On the other hand, let us note that the assumption $x \in \mathcal{N}\mathrm{D}^{1/p}\mathcal{M}$ (resp. $x \in \mathcal{M}\mathrm{D}^{1/p}\mathcal{N}$) is not needed to define the norm of x in $L_p^r(\mathcal{M}, \mathsf{E})$ (resp. $L_p^c(\mathcal{M}, \mathsf{E})$) for $2 \le p \le \infty$. Indeed, given $x \in L_p(\mathcal{M})$ we can define

$$\|x\|_{L_p^r(\mathcal{M}, \mathsf{E})} = \|\mathsf{E}(xx^*)\|_{L_{p/2}(\mathcal{N})}^{1/2} \quad \text{and} \quad \|x\|_{L_p^c(\mathcal{M}, \mathsf{E})} = \|\mathsf{E}(x^*x)\|_{L_{p/2}(\mathcal{N})}^{1/2}.$$

In that case $\mathsf{E}(xx^*)$ and $\mathsf{E}(x^*x)$ are well-defined and

$$\max\left\{\|\mathsf{E}(xx^*)^{1/2}\|_{L_p(\mathcal{N})}, \|\mathsf{E}(x^*x)^{1/2}\|_{L_p(\mathcal{N})}\right\} \le \|x\|_{L_p(\mathcal{M})}.$$

Hence, by the density of $\mathcal{N}\mathrm{D}^{1/p}\mathcal{M}$ and $\mathcal{M}\mathrm{D}^{1/p}\mathcal{N}$ in $L_p(\mathcal{M})$, we obtain the same closure as in (1.1). However, in the case $1 \le p < 2$ the conditional expectation E is no longer continuous on $L_{p/2}(\mathcal{M})$ so that we need this alternative definition.

The duality of $L_p^r(\mathcal{M}, \mathsf{E})$ and $L_p^c(\mathcal{M}, \mathsf{E})$ was studied in [16]. For the moment we just need to know that, given $1 < p < \infty$, the following isometries hold via the anti-linear duality bracket $\langle x, y \rangle = \mathrm{tr}(x^*y)$

$$
\begin{aligned}
L_p^r(\mathcal{M}, \mathsf{E})^* &= L_{p'}^r(\mathcal{M}, \mathsf{E}), \\
L_p^c(\mathcal{M}, \mathsf{E})^* &= L_{p'}^c(\mathcal{M}, \mathsf{E}).
\end{aligned}
\tag{1.2}
$$

LEMMA 1.8. *If $2 \le p \le \infty$ and $1/2 = 1/p + 1/s$, we have*

$$
\begin{aligned}
\|x\|_{L_p^r(\mathcal{M}, \mathsf{E})} &= \sup\left\{\|\alpha x\|_{L_2(\mathcal{M})} \mid \|\alpha\|_{L_s(\mathcal{N})} \le 1\right\}, \\
\|x\|_{L_p^c(\mathcal{M}, \mathsf{E})} &= \sup\left\{\|x\beta\|_{L_2(\mathcal{M})} \mid \|\beta\|_{L_s(\mathcal{N})} \le 1\right\}.
\end{aligned}
$$

PROOF. Given $x \in L_p^r(\mathcal{M}, \mathsf{E})$, the operator $\mathsf{E}(xx^*)$ is positive. Hence

$$
\begin{aligned}
\|x\|_{L_p^r(\mathcal{M}, \mathsf{E})}^2 &= \sup\left\{\mathrm{tr}(a\mathsf{E}(xx^*)) \mid a \ge 0, \|a\|_{L_{(p/2)'}(\mathcal{N})} \le 1\right\} \\
&= \sup\left\{\mathrm{tr}(\alpha \mathsf{E}(xx^*)\alpha^*) \mid \|\alpha^*\alpha\|_{L_{(p/2)'}(\mathcal{N})} \le 1\right\} \\
&= \sup\left\{\mathrm{tr}(\alpha xx^*\alpha^*) \mid \|\alpha^*\alpha\|_{L_{(p/2)'}(\mathcal{N})} \le 1\right\}
\end{aligned}
$$

$$= \sup\Big\{\|\alpha x\|^2_{L_2(\mathcal{M})} \,\big|\, \|\alpha\|_{L_{2(p/2)'}(\mathcal{N})} \leq 1\Big\}.$$

Finally we recall that $s = 2(p/2)'$. The proof for $L_p^c(\mathcal{M}, \mathsf{E})$ is entirely similar. □

As usual, we shall write $L_2^r(\mathcal{M}) = \mathcal{B}(L_2(\mathcal{M}), \mathbb{C})$ and $L_2^c(\mathcal{M}) = \mathcal{B}(\mathbb{C}, L_2(\mathcal{M}))$ to denote the row and column Hilbert spaces over $L_2(\mathcal{M})$. Both spaces embed isometrically in $\mathcal{B}(L_2(\mathcal{M}))$. Hence, they admit a natural o.s.s. Given any index $2 \leq p \leq \infty$, we generalize these spaces as follows. According to the definition of Haagerup L_p spaces, we may consider the contractive inclusions

$$j_r : x \in \mathcal{M} \mapsto \mathrm{D}^{\frac{1}{2}}x \in L_2(\mathcal{M}),$$
$$j_c : x \in \mathcal{M} \mapsto x\mathrm{D}^{\frac{1}{2}} \in L_2(\mathcal{M}).$$

Then we define for $2 \leq p \leq \infty$

(1.3)
$$L_p^r(\mathcal{M}) = \big[j_r(\mathcal{M}), L_2^r(\mathcal{M})\big]_{\frac{2}{p}} \quad \text{with} \quad \|x\|_0 = \|\mathrm{D}^{-1/2}x\|_{\mathcal{M}},$$
$$L_p^c(\mathcal{M}) = \big[j_c(\mathcal{M}), L_2^c(\mathcal{M})\big]_{\frac{2}{p}} \quad \text{with} \quad \|x\|_0 = \|x\mathrm{D}^{-1/2}\|_{\mathcal{M}}.$$

Note that $L_p^r(\mathcal{M}) = L_p(\mathcal{M}) = L_p^c(\mathcal{M})$ as Banach spaces.

Let $\mathcal{N}$ be a von Neumann subalgebra of $\mathcal{M}$. Then, given $2 \leq u, q, v \leq \infty$, we consider the closed ideal $\mathcal{I}_1$ in the Haagerup tensor product $L_u^r(\mathcal{N}) \otimes_h L_q^c(\mathcal{M})$ generated by the differences $x\gamma \otimes y - x \otimes \gamma y$, with $x \in L_u^r(\mathcal{N})$, $y \in L_q^c(\mathcal{M})$ and $\gamma \in \mathcal{N}$. Similarly, we consider the closed ideal $\mathcal{I}_2$ in $L_q^r(\mathcal{M}) \otimes_h L_v^c(\mathcal{N})$ generated by the differences $x\gamma \otimes y - x \otimes \gamma y$, with $x \in L_q^r(\mathcal{M})$, $y \in L_v^c(\mathcal{N})$ and $\gamma \in \mathcal{N}$. Then, we define the amalgamated Haagerup tensor products as the quotients

$$L_u^r(\mathcal{N}) \otimes_{\mathcal{N}, h} L_q^c(\mathcal{M}) = L_u^r(\mathcal{N}) \otimes_h L_q^c(\mathcal{M}) / \mathcal{I}_1,$$
$$L_q^r(\mathcal{M}) \otimes_{\mathcal{N}, h} L_v^c(\mathcal{N}) = L_q^r(\mathcal{M}) \otimes_h L_v^c(\mathcal{N}) / \mathcal{I}_2.$$

Let $a_1, a_2, \ldots, a_m \in L_u(\mathcal{N})$ and $b_1, b_2, \ldots, b_m \in L_q(\mathcal{M})$. Using that the Haagerup tensor product commutes with complex interpolation, it is not difficult to check that the following identities hold

$$\Big\|\sum_{k=1}^m a_k \otimes e_{1k}\Big\|_{L_u^r(\mathcal{N}) \otimes_h R_m} = \Big\|\big(\sum_{k=1}^m a_k a_k^*\big)^{1/2}\Big\|_{L_u(\mathcal{N})},$$
$$\Big\|\sum_{k=1}^m e_{k1} \otimes b_k\Big\|_{C_m \otimes_h L_q^c(\mathcal{M})} = \Big\|\big(\sum_{k=1}^m b_k^* b_k\big)^{1/2}\Big\|_{L_q(\mathcal{M})}.$$

Let us write $\mathcal{S}_r(a) = (\sum_k a_k a_k^*)^{1/2}$ and $\mathcal{S}_c(b) = (\sum_k b_k^* b_k)^{1/2}$. Then, following the definition of the Haagerup tensor product, the norm of x in $L_u^r(\mathcal{N}) \otimes_{\mathcal{N}, h} L_q^c(\mathcal{M})$ can be written as

$$\|x\|_{u \cdot q} = \inf\Big\{\|\mathcal{S}_r(a)\|_{L_u(\mathcal{N})} \|\mathcal{S}_c(b)\|_{L_q(\mathcal{M})} \,\Big|\, x \simeq \sum_{k=1}^m a_k \otimes b_k\Big\},$$

where $\simeq$ means that the difference belongs to $\mathcal{I}_1$. Similarly, given $e_1, e_2, \ldots, e_m$ in $L_q^r(\mathcal{M})$ and $f_1, f_2, \ldots, f_m$ in $L_v^c(\mathcal{N})$, we can write the norm of an element x in the space $L_q^r(\mathcal{M}) \otimes_{\mathcal{N}, h} L_v^c(\mathcal{N})$ as follows

$$\|x\|_{q \cdot v} = \inf\Big\{\|\mathcal{S}_r(e)\|_{L_q(\mathcal{M})} \|\mathcal{S}_c(f)\|_{L_v(\mathcal{N})} \,\Big|\, x \simeq \sum_{k=1}^m e_k \otimes f_k\Big\}.$$

LEMMA 1.9. *The following identities hold*
$$\|x\|_{u \cdot q} = \inf_{x=\alpha y} \|\alpha\|_{L_u(\mathcal{N})} \|y\|_{L_q(\mathcal{M})},$$
$$\|x\|_{q \cdot v} = \inf_{x=y\beta} \|y\|_{L_q(\mathcal{M})} \|\beta\|_{L_v(\mathcal{N})}.$$

PROOF. The upper estimates
$$\|x\|_{u \cdot q} \leq \inf_{x=\alpha y} \|\alpha\|_{L_u(\mathcal{N})} \|y\|_{L_q(\mathcal{M})},$$
$$\|x\|_{q \cdot v} \leq \inf_{x=y\beta} \|y\|_{L_q(\mathcal{M})} \|\beta\|_{L_v(\mathcal{N})},$$

are trivial. We only prove the converse for the first identity since the second identity can be derived in the same way. Let us assume that $\|x\|_{u \cdot q} < 1$. Then we can find $a_1, a_2, \ldots, a_m \in L_u(\mathcal{N})$ and $b_1, b_2, \ldots, b_m \in L_q(\mathcal{M})$ such that
$$x \simeq \sum_{k=1}^{m} a_k \otimes b_k \quad \text{and} \quad \|\mathcal{S}_r(a)\|_{L_u(\mathcal{N})} \|\mathcal{S}_c(b)\|_{L_q(\mathcal{M})} < 1.$$

If D denotes the density associated to φ, we define
$$\widetilde{\mathcal{S}}_r(a) = \Big(\sum_{k=1}^{m} a_k a_k^* + \delta \mathrm{D}^{2/u} \Big)^{1/2} \quad \text{with} \quad \delta > 0.$$

Since $\operatorname{supp} \mathrm{D} = 1$, $\widetilde{\mathcal{S}}_r(a)$ is invertible and we can define α_k by $a_k = \widetilde{\mathcal{S}}_r(a)\alpha_k$ so that
$$\|\mathcal{S}_r(\alpha)\|_{L_\infty(\mathcal{N})} = \Big\| \Big(\sum_{i=1}^{m} \alpha_i \alpha_i^* \Big)^{1/2} \Big\|_{L_\infty(\mathcal{N})} \leq 1.$$

The amalgamation over $\mathcal{N}$ allows us to write
$$\sum_{k=1}^{m} a_k \otimes b_k \simeq \widetilde{\mathcal{S}}_r(a) \otimes \Big(\sum_{k=1}^{m} \alpha_k b_k \Big)$$

for any possible decomposition of x. Then, we have
$$\Big\| \sum_{k=1}^{m} \alpha_k b_k \Big\|_{L_q(\mathcal{M})} \leq \|\mathcal{S}_r(\alpha)\|_{L_\infty(\mathcal{N})} \|\mathcal{S}_c(b)\|_{L_q(\mathcal{M})} \leq \|\mathcal{S}_c(b)\|_{L_q(\mathcal{M})}.$$

Moreover, applying the triangle inequality we obtain
$$\|\widetilde{\mathcal{S}}_r(a)\|_{L_u(\mathcal{N})}^2 = \Big\| \sum_{k=1}^{m} a_k a_k^* + \delta \mathrm{D}^{2/u} \Big\|_{L_{u/2}(\mathcal{N})} \leq \|\mathcal{S}_r(a)\|_{L_u(\mathcal{N})}^2 + \delta$$

In summary, we have
$$\inf_{x=\alpha y} \|\alpha\|_{L_u(\mathcal{N})} \|y\|_{L_q(\mathcal{M})} \leq \sqrt{\|\mathcal{S}_r(a)\|_u^2 + \delta} \, \|\mathcal{S}_c(b)\|_{L_q(\mathcal{M})}.$$

This holds for any decomposition $x \simeq \sum_k a_k \otimes b_k$. We conclude letting $\delta \to 0$. □

LEMMA 1.10. *Let $1 \leq p \leq \infty$ and let $\mathcal{M}$ be a von Neumann algebra. Assume that α, β are elements of $L_p(\mathcal{M})$, β is positive and $\alpha \alpha^* \leq C \beta^2$. Then, there exists $w \in \mathcal{M}$ such that the identity $\alpha = \beta w$ holds.*

PROOF. We first construct w in the crossed product von Neumann algebra $\mathcal{M} \rtimes_\sigma \mathbb{R}$. Indeed, multiplying at both sides of $\alpha\alpha^* \leq C\beta^2$ by $1_{(\lambda,\infty)}(\beta)\beta^{-1}$ for each positive λ, we conclude that $\beta^{-1}\alpha\alpha^*\beta^{-1} \leq C$ and hence $\beta^{-1}\alpha$ is in $\mathcal{M} \rtimes_\sigma \mathbb{R}$. Now, using the dual automorphism group, we have
$$\hat{\sigma}_t(\beta^{-1}\alpha) = \hat{\sigma}_t(\beta^{-1})\hat{\sigma}_t(\alpha) = \hat{\sigma}_t(\beta)^{-1}\hat{\sigma}_t(\alpha) = e^{t/p}\beta^{-1}e^{-t/p}\alpha = \beta^{-1}\alpha,$$
which means that $\beta^{-1}\alpha \in \mathcal{M}$. This completes the proof. $\square$

PROPOSITION 1.11. *Let $2 \leq p < \infty$ and $2 < s \leq \infty$ related by $1/2 = 1/p + 1/s$. Then, we have the following isometries*
$$L_{p'}^r(\mathcal{M}, \mathsf{E}) = L_p^r(\mathcal{M}, \mathsf{E})^* = L_s^r(\mathcal{N}) \otimes_{\mathcal{N},h} L_2^c(\mathcal{M}),$$
$$L_{p'}^c(\mathcal{M}, \mathsf{E}) = L_p^c(\mathcal{M}, \mathsf{E})^* = L_2^r(\mathcal{M}) \otimes_{\mathcal{N},h} L_s^c(\mathcal{N}).$$

PROOF. As we have already said, the isometries
$$L_{p'}^r(\mathcal{M}, \mathsf{E}) = L_p^r(\mathcal{M}, \mathsf{E})^*,$$
$$L_{p'}^c(\mathcal{M}, \mathsf{E}) = L_p^c(\mathcal{M}, \mathsf{E})^*,$$
were proved in [**16**]. Now let us consider an element $x \in L_s^r(\mathcal{N}) \otimes_{\mathcal{N},h} L_2^c(\mathcal{M})$ and let $x = \alpha y$ be a decomposition with $\alpha \in L_s(\mathcal{N})$ and $y \in L_2(\mathcal{M})$. Then, (1.2) and Lemma 1.8 provide the following inequality
$$\|x\|_{L_{p'}^r(\mathcal{M},\mathsf{E})} = \sup\left\{\mathrm{tr}(\alpha^* z y^*) \,\Big|\, \sup_{\|\gamma\|_{L_s(\mathcal{N})} \leq 1} \|\gamma z\|_{L_2(\mathcal{M})} \leq 1\right\} \leq \|\alpha\|_{L_s(\mathcal{N})}\|y\|_{L_2(\mathcal{M})}.$$
According to Lemma 1.9, this proves that
$$id : x \in L_s^r(\mathcal{N}) \otimes_{\mathcal{N},h} L_2^c(\mathcal{M}) \mapsto x \in L_{p'}^r(\mathcal{M}, \mathsf{E})$$
is a contraction. Reciprocally, let us consider an element of the form $x = \xi \mathrm{D}^{1/p'} a$ with $\xi \in \mathcal{N}$ and $a \in \mathcal{M}$. Then, given any $\delta > 0$, we consider the decomposition $x = \alpha_\delta y_\delta$ with α_δ and y_δ given by
$$\alpha_\delta = \left(\xi \mathrm{D}^{\frac{1}{p'}} \mathsf{E}(aa^* + \delta 1) \mathrm{D}^{\frac{1}{p'}} \xi^*\right)^{\gamma/2} \quad \text{and} \quad y_\delta = \alpha_\delta^{-1} x$$
with
$$1 - \gamma = \frac{p'}{2} = \frac{s\gamma}{2}.$$
Let us note that y_δ is a well-defined element of $L_2(\mathcal{M})$. Indeed, if we set
$$A_\delta = \left(\xi \mathrm{D}^{\frac{1}{p'}} \mathsf{E}(aa^* + \delta 1) \mathrm{D}^{\frac{1}{p'}} \xi^*\right)^{1/2},$$
it is clear that
$$xx^* \leq \delta^{-1} \|a\|_{\mathcal{M}}^2 A_\delta^2.$$
By Lemma 1.10, we know that $x = A_\delta w$ for some $w \in \mathcal{M}$. In particular, this gives $y_\delta = A_\delta^{1-\gamma} w$ and since $1 - \gamma = p'/2$ we obtain an element in $L_2(\mathcal{M})$. Let us now estimate the norms of α_δ and y_δ. We have
$$\|\alpha_\delta\|_{L_s(\mathcal{N})} = \left\|\xi \mathrm{D}^{\frac{1}{p'}} \mathsf{E}(aa^* + \delta 1) \mathrm{D}^{\frac{1}{p'}} \xi^*\right\|_{L_{s\gamma/2}(\mathcal{N})}^{\gamma/2} = \left\|\xi \mathrm{D}^{\frac{1}{p'}} \mathsf{E}(aa^* + \delta 1) \mathrm{D}^{\frac{1}{p'}} \xi^*\right\|_{L_{p'/2}(\mathcal{N})}^{\gamma/2}$$
and
$$\|y_\delta\|_{L_2(\mathcal{M})} = \mathrm{tr}\left(xx^*\left[\xi \mathrm{D}^{\frac{1}{p'}} \mathsf{E}(aa^* + \delta 1) \mathrm{D}^{\frac{1}{p'}} \xi^*\right]^{-\gamma}\right)^{1/2}$$
$$= \mathrm{tr}\left(\xi \mathrm{D}^{\frac{1}{p'}} \mathsf{E}(aa^*) \mathrm{D}^{\frac{1}{p'}} \xi^* \left[\xi \mathrm{D}^{\frac{1}{p'}} \mathsf{E}(aa^* + \delta 1) \mathrm{D}^{\frac{1}{p'}} \xi^*\right]^{-\gamma}\right)^{1/2}$$

$$\leq \operatorname{tr}\Big(\xi \mathrm{D}^{\frac{1}{p'}}\mathsf{E}(aa^*+\delta 1)\mathrm{D}^{\frac{1}{p'}}\xi^*\big[\xi \mathrm{D}^{\frac{1}{p'}}\mathsf{E}(aa^*+\delta 1)\mathrm{D}^{\frac{1}{p'}}\xi^*\big]^{-\gamma}\Big)^{1/2}.$$

In particular,
$$\|y_\delta\|_{L_2(\mathcal{M})} \leq \big\|\xi \mathrm{D}^{\frac{1}{p'}}\mathsf{E}(aa^*+\delta 1)\mathrm{D}^{\frac{1}{p'}}\xi^*\big\|_{L_{p'/2}(\mathcal{N})}^{1/2-\gamma/2}.$$

We finally conclude
$$\inf_{x=\alpha y} \|\alpha\|_{L_s(\mathcal{N})}\|y\|_{L_2(\mathcal{M})} \leq \lim_{\delta\to 0} \|\alpha_\delta\|_{L_s(\mathcal{N})}\|y_\delta\|_{L_2(\mathcal{M})} \leq \|x\|_{L^r_{p'}(\mathcal{M},\mathsf{E})}.$$

Note that the case $p=2$ degenerates since we obtain $\gamma = 0$ and $s = \infty$. However, this is a trivial case since it suffices to consider the decomposition given by $\alpha = 1$ and $y = x$. In summary, we have seen that the norms of $L^r_s(\mathcal{N}) \otimes_{\mathcal{N},h} L^c_2(\mathcal{M})$ and of $L^r_{p'}(\mathcal{M},\mathsf{E})$ coincide on
$$\mathcal{A}_{p'} = \mathcal{N}\mathrm{D}^{1/p'}\mathcal{M}.$$
Moreover, $\mathcal{A}_{p'} = \mathcal{N}\mathrm{D}^{1/s}\mathrm{D}^{1/2}\mathcal{M}$ is clearly dense in $L^r_s(\mathcal{N}) \otimes_{\mathcal{N},h} L^c_2(\mathcal{M})$. Therefore, since the density of $\mathcal{A}_{p'}$ in $L^r_{p'}(\mathcal{M},\mathsf{E})$ follows by definition, we obtain the desired isometry. The arguments for the column case are exactly the same. $\square$

REMARK 1.12. Proposition 1.11 provides an *o.s.s.* for $L^r_p(\mathcal{M},\mathsf{E})$ and $L^c_p(\mathcal{M},\mathsf{E})$ when $1 < p \leq 2$. Moreover, by anti-linear duality we also obtain a natural *o.s.s.* for $L^r_p(\mathcal{M},\mathsf{E})$ and $L^c_p(\mathcal{M},\mathsf{E})$ when $2 \leq p < \infty$. However, we shall be interested only on the Banach space structure of these spaces rather than the operator space one. Therefore, we shall use the simpler notation
$$L_u(\mathcal{N})L_q(\mathcal{M}) \quad \text{and} \quad L_q(\mathcal{M})L_v(\mathcal{N})$$
to denote the underlying Banach space of $L^r_u(\mathcal{N})\otimes_{\mathcal{N},h}L^c_q(\mathcal{M})$ or $L^r_q(\mathcal{M})\otimes_{\mathcal{N},h}L^c_v(\mathcal{N})$.

We conclude by giving one more characterization of the norm of $L^r_p(\mathcal{M},\mathsf{E})$ and $L^c_p(\mathcal{M},\mathsf{E})$ for $2 \leq p \leq \infty$. To that aim, we shall denote by $\mathsf{Lm}_\mathcal{N}(\mathrm{X}_1,\mathrm{X}_2)$ and $\mathsf{Rm}_\mathcal{N}(\mathrm{X}_1,\mathrm{X}_2)$ the spaces of left and right $\mathcal{N}$-module maps between X_1 and X_2.

PROPOSITION 1.13. *Let $\mathcal{N}$ be a von Neumann subalgebra of $\mathcal{M}$. Then, given any three indices $2 \leq u,q,v < \infty$, we have the following isometric isomorphisms*
$$\begin{aligned}\big(L_u(\mathcal{N})L_q(\mathcal{M})\big)^* &= \mathsf{Rm}_\mathcal{N}(L_u(\mathcal{N}), L_{q'}(\mathcal{M})) = \mathsf{Lm}_\mathcal{N}(L_q(\mathcal{M}), L_{u'}(\mathcal{N})),\\ \big(L_q(\mathcal{M})L_v(\mathcal{N})\big)^* &= \mathsf{Lm}_\mathcal{N}(L_v(\mathcal{N}), L_{q'}(\mathcal{M})) = \mathsf{Rm}_\mathcal{N}(L_q(\mathcal{M}), L_{v'}(\mathcal{N})).\end{aligned}$$

PROOF. The arguments we shall be using hold for both isometries. Hence, we only prove the first one. Let us consider a linear functional $\Phi: L_u(\mathcal{N})L_q(\mathcal{M}) \to \mathbb{C}$ and let Ψ denote the associated bilinear map
$$\Psi: L_u(\mathcal{N}) \times L_q(\mathcal{M}) \to \mathbb{C},$$
defined by $\Psi(\alpha,x) = \Phi(\alpha \otimes x)$. Clearly, we have
$$\|\Psi\| = \sup\Big\{\big|\Phi(\alpha\otimes x)\big|\,\big|\,\|\alpha\|_{L_u(\mathcal{N})},\|x\|_{L_q(\mathcal{M})} \leq 1\Big\} = \|\Phi\|.$$
On the other hand, we use the isometry
$$\mathcal{B}(L_u(\mathcal{N}) \times L_q(\mathcal{M}),\mathbb{C}) = \mathcal{B}(L_u(\mathcal{N}), L_{q'}(\mathcal{M}))$$
defined by $\Psi(\alpha,x) = \operatorname{tr}(\mathrm{T}(\alpha)x)$. By the amalgamation over $\mathcal{N}$, any such linear map T is a right $\mathcal{N}$-module map (i.e. $\mathrm{T}(\alpha\beta) = \mathrm{T}(\alpha)\beta$ for all $\beta \in \mathcal{N}$). Indeed, given $\beta \in \mathcal{N}$, we have
$$\operatorname{tr}\big(\mathrm{T}(\alpha\beta)x\big) = \Phi(\alpha\beta \otimes x) = \Phi(\alpha \otimes \beta x) = \operatorname{tr}\big(\mathrm{T}(\alpha)\beta x\big)$$

for all $x \in L_q(\mathcal{M})$, which implies our claim. Reciprocally, if $T : L_u(\mathcal{N}) \to L_{q'}(\mathcal{M})$ is a right $\mathcal{N}$-module map, we know by the same argument that T arises from a bilinear map Ψ satisfying $\Psi(\alpha\beta, x) = \Psi(\alpha, \beta x)$, which corresponds to a linear functional Φ on $L_u(\mathcal{N})L_q(\mathcal{M})$ with the same norm. This proves the first identity. For the second we proceed in a similar way by using the adjoint map T* instead of T. □

REMARK 1.14. Note that the proof given above still works when we only require one of the two indices to be finite. In particular, we cover all combinations appearing for $L_p^r(\mathcal{M}, \mathsf{E})$ and $L_p^c(\mathcal{M}, \mathsf{E})$ since in that case the $\mathcal{M}$-index is always 2.

CHAPTER 2

Amalgamated L_p spaces

Let $\mathcal{M}$ be a von Neumann algebra equipped with a *n.f.* state φ and let us consider a von Neumann subalgebra $\mathcal{N}$ of $\mathcal{M}$. In what follows we shall work with indices (u, q, v) satisfying the following property

(2.1) $\quad 1 \leq q \leq \infty \quad \text{and} \quad 2 \leq u, v \leq \infty \quad \text{and} \quad \dfrac{1}{u} + \dfrac{1}{q} + \dfrac{1}{v} = \dfrac{1}{p} \leq 1.$

Let us define $\mathcal{N}_u L_q(\mathcal{M}) \mathcal{N}_v$ to be the space $L_q(\mathcal{M})$ equipped with

$$|||x|||_{u \cdot q \cdot v} = \inf_{x = \alpha y \beta} \left\{ \left\| \mathrm{D}^{\frac{1}{u}} \alpha \right\|_{L_u(\mathcal{N})} \|y\|_{L_q(\mathcal{M})} \left\| \beta \mathrm{D}^{\frac{1}{v}} \right\|_{L_v(\mathcal{N})} \,\bigg|\, \alpha, \beta \in \mathcal{N}, y \in L_q(\mathcal{M}) \right\}.$$

In the sequel it will be quite useful to have a geometric representation of the indices (u, q, v) satisfying property (2.1). To that aim, we consider the variables $(1/u, 1/v, 1/q)$ in the Euclidean space $\mathbb{R}^3$. Then, we note that the points satisfying (2.1) are given by the intersection of the simplex $0 \leq 1/u + 1/v + 1/q \leq 1$ with the prism $0 \leq \min(1/u, 1/v) \leq \max(1/u, 1/v) \leq 1/2$. This gives rise to a solid K sketched below.

Now, let (u, q, v) be any three indices satisfying property (2.1). Then we define the *amalgamated L_p space* $L_u(\mathcal{N}) L_q(\mathcal{M}) L_v(\mathcal{N})$ as the set of elements $x \in L_p(\mathcal{M})$, with $1/p = 1/u + 1/q + 1/v$, such that there exists a factorization $x = ayb$ with $a \in L_u(\mathcal{N})$, $y \in L_q(\mathcal{M})$ and $b \in L_v(\mathcal{N})$. For any such element x, we define

$$\|x\|_{u \cdot q \cdot v} = \inf \left\{ \|a\|_{L_u(\mathcal{N})} \|y\|_{L_q(\mathcal{M})} \|b\|_{L_v(\mathcal{N})} \,\bigg|\, x = ayb \right\}.$$

Chapters 2, 3 and 4 are devoted to prove that $L_u(\mathcal{N}) L_q(\mathcal{M}) L_v(\mathcal{N})$ is a Banach space for any (u, q, v) satisfying (2.1) and to study the complex interpolation and duality of such spaces. In the process, it will be quite relevant to know that the spaces

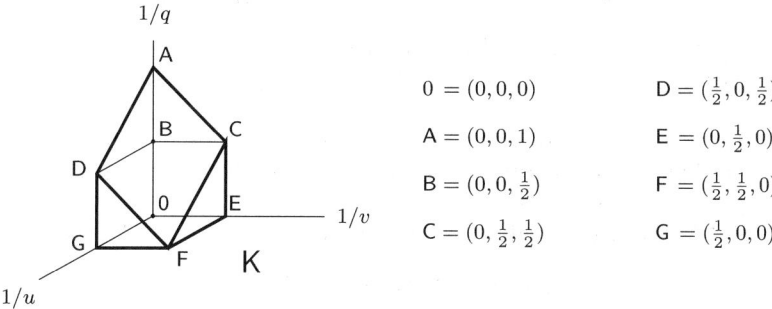

FIGURE I: THE SOLID K.

$\mathcal{N}_u L_q(\mathcal{M})\mathcal{N}_v$ embed isometrically in $L_u(\mathcal{N})L_q(\mathcal{M})L_v(\mathcal{N})$ as dense subspaces. This will be essentially the aim of this chapter.

REMARK 2.1. In order to justify the restriction $2 \leq u, v \leq \infty$ on (2.1), let us show how the triangle inequality might fail when this restriction is not satisfied. Let us take $\mathcal{M} = \ell_\infty(\mathcal{N})$, then we claim that

$$L_w(\mathcal{N})L_\infty(\mathcal{M})L_\infty(\mathcal{N}) \quad \text{and} \quad L_\infty(\mathcal{N})L_\infty(\mathcal{M})L_w(\mathcal{N})$$

are not normed for $1 \leq w < 2$. Indeed, given a sequence $x = (x_n)_{n \geq 1}$ in $L_\infty(\mathcal{M})$ and elements $a, b \in L_w(\mathcal{N})$, we have by definition the following inequalities

$$\|ax\|_{w \cdot \infty \cdot \infty} \leq \|a\|_{L_w(\mathcal{N})} \|x\|_{L_\infty(\mathcal{M})} \quad \text{and} \quad \|xb\|_{\infty \cdot \infty \cdot w} \leq \|x\|_{L_\infty(\mathcal{M})} \|b\|_{L_w(\mathcal{N})}.$$

In particular, if these spaces were normed they should contain contractively the projective tensor product $L_w(\mathcal{N}) \otimes_\pi L_\infty(\mathcal{M})$. However, by the noncommutative Menchoff-Rademacher inequality in [**7**], we may find $(z_n) \in L_w(\mathcal{N}) \otimes_\pi L_\infty(\mathcal{M})$ of the form

$$z_n = \sum_{k=1}^n \varepsilon_k \frac{x_n}{1 + \log k} \quad \text{with} \quad x_k \in L_w(\mathcal{N})$$

and such that $(z_n) \notin L_w(\mathcal{N})L_\infty(\mathcal{M})L_\infty(\mathcal{N})$. Moreover, taking adjoints we also may find a sequence (z_n) such that $(z_n) \in L_w(\mathcal{N}) \otimes_\pi L_\infty(\mathcal{M}) \setminus L_\infty(\mathcal{N})L_\infty(\mathcal{M})L_w(\mathcal{N})$.

EXAMPLE 2.2. As noted in the Introduction, several noncommutative function spaces arise as particular cases of our notion of amalgamated L_p space. Let us mention four particularly relevant examples:

(a) The noncommutative L_p spaces arise as

$$L_q(\mathcal{M}) = L_\infty(\mathcal{N})L_q(\mathcal{M})L_\infty(\mathcal{N}).$$

Note that these spaces are represented by the segment 0A in Figure I.

(b) If $p \leq q$ and $1/r = 1/p - 1/q$, the spaces $L_p(\mathcal{N}_1; L_q(\mathcal{N}_2))$ arise as

$$L_{2r}(\mathcal{N}_1)L_q(\mathcal{N}_1 \bar{\otimes} \mathcal{N}_2)L_{2r}(\mathcal{N}_1).$$

When the index p is fixed, these spaces are represented in Figure I as segments parallel to the upper face ACDF whose projection into the plane xy goes in the direction of the diagonal.

(c) By Proposition 1.11, $L_p^r(\mathcal{M}, \mathsf{E})$ and $L_p^c(\mathcal{M}, \mathsf{E})$ ($1 \leq p \leq 2$) arise as

$$L_p^r(\mathcal{M}, \mathsf{E}) = L_s(\mathcal{N})L_2(\mathcal{M})L_\infty(\mathcal{N}) \quad \text{with} \quad 1/p = 1/2 + 1/s,$$
$$L_p^c(\mathcal{M}, \mathsf{E}) = L_\infty(\mathcal{N})L_2(\mathcal{M})L_s(\mathcal{N}) \quad \text{with} \quad 1/p = 1/2 + 1/s.$$

These spaces are represented by the segments BD and BC in Figure I. As we shall see in Chapter 4, the spaces $L_p(\mathcal{M}; R_p^n)$ and $L_p(\mathcal{M}; C_p^n)$ arise as particular cases of the latter ones.

(d) If $\mathcal{M} = \mathcal{N}$, the so-called asymmetric noncommutative L_p spaces arise as

$$L_{(u,v)}(\mathcal{M}) = L_u(\mathcal{M})L_\infty(\mathcal{M})L_v(\mathcal{M}).$$

These spaces are represented by the square 0EFG in Figure I. The reader is refereed to [**22**] and Chapter 7 below for a detailed exposition of the main properties of these spaces and their vector-valued analogues.

2.1. Haagerup's construction

We now briefly sketch a well-known unpublished result due to Haagerup [12] which will be essential for our further purposes. Let us consider a σ-finite von Neumann algebra $\mathcal{M}$ equipped with a distinguished n.f. state φ and let us define the discrete multiplicative group

$$G = \bigcup_{n \in \mathbb{N}} 2^{-n}\mathbb{Z}.$$

Then we can construct the crossed product $\mathcal{R} = \mathcal{M} \rtimes_\sigma G$ in the same way we constructed the crossed product $\mathcal{M} \rtimes_\sigma \mathbb{R}$. That is, if $\mathcal{H}$ is the Hilbert space provided by the GNS construction applied to φ, $\mathcal{R}$ is generated by $\pi : \mathcal{M} \to \mathcal{B}(L_2(G; \mathcal{H}))$ and $\lambda : G \to \mathcal{B}(L_2(G; \mathcal{H}))$ where

$$\big(\pi(x)\xi\big)(g) = \sigma_{-g}(x)\xi(g) \quad \text{and} \quad \big(\lambda(h)\xi\big)(g) = \xi(g - h).$$

By the faithfulness of π we are allowed to identify $\mathcal{M}$ with $\pi(\mathcal{M})$. Then, a generic element in $\mathcal{R}$ has the form $\sum_g x_g \lambda(g)$ with $x_g \in \mathcal{M}$ and we have the conditional expectation

$$\mathsf{E}_\mathcal{M} : \sum_{g \in G} x_g \lambda(g) \in \mathcal{R} \mapsto x_0 \in \mathcal{M}.$$

This gives rise to the state

$$\widehat{\varphi} : \sum_{g \in G} x_g \lambda(g) \in \mathcal{R} \mapsto \varphi \circ \mathsf{E}_\mathcal{M}\Big(\sum_{g \in G} x_g \lambda(g)\Big) = \varphi(x_0) \in \mathbb{C}.$$

According to [12], $\mathcal{R}$ is the closure of a union of finite von Neumann algebras

$$\bigcup_{k \geq 1} \mathcal{M}_k$$

where $(\mathcal{M}_k)_{k \geq 1}$ is directed by inclusion $\mathcal{M}_1 \subset \mathcal{M}_2 \subset \ldots$ and each $\mathcal{M}_k$ satisfies

(2.2) $$c_1(k) 1_{\mathcal{M}_k} \leq D_{\varphi_k} \leq c_2(k) 1_{\mathcal{M}_k}$$

for some constants $0 < c_1(k) \leq c_2(k) < \infty$. Here φ_k denotes the restriction of $\widehat{\varphi}$ to $\mathcal{M}_k$ and D_{φ_k} stands for the corresponding density. Moreover, it also follows from [12] that we can find for any integer $k \geq 1$ conditional expectations

$$\mathcal{E}_k : \mathcal{R} \to \mathcal{M}_k$$

such that the following limit holds for every $\hat{x} \in \mathcal{R}$, any $1 \leq p < \infty$ and all $0 \leq \eta \leq 1$

(2.3) $$\Big\| D_{\widehat{\varphi}}^{(1-\eta)/p} \big(\hat{x} - \mathcal{E}_k(\hat{x})\big) D_{\widehat{\varphi}}^{\eta/p} \Big\|_{L_p(\mathcal{R})} \longrightarrow 0 \quad \text{as} \quad k \to \infty.$$

REMARK 2.3. The σ-finiteness assumption might be dropped if we replaced sequences of finite von Neumann algebras by nets. However, it suffices for our aims to consider the σ-finite case.

In the following result we provide the first application of this construction. As usual, we write $\mathcal{S}$ for the strip of complex numbers $z \in \mathbb{C}$ with $0 < \mathrm{Re}(z) < 1$ where we consider the decomposition $\partial \mathcal{S} = \partial_0 \cup \partial_1$ of its boundary into the sets

$$\partial_0 = \big\{z \in \mathbb{C} \,\big|\, \mathrm{Re}(z) = 0\big\} \quad \text{and} \quad \partial_1 = \big\{z \in \mathbb{C} \,\big|\, \mathrm{Re}(z) = 1\big\}.$$

Let X be a Banach space and let x be an element of X. Then, given $0 < \theta < 1$, we write $\mathcal{A}(X, \theta, x)$ for the set of bounded analytic functions $f : \mathcal{S} \to X$ (i.e. bounded and continuous on $\mathcal{S} \cup \partial \mathcal{S}$ and analytic on $\mathcal{S}$) which satisfy $f(\theta) = x$.

LEMMA 2.4. *Let x be an element in a von Neumann algebra $\mathcal{M}$. Then, given $0 \leq \eta \leq 1$ and $1/p_\theta = (1-\theta)/p$ for some $0 < \theta < 1$ and some $1 \leq p < \infty$, we have*

$$\left\|\mathrm{D}^{\frac{1-\eta}{p_\theta}} x \mathrm{D}^{\frac{\eta}{p_\theta}}\right\|_{L_{p_\theta}(\mathcal{M})} = \inf \left\{ \max \left\{ \sup_{z \in \partial_0} \left\|\mathrm{D}^{\frac{1-\eta}{p}} f(z) \mathrm{D}^{\frac{\eta}{p}}\right\|_{L_p(\mathcal{M})}, \sup_{z \in \partial_1} \|f(z)\|_{L_\infty(\mathcal{M})} \right\} \right\},$$

where the infimum runs over all analytic functions $f : \mathcal{S} \to \mathcal{M}$ in the set $\mathcal{A}(\mathcal{M}, \theta, x)$.

PROOF. The upper estimate follows by Kosaki's interpolation. To prove the lower estimate, we assume by homogeneity that the norm on the left is 1. Then we begin with the case where $\mathcal{M}$ is a finite von Neumann algebra equipped with a n.f. state φ_0 such that the density D_{φ_0} satisfies (2.2). That is,

$$c_1 1_\mathcal{M} \leq \mathrm{D}_{\varphi_0} \leq c_2 1_\mathcal{M}$$

for some positive numbers $0 < c_1 \leq c_2 < \infty$. Note that our assumption implies $\mathrm{D}_{\varphi_0} \in \mathcal{M}$. In particular, the function $z \to \mathrm{D}_{\varphi_0}^{\lambda z}$ is analytic for any scalar $\lambda \in \mathbb{C}$. Then it is easy to find an optimal function. Indeed, since

$$x(\eta, \theta) = \mathrm{D}_{\varphi_0}^{(1-\eta)/p_\theta} x \, \mathrm{D}_{\varphi_0}^{\eta/p_\theta} \in \mathcal{M}$$

we find by polar decomposition a partial isometry ω such that $x(\eta, \theta) = \omega |x(\eta, \theta)|$. Our optimal function is defined as

$$f(z) = \mathrm{D}_{\varphi_0}^{-\frac{(1-\eta)(1-z)}{p}} \omega \, |x(\eta, \theta)|^{\frac{p_\theta(1-z)}{p}} \mathrm{D}_{\varphi_0}^{-\frac{\eta(1-z)}{p}}.$$

Our assumption on D_{φ_0} implies the boundedness and analyticity of f. On the other hand, it is easy to check that $f(\theta) = x$ so that $f \in \mathcal{A}(\mathcal{M}, \theta, x)$. Then, recalling that both $|x(\eta, \theta)|^{i\lambda}$ and $\mathrm{D}_{\varphi_0}^{i\lambda}$ are unitaries for any $\lambda \in \mathbb{R}$, we have

$$\sup_{z \in \partial_0} \left\|\mathrm{D}_{\varphi_0}^{\frac{1-\eta}{p}} f(z) \mathrm{D}_{\varphi_0}^{\frac{\eta}{p}}\right\|_{L_p(\mathcal{M})} = \sup_{z \in \partial_0} \left\|\mathrm{D}_{\varphi_0}^{\frac{(1-\eta)z}{p}} \omega |x(\eta, \theta)|^{\frac{p_\theta}{p}} |x(\eta,\theta)|^{-\frac{p_\theta z}{p}} \mathrm{D}_{\varphi_0}^{\frac{\eta z}{p}}\right\|_{L_p(\mathcal{M})}$$

$$\leq \left\||x(\eta,\theta)|^{\frac{p_\theta}{p}}\right\|_{L_p(\mathcal{M})} = \left\|\mathrm{D}_{\varphi_0}^{\frac{1-\eta}{p_\theta}} x \mathrm{D}_{\varphi_0}^{\frac{\eta}{p_\theta}}\right\|_{L_{p_\theta}(\mathcal{M})}^{\frac{1}{1-\theta}} = 1.$$

Similarly, we have on ∂_1

$$\sup_{z \in \partial_1} \|f(z)\|_{L_\infty(\mathcal{M})} = \sup_{z \in \partial_1} \left\|\mathrm{D}_{\varphi_0}^{\frac{i(1-\eta)\mathrm{Im}\,z}{p}} \omega |x(\eta,\theta)|^{-\frac{ip_\theta \mathrm{Im}\,z}{p}} \mathrm{D}_{\varphi_0}^{\frac{i\eta\mathrm{Im}\,z}{p}}\right\|_{L_\infty(\mathcal{M})} \leq 1.$$

This completes the proof for finite von Neumann algebras satisfying (2.2). In the case of a general von Neumann algebra $\mathcal{M}$ we use the Haagerup construction sketched above. Let us introduce the shorter notation

$$\mathsf{N}_1(\mathcal{M}, \varphi, x) = \left\|\mathrm{D}^{\frac{1-\eta}{p_\theta}} x \mathrm{D}^{\frac{\eta}{p_\theta}}\right\|_{L_{p_\theta}(\mathcal{M})},$$

$$\mathsf{N}_2(\mathcal{M}, \varphi, x) = \inf \left\{ \max \left\{ \sup_{z \in \partial_0} \left\|\mathrm{D}^{\frac{1-\eta}{p}} f(z) \mathrm{D}^{\frac{\eta}{p}}\right\|_{L_p(\mathcal{M})}, \sup_{z \in \partial_1} \|f(z)\|_{L_\infty(\mathcal{M})} \right\} \right\}.$$

We are interested in proving $\mathsf{N}_1(\mathcal{M}, \varphi, x) \geq \mathsf{N}_2(\mathcal{M}, \varphi, x)$. Combining property (2.2) with the first part of this proof, we deduce that any $x \in \mathcal{M}$ satisfies the following identities

(2.4) $\quad \mathsf{N}_1(\mathcal{M}_k, \varphi_k, \mathcal{E}_k(x)) = \mathsf{N}_2(\mathcal{M}_k, \varphi_k, \mathcal{E}_k(x)) \quad$ for all $\quad k \geq 1$.

By the triangle inequality

$$\mathsf{N}_2(\mathcal{M}, \varphi, x) \leq \limsup_{k \to \infty} \mathsf{N}_2(\mathcal{M}, \varphi, \mathsf{E}_\mathcal{M}(\mathcal{E}_k(x))) + \mathsf{N}_2(\mathcal{M}, \varphi, x - \mathsf{E}_\mathcal{M}(\mathcal{E}_k(x))).$$

Applying the contractivity of $\mathsf{E}_\mathcal{M}$, (2.3) and (2.4)

$$\begin{aligned}
\limsup_{k\to\infty} \mathsf{N}_2(\mathcal{M},\varphi,\mathsf{E}_\mathcal{M}(\mathcal{E}_k(x))) &\leq \limsup_{k\to\infty} \mathsf{N}_2(\mathcal{M}_k,\varphi_k,\mathcal{E}_k(x)) \\
&= \limsup_{k\to\infty} \mathsf{N}_1(\mathcal{M}_k,\varphi_k,\mathcal{E}_k(x)) \\
&= \mathsf{N}_1(\mathcal{M},\varphi,x)
\end{aligned}$$

It remains to see that $\mathsf{N}_2(\mathcal{M},\varphi,x - \mathsf{E}_\mathcal{M}(\mathcal{E}_k(x)))$ is arbitrary small. First we note that

$$\mathsf{N}_2(\mathcal{M},\varphi,x - \mathsf{E}_\mathcal{M}(\mathcal{E}_k(x))) = \mathsf{N}_2(\mathcal{M},\varphi,\mathsf{E}_\mathcal{M}(x - \mathcal{E}_k(x))) \leq \mathsf{N}_2(\mathcal{R},\widehat{\varphi},x - \mathcal{E}_k(x)).$$

Then we consider the bounded analytic functions

$$f_k : z \in \mathcal{S} \mapsto \mathsf{m}_{0k}^{1-z} \mathsf{m}_{1k}^{z} (x - \mathcal{E}_k(x)) \in \mathcal{R}$$

with the constants $\mathsf{m}_{0k}, \mathsf{m}_{1k}$ defined by

$$\begin{aligned}
\mathsf{m}_{0k} &= \left\| \mathrm{D}_{\widehat{\varphi}}^{(1-\eta)/p} (x - \mathcal{E}_k(x)) \mathrm{D}_{\widehat{\varphi}}^{\eta/p} \right\|_{L_p(\mathcal{R})}^{-1/2}, \\
\mathsf{m}_{1k} &= \left\| \mathrm{D}_{\widehat{\varphi}}^{(1-\eta)/p} (x - \mathcal{E}_k(x)) \mathrm{D}_{\widehat{\varphi}}^{\eta/p} \right\|_{L_p(\mathcal{R})}^{(1-\theta)/2\theta}.
\end{aligned}$$

Note that $f_k \in \mathcal{A}(\mathcal{R},\theta,x - \mathcal{E}_k(x))$ since $\mathsf{m}_{0k}^{1-\theta} \mathsf{m}_{1k}^\theta = 1$. On the other hand,

$$\sup_{z \in \partial_0} \left\| \mathrm{D}_{\widehat{\varphi}}^{(1-\eta)/p} f_k(z) \mathrm{D}_{\widehat{\varphi}}^{\eta/p} \right\|_{L_p(\mathcal{R})} = \mathsf{m}_{0k} \left\| \mathrm{D}_{\widehat{\varphi}}^{(1-\eta)/p} (x - \mathcal{E}_k(x)) \mathrm{D}_{\widehat{\varphi}}^{\eta/p} \right\|_{L_p(\mathcal{R})}.$$

Similarly, we have on ∂_1

$$\sup_{z \in \partial_1} \|f_k(z)\|_{L_\infty(\mathcal{R})} = \mathsf{m}_{1k} \|x - \mathcal{E}_k(x)\|_{L_\infty(\mathcal{R})} \leq 2\mathsf{m}_{1k} \|x\|_\mathcal{M}.$$

Therefore, the proof is completed since from (2.3) both terms tend to 0 with k. □

2.2. Triangle inequality on $\partial_\infty \mathsf{K}$

Let $\partial_\infty \mathsf{K}$ be the subset of the boundary of K given by the intersection of K with the coordinate planes (i.e. the union of the plane regions 0ACE, 0ADG and 0EFG). This set will appear repeatedly in what follows. Note that the indices (u,q,v) which are represented in Figure I by the points of $\partial_\infty \mathsf{K}$ are those satisfying (2.1) and

(2.5) $$\min\left\{1/u, 1/q, 1/v\right\} = 0.$$

In the following result, we apply a complex interpolation trick based on the operator-valued version of Szegö's classical factorization theorem. This result is due to Devinatz [8]. A precise statement of Devinatz's theorem adapted to our aims can be found in Pisier's paper [44], where he also applies it in the context of complex interpolation. We note in passing that a more general result (combining previous results due to Devinatz, Helson & Lowdenslager, Sarason and Wiener & Masani) can be found in the survey [52], see Corollary 8.2. This interpolation technique will be used frequently in the sequel.

LEMMA 2.5. *If* $(1/u, 1/v, 1/q) \in \partial_\infty \mathsf{K}$, $\mathcal{N}_u L_q(\mathcal{M}) \mathcal{N}_v$ *is a normed space.*

PROOF. Assuming (2.1), Hölder inequality gives for $1/p = 1/u + 1/q + 1/v$

$$\left\|D^{\frac{1}{u}} x D^{\frac{1}{v}}\right\|_{L_p(\mathcal{M})} \leq |||x|||_{u \cdot q \cdot v} \qquad \text{for all} \qquad x \in \mathcal{N}_u L_q(\mathcal{M}) \mathcal{N}_v.$$

Therefore, $|||x|||_{u \cdot q \cdot v} = 0$ implies $x = 0$. Since the homogeneity over $\mathbb{R}_+$ is clear, it remains to show that the triangle inequality holds in this case. Our assumption $(1/u, 1/v, 1/q) \in \partial_\infty \mathsf{K}$ reduces the possibilities to those in which at least one of the indices is infinite. When $1/q = 0$, Pisier's factorization argument (*c.f.* Lemma 3.5 in [**46**]) suffices to obtain the triangle inequality. Indeed, it can be checked that this factorization provides the estimate

$$|||x_1 + x_2|||_{u \cdot q \cdot v} \leq 2^{1/q} \big(|||x_1|||_{u \cdot q \cdot v} + |||x_2|||_{u \cdot q \cdot v} \big).$$

Hence, it remains to consider the cases $1/u = 0$ and $1/v = 0$. Both can be treated with the same arguments so that we only show the triangle inequality for $1/u = 0$. In that case, the left term $\mathcal{N}_\infty$ is irrelevant in $\mathcal{N}_\infty L_q(\mathcal{M}) \mathcal{N}_v$ so that we shall ignore it in what follows. We need to consider two different cases.

CASE I. We first assume that $1 \leq q \leq 2$. Let $x_1, x_2, \ldots, x_m$ be a finite sequence of vectors in $L_q(\mathcal{M}) \mathcal{N}_v$ satisfying $|||x_k|||_{q \cdot v} < 1$ and let us also consider a finite sequence of positive numbers $\lambda_1, \lambda_2, \ldots, \lambda_m$ with $\sum_k \lambda_k = 1$. Then, it clearly suffices to see that

$$\left\| \left\| \sum_{k=1}^m \lambda_k x_k \right\| \right\|_{q \cdot v} \leq 1.$$

By hypothesis, we may assume that $x_k = y_k \beta_k$, with

$$\max \left\{ \|y_k\|_{L_q(\mathcal{M})}, \left\|\beta_k D^{\frac{1}{v}}\right\|_{L_v(\mathcal{N})} \right\} < 1.$$

On the other hand, since $1 \leq q \leq 2$ we can consider $2 \leq q_1 \leq \infty$ defined by

$$\frac{1}{q_1} + \frac{1}{2} = \frac{1}{p} = \frac{1}{q} + \frac{1}{v}.$$

It is not difficult to check that

$$\begin{aligned} L_q(\mathcal{M}) &= \big[L_p(\mathcal{M}), L_{q_1}(\mathcal{M})\big]_{2/v}, \\ L_v(\mathcal{N}) &= \big[L_\infty(\mathcal{N}), L_2(\mathcal{N})\big]_{2/v}. \end{aligned}$$

Then, by the complex interpolation method we can find bounded analytic functions

$$f_{1k} : \mathcal{S} \to L_p(\mathcal{M}) + L_{q_1}(\mathcal{M})$$

satisfying $f_{1k}(2/v) = y_k$ for all $k = 1, 2, \ldots, m$ and

(2.6)
$$\begin{aligned} \sup_{z \in \partial_0} \|f_{1k}(z)\|_{L_p(\mathcal{M})} &< 1, \\ \sup_{z \in \partial_1} \|f_{1k}(z)\|_{L_{q_1}(\mathcal{M})} &< 1. \end{aligned}$$

Similarly, Lemma 2.4 provides us with bounded analytic functions

$$f_{2k} : \mathcal{S} \to \mathcal{N}$$

satisfying $f_{2k}(2/v) = \beta_k$ for all $k = 1, 2, \ldots, m$ and

$$\begin{aligned} \sup_{z \in \partial_0} \|f_{2k}(z)\|_{L_\infty(\mathcal{N})} &< 1, \\ \sup_{z \in \partial_1} \left\|f_{2k}(z) D^{\frac{1}{2}}\right\|_{L_2(\mathcal{N})} &< 1. \end{aligned}$$

2.2. TRIANGLE INEQUALITY ON $\partial_\infty \mathsf{K}$

Given $\delta > 0$, we define the following function on $\partial \mathcal{S}$

$$W(z) = \begin{cases} 1 & \text{if } z \in \partial_0, \\ \delta 1 + \sum_k \lambda_k f_{2k}(z)^* f_{2k}(z) & \text{if } z \in \partial_1. \end{cases}$$

According to Devinatz's factorization theorem [44], there exists a bounded analytic function $w : \mathcal{S} \to \mathcal{N}$ with bounded analytic inverse and satisfying the following identity on $\partial \mathcal{S}$

$$w(z)^* w(z) = W(z).$$

Let us consider the bounded analytic function

$$g(z) = \Big(\sum_{k=1}^m \lambda_k \, f_{1k}(z) f_{2k}(z)\Big) w^{-1}(z).$$

Then, since $w(z)^* w(z) \equiv 1$ on ∂_0, we have

$$\sup_{z \in \partial_0} \|g(z)\|_{L_p(\mathcal{M})} \leq \sup_{z \in \partial_0} \sum_{k=1}^m \lambda_k \|f_{1k}(z)\|_{L_p(\mathcal{M})} \|f_{2k}(z) w^{-1}(z)\|_{L_\infty(\mathcal{N})} < 1.$$

On the other hand, we can write

$$g(z) = \sum_{k=1}^m \sqrt{\lambda_k} \, f_{1k}(z) \gamma_k(z) \quad \text{with} \quad \sqrt{\lambda_k} f_{2k}(z) = \gamma_k(z) w(z).$$

Note that, according to the definition of w, we have for any $z \in \partial_1$

$$\sum_{k=1}^m \gamma_k(z)^* \gamma_k(z) \leq 1. \tag{2.7}$$

Therefore, the following estimate follows from (2.6) and (2.7)

$$\begin{aligned}
\sup_{z \in \partial_1} \|g(z)\|_{L_{q_1}(\mathcal{M})} &\leq \sup_{z \in \partial_1} \Big\|\Big(\sum_{k=1}^m \lambda_k f_{1k}(z) f_{1k}(z)^*\Big)^{1/2}\Big\|_{L_{q_1}(\mathcal{M})} \\
&\quad \times \sup_{z \in \partial_1} \Big\|\Big(\sum_{k=1}^m \gamma_k(z)^* \gamma_k(z)\Big)^{1/2}\Big\|_{L_\infty(\mathcal{N})} \\
&\leq \sup_{z \in \partial_1} \Big(\sum_{k=1}^m \lambda_k \|f_{1k}(z)\|_{L_{q_1}(\mathcal{M})}^2\Big)^{1/2} < 1.
\end{aligned}$$

Note that the last estimate uses the triangle inequality on $L_{q_1/2}(\mathcal{M})$ and that we are allowed to do so since $q_1 \geq 2$. Combining the estimates obtained so far and applying Kosaki's interpolation we obtain

$$\|g(2/v)\|_{L_q(\mathcal{M})} < 1.$$

On the other hand, we have

$$\begin{aligned}
\sup_{z \in \partial_0} \|w(z)\|_{L_\infty(\mathcal{N})} &= \sup_{z \in \partial_0} \|w(z)^* w(z)\|_{L_\infty(\mathcal{N})}^{1/2} = 1, \\
\sup_{z \in \partial_1} \|w(z) D^{\frac{1}{2}}\|_{L_2(\mathcal{N})}^2 &= \sup_{z \in \partial_1} \|D^{\frac{1}{2}} w(z)^* w(z) D^{\frac{1}{2}}\|_{L_1(\mathcal{N})} \\
&\leq \sup_{z \in \partial_1} \sum_{k=1}^m \lambda_k \|f_{2k}(z) D^{\frac{1}{2}}\|_{L_2(\mathcal{N})}^2 + \delta < 1 + \delta.
\end{aligned}$$

Again, Kosaki's interpolation provides the estimate
$$\|w(2/v)D^{\frac{1}{v}}\|_{L_v(\mathcal{N})} \leq 1+\delta.$$
In summary, recalling that
$$\sum_{k=1}^{m} \lambda_k x_k = \sum_{k=1}^{m} \lambda_k f_{1k}(2/v) f_{2k}(2/v) = g(2/v) w(2/v),$$
we obtain from our previous estimates that
$$\Big\|\Big|\Big\|\sum_{k=1}^{m} \lambda_k x_k \Big\|\Big|\Big\|_{q\cdot v} < 1+\delta.$$
Thus, the triangle inequality follows by letting $\delta \to 0$ in the expression above.

CASE II. It remains to consider the case $2 \leq q \leq \infty$. This case is simpler. Indeed, given any family of vectors $x_1, x_2, \ldots, x_m$ and scalars $\lambda_1, \lambda_2, \ldots, \lambda_m$ as above (with $x_k = y_k \beta_k$ and y_k, β_k satisfying the same inequalities), we define for $\delta > 0$
$$\mathcal{S}_\beta = \Big(\sum_{k=1}^{m} \lambda_k \beta_k^* \beta_k + \delta 1\Big)^{1/2} \qquad \text{so that} \qquad \beta_k = b_k \mathcal{S}_\beta,$$
for some $b_1, b_2, \ldots, b_m \in \mathcal{N}$. Then we have a factorization
$$\sum_{k=1}^{m} \lambda_k x_k = \Big(\sum_{k=1}^{m} \lambda_k y_k b_k\Big) \mathcal{S}_\beta.$$
Now, since $2 \leq q \leq \infty$, we have triangle inequality in $L_{q/2}(\mathcal{M})$ and
$$\Big\|\sum_{k=1}^{m} \lambda_k y_k b_k\Big\|_{L_q(\mathcal{M})} \leq \Big\|\Big(\sum_{k=1}^{m} \lambda_k y_k y_k^*\Big)^{1/2}\Big\|_{L_q(\mathcal{M})} \Big\|\Big(\sum_{k=1}^{m} \lambda_k b_k^* b_k\Big)^{1/2}\Big\|_{L_\infty(\mathcal{N})}$$
$$\leq \Big(\sum_{k=1}^{m} \lambda_k \|y_k\|^2_{L_q(\mathcal{M})}\Big)^{1/2} \Big\|\mathcal{S}_\beta^{-1}\Big(\sum_{k=1}^{m} \lambda_k \beta_k^* \beta_k\Big)\mathcal{S}_\beta^{-1}\Big\|^{1/2}_{L_\infty(N)},$$
which is clearly bounded by 1. On the other hand,
$$\|\mathcal{S}_\beta D^{\frac{1}{v}}\|^2_{L_v(\mathcal{N})} = \Big\|D^{\frac{1}{v}}\Big(\sum_{k=1}^{m} \lambda_k \beta_k^* \beta_k + \delta 1\Big) D^{\frac{1}{v}}\Big\|_{L_{v/2}(\mathcal{N})}$$
$$\leq \sum_{k=1}^{m} \lambda_k \|\beta_k D^{\frac{1}{v}}\|^2_{L_v(\mathcal{N})} + \delta < 1+\delta.$$
Therefore, the triangle inequality follows one more time by letting $\delta \to 0$. □

The following will be a key point in the proof of Theorem 3.2, the main result in Chapter 3. Note that we state it under the assumption that $\mathcal{N}_u L_q(\mathcal{M}) \mathcal{N}_v$ is a normed space and that for the moment we only know it (from Lemma 2.5) whenever $(1/u, 1/v, 1/q) \in \partial_\infty \mathsf{K}$. However, since eventually we shall need to apply this result for any point in K, we state it in full generality.

PROPOSITION 2.6. *If $\mathcal{N}_u L_q(\mathcal{M}) \mathcal{N}_v$ is a normed space, we have*
  i) *The following map is an isometry*
$$j_{u,v} : x \in \mathcal{N}_u L_q(\mathcal{M}) \mathcal{N}_v \mapsto D^{\frac{1}{u}} x D^{\frac{1}{v}} \in L_u(\mathcal{N}) L_q(\mathcal{M}) L_v(\mathcal{N}).$$

ii) $L_u(\mathcal{N})L_q(\mathcal{M})L_v(\mathcal{N})$ is a Banach space which completes $\mathcal{N}_u L_q(\mathcal{M})\mathcal{N}_v$.

PROOF. It is clear that
$$\|j_{u,v}(x)\|_{u \cdot q \cdot v} \leq \||x|\|_{u \cdot q \cdot v} \qquad \text{for all} \qquad x \in \mathcal{N}_u L_q(\mathcal{M})\mathcal{N}_v.$$
Let us see that the reverse inequality holds. Assume it does not hold, then we can find $x_0 \in \mathcal{N}_u L_q(\mathcal{M})\mathcal{N}_v$ such that $\||x_0|\|_{u \cdot q \cdot v} > 1$ and $\|j_{u,v}(x_0)\|_{u \cdot q \cdot v} < 1$. By the Hahn-Banach theorem (here we use the assumption that the space $\mathcal{N}_u L_q(\mathcal{M})\mathcal{N}_v$ is normed) there exists a norm one functional
$$\varphi : \mathcal{N}_u L_q(\mathcal{M})\mathcal{N}_v \to \mathbb{C}$$
satisfying

(a) $\varphi(x_0) = \||x_0|\|_{u \cdot q \cdot v} > 1$.

(b) $|\varphi(\alpha y \beta)| \leq \|\mathrm{D}^{\frac{1}{u}}\alpha\|_{L_u(\mathcal{N})} \|y\|_{L_q(\mathcal{M})} \|\beta \mathrm{D}^{\frac{1}{v}}\|_{L_v(\mathcal{N})}$.

Note that any $y \in L_q(\mathcal{M})$ provides a densely defined bilinear map
$$\Phi_y : \left(\mathrm{D}^{\frac{1}{u}}\alpha, \beta \mathrm{D}^{\frac{1}{v}}\right) \in L_u(\mathcal{N}) \times L_v(\mathcal{N}) \mapsto \varphi(\alpha y \beta) \in \mathbb{C},$$
which satisfies $\|\Phi_y\| \leq \|y\|_q$ and
$$\Phi_{n_1 y n_2}\left(\mathrm{D}^{\frac{1}{u}}\alpha, \beta \mathrm{D}^{\frac{1}{v}}\right) = \Phi_y\left(\mathrm{D}^{\frac{1}{u}}\alpha n_1, n_2 \beta \mathrm{D}^{\frac{1}{v}}\right) \quad \text{for all} \quad n_1, n_2 \in \mathcal{N}.$$
On the other hand, since $\|j_{u,v}(x_0)\|_{u \cdot q \cdot v} < 1$ we must have $j_{u,v}(x_0) = a_0 y_0 b_0$ with
$$\max\left\{\|a_0\|_{L_u(\mathcal{N})}, \|y_0\|_{L_q(\mathcal{M})}, \|b_0\|_{L_v(\mathcal{N})}\right\} < 1.$$
If we consider the invertible elements
$$a = \left(a_0 a_0^* + \delta \mathrm{D}^{2/u}\right)^{1/2} \qquad \text{and} \qquad b = \left(b_0^* b_0 + \delta \mathrm{D}^{2/v}\right)^{1/2},$$
we can write $j_{u,v}(x_0) = aa^{-1}a_0 y_0 b_0 b^{-1} b = ayb$. Moreover, for $\delta > 0$ small enough
$$\max\left\{\|a\|_{L_u(\mathcal{N})}, \|y\|_{L_q(\mathcal{M})}, \|b\|_{L_v(\mathcal{N})}\right\} < 1.$$
Since $a^2 \geq \delta \mathrm{D}^{2/u}$ and $b^2 \geq \delta \mathrm{D}^{2/v}$, there exist bounded elements $\alpha, \beta \in \mathcal{N}$ with
$$\mathrm{D}^{1/u} = a\alpha \qquad \text{and} \qquad \mathrm{D}^{1/v} = \beta b.$$
Let us denote by e the left support of α and by f the right support of β. We note that the right support of α and the left support of β is 1. Then, we use polar decomposition to find strictly positive densities $d_1 \in e\mathcal{N}e$ and $d_2 \in f\mathcal{N}f$ and partial isometries w_1 and w_2 such that
$$\alpha = d_1 w_1 \qquad \text{and} \qquad \beta = w_2 d_2.$$
Note that $w_1^* w_1 = 1$, $w_1 w_1^* = e$, $w_2^* w_2 = f$ and $w_2 w_2^* = 1$. Then we observe that
$$ayb = j_{u,v}(x_0) = \mathrm{D}^{1/u} x_0 \mathrm{D}^{1/v} = a\alpha x_0 \beta b \Rightarrow y = \alpha x_0 \beta.$$
This yields
$$eyf = \alpha x_0 \beta = d_1 w_1 x_0 w_2 d_2.$$
In particular, taking spectral projections $e_n = 1_{[\frac{1}{n}, n]}(d_1)$ and $f_n = 1_{[\frac{1}{n}, n]}(d_2)$
$$w_1^* e_n d_1^{-1} eyf d_2^{-1} f_n w_2^* = w_1^* e_n w_1 x_0 w_2 f_n w_2^*.$$
This implies that
$$\left|\varphi(w_1^* e_n w_1 x_0 w_2 f_n w_2^*)\right| = \left|\Phi_{w_1^* e_n w_1 x_0 w_2 f_n w_2^*}\left(\mathrm{D}^{\frac{1}{u}}, \mathrm{D}^{\frac{1}{v}}\right)\right|$$

$$\begin{aligned}
&= \left|\Phi_{w_1^* e_n d_1^{-1} eyf d_2^{-1} f_n w_2^*}(\mathrm{D}^{\frac{1}{u}}, \mathrm{D}^{\frac{1}{v}})\right| \\
&= \left|\Phi_{w_1^* e_n d_1^{-1} eyf d_2^{-1} f_n w_2^*}(ad_1 w_1, w_2 d_2 b)\right| \\
&= \left|\Phi_{eyf}(ae_n, f_n b)\right| \\
&\leq \|ae_n\|_u \|eyf\|_q \|f_n b\|_v \\
&\leq \|a\|_u \|y\|_q \|b\|_v < 1.
\end{aligned}$$

On the other hand, $(w_1^* e_n w_1)$ (resp. $(w_2 f_n w_2^*)$) converges to $w_1^* w_1 = 1$ (resp. $w_2 w_2^* = 1$) strongly. By Lemma 2.3 in [**16**], this implies that $(\mathrm{D}^{1/u} w_1^* e_n w_1)$ (resp. $(w_2 f_n w_2^* \mathrm{D}^{1/v})$) converges to $\mathrm{D}^{1/u}$ (resp. $\mathrm{D}^{1/v}$) in the norm of $L_u(\mathcal{N})$ (resp. $L_v(\mathcal{N})$). This combined with the continuity of Φ_{x_0} gives

$$\begin{aligned}
|\varphi(x_0)| &= \lim_{n\to\infty} \left|\Phi_{x_0}(\mathrm{D}^{\frac{1}{u}} w_1^* e_n w_1, w_2 f_n w_2^* \mathrm{D}^{\frac{1}{v}})\right| \\
&= \lim_{n\to\infty} \left|\varphi(w_1^* e_n w_1 x_0 w_2 f_n w_2^*)\right| \leq 1.
\end{aligned}$$

This contradicts condition (a). Therefore, the map $j_{u,v}$ defines an isometry and the proof of i) is completed. Next, we see that $\|\ \|_{u \cdot q \cdot v}$ is a norm 'outside' the space $\mathcal{N}_u L_q(\mathcal{M})\mathcal{N}_v$. The homogeneity over $\mathbb{R}_+$ is clear and the positive definiteness follows as in Lemma 2.5. Thus, it suffices to show that the triangle inequality holds. As we did above, we begin by taking a family $x_1, x_2, \ldots, x_m$ of elements in $L_u(\mathcal{N}) L_q(\mathcal{M}) L_v(\mathcal{N})$ satisfying $\|x_k\|_{u \cdot q \cdot v} < 1$ and a collection of positive numbers $\lambda_1, \lambda_2, \ldots, \lambda_m$ with $\sum_k \lambda_k = 1$. By hypothesis, we may assume that $x_k = a_k y_k b_k$ with

$$\max\left\{\|a_k\|_{L_u(\mathcal{N})}, \|y_k\|_{L_q(\mathcal{M})}, \|b_k\|_{L_v(\mathcal{N})}\right\} < 1.$$

Given any $\xi > 1$ and by the density of $\mathrm{D}^{\frac{1}{u}}\mathcal{N}$ (resp. $\mathcal{N}\mathrm{D}^{\frac{1}{v}}$) in $L_u(\mathcal{N})$ (resp. $L_v(\mathcal{N})$), it is not difficult to check that both a_k and b_k can be written in the following way

$$a_k = \sum_{i=0}^{\infty} \mathrm{D}^{\frac{1}{u}} a_{ik} \quad \text{with} \quad \left\|\mathrm{D}^{\frac{1}{u}} a_{ik}\right\|_{L_u(\mathcal{N})} \leq \xi^{-i},$$

$$b_k = \sum_{j=0}^{\infty} b_{jk} \mathrm{D}^{\frac{1}{v}} \quad \text{with} \quad \left\|b_{jk} \mathrm{D}^{\frac{1}{v}}\right\|_{L_v(\mathcal{N})} \leq \xi^{-j}.$$

This gives rise to

$$\sum_{k=1}^{m} \lambda_k x_k = \sum_{i,j=0}^{\infty} \mathrm{D}^{\frac{1}{u}} \Big(\sum_{k=1}^{m} \lambda_k a_{ik} y_k b_{jk}\Big) \mathrm{D}^{\frac{1}{v}}.$$

Assuming the triangle inequality on $\mathcal{N}_u L_q(\mathcal{M})\mathcal{N}_v$, we can write

$$\mathrm{D}^{\frac{1}{u}} \Big(\sum_{k=1}^{m} \lambda_k a_{ik} y_k b_{jk}\Big) \mathrm{D}^{\frac{1}{v}} = \mathsf{A}_{ij} \mathsf{Y}_{ij} \mathsf{B}_{ij}$$

where

$$\|\mathsf{A}_{ij}\|_{L_u(\mathcal{N})} \leq \xi^{-\frac{i+j}{4}}, \quad \|\mathsf{Y}_{ij}\|_{L_q(\mathcal{M})} \leq \xi^{-\frac{i+j}{2}}, \quad \|\mathsf{B}_{ij}\|_{L_v(\mathcal{N})} \leq \xi^{-\frac{i+j}{4}}.$$

Then, we define for $\delta > 0$

$$\mathcal{S}_r(\mathsf{A}) = \Big(\sum_{i,j=0}^{\infty} \mathsf{A}_{ij} \mathsf{A}_{ij}^* + \delta \mathrm{D}^{\frac{2}{u}}\Big)^{1/2} \quad \text{and} \quad \mathcal{S}_c(\mathsf{B}) = \Big(\sum_{i,j=0}^{\infty} \mathsf{B}_{ij}^* \mathsf{B}_{ij} + \delta \mathrm{D}^{\frac{2}{v}}\Big)^{1/2},$$

so that there exist $\alpha_{ij}, \beta_{ij} \in \mathcal{N}$ given by $\mathcal{S}_r(\mathsf{A})\alpha_{ij} = \mathsf{A}_{ij}$ and $\beta_{ij}\mathcal{S}_c(\mathsf{B}) = \mathsf{B}_{ij}$. Hence,

$$\sum_{k=1}^{m} \lambda_k x_k = \mathcal{S}_r(\mathsf{A})\Big(\sum_{i,j=0}^{\infty} \alpha_{ij}\mathsf{Y}_{ij}\beta_{ij}\Big)\mathcal{S}_c(\mathsf{B}).$$

Moreover, we have

$$\big\|\mathcal{S}_r(\mathsf{A})\big\|_{L_u(\mathcal{N})} \le \Big(\delta + \sum_{i,j=0}^{\infty} \|\mathsf{A}_{ij}\|_{L_u(\mathcal{N})}^2\Big)^{1/2} \le \sqrt{\delta} + \frac{1}{1 - 1/\sqrt{\xi}}.$$

The same estimate applies to $\mathcal{S}_c(\mathsf{B})$. The middle term satisfies

$$\Big\|\sum_{i,j=0}^{\infty} \alpha_{ij}\mathsf{Y}_{ij}\beta_{ij}\Big\|_q \le \Big\|\Big(\sum_{i,j=0}^{\infty} \alpha_{ij}\alpha_{ij}^*\Big)^{\frac{1}{2}}\Big\|_{\infty} \Big(\sum_{i,j=0}^{\infty} \|\mathsf{Y}_{ij}\|_q^q\Big)^{\frac{1}{q}} \Big\|\Big(\sum_{i,j=0}^{\infty} \beta_{ij}^*\beta_{ij}\Big)^{\frac{1}{2}}\Big\|_{\infty}$$

$$\le \Big(\sum_{i,j=0}^{\infty} \|\mathsf{Y}_{ij}\|_q^q\Big)^{\frac{1}{q}} \le \Big(\frac{1}{1 - 1/\xi^{q/2}}\Big)^{2/q}.$$

In summary,

$$\Big\|\sum_{k=1}^{m} \lambda_k x_k\Big\|_{u \cdot q \cdot v} \le \big\|\mathcal{S}_r(\mathsf{A})\big\|_{L_u(\mathcal{N})} \Big\|\sum_{i,j=0}^{\infty} \alpha_{ij}\mathsf{Y}_{ij}\beta_{ij}\Big\|_{L_q(\mathcal{M})} \big\|\mathcal{S}_c(\mathsf{B})\big\|_{L_v(\mathcal{N})} \longrightarrow 1$$

as $\delta \to 0$ and $\xi \to \infty$. This proves the triangle inequality. To prove completeness we use again a geometric series argument. Let $x_1, x_2, \ldots$ be a countable family of elements in $L_u(\mathcal{N})L_q(\mathcal{M})L_v(\mathcal{N})$ with $\|x_k\|_{u \cdot q \cdot v} < 4^{-k}$ for any integer $k \ge 1$. That is, we have $x_k = 4^{-k} a_k y_k b_k$ with

$$\max\Big\{\|a_k\|_{L_u(\mathcal{N})}, \|y_k\|_{L_q(\mathcal{M})}, \|b_k\|_{L_v(\mathcal{N})}\Big\} < 1.$$

Then, by a well known characterization of completeness, it suffices to see that the sum $\sum_k x_k$ belongs to $L_u(\mathcal{N})L_q(\mathcal{M})L_v(\mathcal{N})$. We use again the same factorization trick

$$\sum_{k=1}^{\infty} x_k = \mathcal{S}_r(a)\Big(\sum_{k=1}^{\infty} 2^{-k} \alpha_k y_k \beta_k\Big)\mathcal{S}_c(b)$$

where

$$\mathcal{S}_r(a) = \Big(\sum_{k=1}^{\infty} 2^{-k} a_k a_k^* + \delta \mathrm{D}^{\frac{2}{u}}\Big)^{1/2} \quad \text{and} \quad \mathcal{S}_r(a)\alpha_k = 2^{-k/2} a_k,$$

$$\mathcal{S}_c(b) = \Big(\sum_{k=1}^{\infty} 2^{-k} b_k^* b_k + \delta \mathrm{D}^{\frac{2}{v}}\Big)^{1/2} \quad \text{and} \quad \beta_k \mathcal{S}_c(b) = 2^{-k/2} b_k.$$

In particular, we just need to show that $\mathcal{S}_r(a) \in L_u(\mathcal{N})$, $\mathcal{S}_c(b) \in L_v(\mathcal{N})$ and the middle term belongs to $L_q(\mathcal{M})$. However, this follows again by applying the same estimates as above, details are left to the reader. To conclude, we just need to show that $\mathcal{N}_u L_q(\mathcal{M})\mathcal{N}_v$ is dense in $L_u(\mathcal{N})L_q(\mathcal{M})L_v(\mathcal{N})$. However, this follows easily from the density of $\mathrm{D}^{1/u}\mathcal{N}$ (resp. $\mathcal{N}\mathrm{D}^{1/v}$) in $L_u(\mathcal{N})$ (resp. $L_v(\mathcal{N})$) and the triangle inequality proved above. $\square$

2.3. A metric structure on the solid K

At this point, we are not able to prove that $\mathcal{N}_u L_q(\mathcal{M})\mathcal{N}_v$ is a normed space for any point $(1/u, 1/v, 1/q)$ in the solid K. However, we need at least to know that we have a metric space structure. In fact, we shall prove that $||| \ |||_{u \cdot q \cdot v}$ is always a γ-norm for some $0 < \gamma \leq 1$.

LEMMA 2.7. *If $(1/u, 1/v, 1/q) \in \mathsf{K}$, there exists $0 < \gamma \leq 1$ such that*
$$|||x_1 + x_2|||_{u \cdot q \cdot v}^\gamma \leq |||x_1|||_{u \cdot q \cdot v}^\gamma + |||x_2|||_{u \cdot q \cdot v}^\gamma \quad \text{for all} \quad x_1, x_2 \in \mathcal{N}_u L_q(\mathcal{M})\mathcal{N}_v.$$

PROOF. According to Lemma 2.5, we can assume in what follows that
$$(1/u, 1/v, 1/q) \in \mathsf{K} \setminus \partial_\infty \mathsf{K}.$$

As we did to prove the triangle inequality, let $x_1, x_2, \ldots, x_m \in \mathcal{N}_u L_q(\mathcal{M})\mathcal{N}_v$ be a sequence of vectors satisfying $|||x_k|||_{u \cdot q \cdot v} < 1$ and let $\lambda_1, \lambda_2, \ldots, \lambda_m \in \mathbb{R}_+$ with sum $\sum_k \lambda_k = 1$. Then it suffices to show that
$$\Big|\Big|\Big|\sum_{k=1}^m \lambda_k x_k\Big|\Big|\Big|_{u \cdot q \cdot v}^\gamma \leq \sum_{k=1}^m \lambda_k^\gamma \quad \text{for some} \quad 0 < \gamma \leq 1.$$

By hypothesis, we may assume that $x_k = \alpha_k y_k \beta_k$, with
$$\max\Big\{\big\|D^{\frac{1}{u}}\alpha_k\big\|_{L_u(\mathcal{N})}, \|y_k\|_{L_q(\mathcal{M})}, \big\|\beta_k D^{\frac{1}{v}}\big\|_{L_v(\mathcal{N})}\Big\} < 1.$$

By Figure I we can always find $1 \leq u_1, q_0, v_1 \leq \infty$ and $0 < \theta < 1$ such that
  (a) We have $1/q_0 = 1/u + 1/q + 1/v = 1/u_1 + 1/v_1$.
  (b) We have $(1/u, 1/v, 1/q) = (0, 0, (1-\theta)/q_0) + (\theta/u_1, \theta/v_1, 0)$.

Indeed, we first consider the plane P parallel to ACDF containing $(1/u, 1/v, 1/q)$. The point $(0, 0, 1/q_0)$ is the intersection of P with the segment 0A. Then, we consider the line L passing through $(0, 0, 1/q_0)$ and $(1/u, 1/v, 1/q)$. Then, the point $(1/u_1, 1/v_1, 0)$ is the intersection of L with the coordinate plane $z = 0$. Note that the point $(1/u_1, 1/v_1, 0)$ is not necessarily in K. However, according to (a) it always satisfies $1/u_1 + 1/v_1 \leq 1$. Note also that, since we have excluded the points in $\partial_\infty \mathsf{K}$ at the beginning of this proof, we can always assume that $0 < \theta < 1$. Then it follows from (b) that
$$\begin{aligned} L_u(\mathcal{N}) &= \big[L_\infty(\mathcal{N}), L_{u_1}(\mathcal{N})\big]_\theta, \\ L_q(\mathcal{M}) &= \big[L_{q_0}(\mathcal{M}), L_\infty(\mathcal{M})\big]_\theta, \\ L_v(\mathcal{N}) &= \big[L_\infty(\mathcal{N}), L_{v_1}(\mathcal{N})\big]_\theta. \end{aligned}$$

By the complex interpolation method, we can find bounded analytic functions
$$f_{2k} : \mathcal{S} \to L_{q_0}(\mathcal{M}) + L_\infty(\mathcal{M})$$

satisfying $f_{2k}(\theta) = y_k$ for $k = 1, 2, \ldots, m$ and such that

(2.8) $$\max\Big\{\sup_{z \in \partial_0} \|f_{2k}(z)\|_{L_{q_0}(\mathcal{M})}, \sup_{z \in \partial_1} \|f_{2k}(z)\|_{L_\infty(\mathcal{M})}\Big\} < 1.$$

Similarly, by Lemma 2.4 we also have bounded analytic functions
$$f_{1k}, f_{3k} : \mathcal{S} \to \mathcal{N}$$

for any $k = 1, 2, \ldots, m$ satisfying $(f_{1k}(\theta), f_{3k}(\theta)) = (\alpha_k, \beta_k)$ and

(2.9)
$$\sup_{z \in \partial_0} \max\left\{ \|f_{1k}(z)\|_{L_\infty(\mathcal{N})}, \|f_{3k}(z)\|_{L_\infty(\mathcal{N})} \right\} < 1,$$
$$\sup_{z \in \partial_1} \max\left\{ \|D^{\frac{1}{u_1}} f_{1k}(z)\|_{L_{u_1}(\mathcal{N})}, \|f_{3k}(z) D^{\frac{1}{v_1}}\|_{L_{v_1}(\mathcal{N})} \right\} < 1.$$

Given $\delta > 0$, we consider the following functions on the boundary

$$W_1(z) = \begin{cases} 1 & \text{if } z \in \partial_0, \\ \delta 1 + \sum_k \lambda_k f_{1k}(z) f_{1k}(z)^* & \text{if } z \in \partial_1, \end{cases}$$

$$W_3(z) = \begin{cases} 1 & \text{if } z \in \partial_0, \\ \delta 1 + \sum_k \lambda_k f_{3k}(z)^* f_{3k}(z) & \text{if } z \in \partial_1. \end{cases}$$

According to Devinatz's factorization theorem [**44**], there exist bounded analytic functions $w_1, w_3 : \mathcal{S} \to \mathcal{N}$ with bounded analytic inverse and satisfying the following identities on $\partial \mathcal{S}$

$$w_1(z) w_1(z)^* = W_1(z),$$
$$w_3(z)^* w_3(z) = W_3(z).$$

Then we can write

$$\sum_{k=1}^m \lambda_k x_k = w_1(\theta) \Big[w_1^{-1}(\theta) \Big(\sum_{k=1}^m \lambda_k f_{1k}(\theta) f_{2k}(\theta) f_{3k}(\theta) \Big) w_3^{-1}(\theta) \Big] w_3(\theta).$$

Let us estimate the norms of the three factors above. First we clearly have

$$\sup_{z \in \partial_0} \|w_1(z)\|_{L_\infty(\mathcal{N})} = \sup_{z \in \partial_0} \|w_1(z) w_1(z)^*\|_{L_\infty(\mathcal{N})}^{1/2} = 1,$$
$$\sup_{z \in \partial_0} \|w_3(z)\|_{L_\infty(\mathcal{N})} = \sup_{z \in \partial_0} \|w_3(z)^* w_3(z)\|_{L_\infty(\mathcal{N})}^{1/2} = 1.$$

On the other hand, since $L_p(\mathcal{N})$ is always a $\min(p,1)$-normed space

$$\sup_{z \in \partial_1} \|D^{\frac{1}{u_1}} w_1(z)\|_{L_{u_1}(\mathcal{N})}^{u_1} = \sup_{z \in \partial_1} \|D^{\frac{1}{u_1}} w_1(z) w_1(z)^* D^{\frac{1}{u_1}}\|_{L_{u_1/2}(\mathcal{N})}^{u_1/2}$$
$$\leq \sup_{z \in \partial_1} \begin{cases} \left(\sum_{k=1}^m \lambda_k \|D^{\frac{1}{u_1}} f_{1k}(z)\|_{L_{u_1}(\mathcal{N})}^2 + \delta \right)^{u_1/2}, & \text{if } u_1 \geq 2, \\ \sum_{k=1}^m \lambda_k^{u_1/2} \|D^{\frac{1}{u_1}} f_{1k}(z)\|_{L_{u_1}(\mathcal{N})}^{u_1} + \delta^{u_1/2}, & \text{if } u_1 < 2. \end{cases}$$
$$< \max\left\{ (1+\delta)^{u_1/2}, \delta^{u_1/2} + \sum_{k=1}^m \lambda_k^{u_1/2} \right\}.$$

Similarly, we have

$$\sup_{z \in \partial_1} \|w_3(z) D^{\frac{1}{v_1}}\|_{L_{v_1}(\mathcal{N})}^{v_1} < \max\left\{ (1+\delta)^{v_1/2}, \delta^{v_1/2} + \sum_{k=1}^m \lambda_k^{v_1/2} \right\}.$$

In summary, we obtain by Kosaki's interpolation

$$\|D^{\frac{1}{u}} w_1(\theta)\|_{L_u(\mathcal{N})} < \max\left\{ \sqrt{1+\delta}, \Big(\delta^{u_1/2} + \sum_{k=1}^m \lambda_k^{u_1/2} \Big)^{1/u_1} \right\},$$

$$\|w_3(\theta) D^{\frac{1}{v}}\|_{L_v(\mathcal{N})} < \max\left\{ \sqrt{1+\delta}, \Big(\delta^{v_1/2} + \sum_{k=1}^m \lambda_k^{v_1/2} \Big)^{1/v_1} \right\}.$$

As we have pointed out above, we have $1/u_1 + 1/v_1 \le 1$. In particular, $u_1/2$ and $v_1/2$ can not be simultaneously less than 1. Thus, at least one of the two sums above must be less or equal than 1. This means that taking
$$\gamma = \min\{1, u_1/2, v_1/2\},$$
we deduce
$$\lim_{\delta \to 0} \left(\left\| D^{\frac{1}{u}} w_1(\theta) \right\|_{L_u(\mathcal{N})} \left\| w_3(\theta) D^{\frac{1}{v}} \right\|_{L_v(\mathcal{N})} \right) < \left(\sum_{k=1}^m \lambda_k^\gamma \right)^{1/2\gamma} \le \left(\sum_{k=1}^m \lambda_k^\gamma \right)^{1/\gamma}.$$
In particular, it suffices to see that
$$\left\| w_1^{-1}(\theta) \left(\sum_{k=1}^m \lambda_k f_{1k}(\theta) f_{2k}(\theta) f_{3k}(\theta) \right) w_3^{-1}(\theta) \right\|_{L_q(\mathcal{M})} \le 1.$$
Let us consider the bounded analytic function
$$g(z) = w_1^{-1}(z) \left(\sum_{k=1}^m \lambda_k f_{1k}(z) f_{2k}(z) f_{3k}(z) \right) w_3^{-1}(z).$$
Since $w_1(z) w_1(z)^* = 1 = w_3(z)^* w_3(z)$ for all $z \in \partial_0$, we clearly have
$$\sup_{z \in \partial_0} \|g(z)\|_{L_{q_0}(\mathcal{M})} = \sup_{z \in \partial_0} \left\| \sum_{k=1}^m \lambda_k f_{1k}(z) f_{2k}(z) f_{3k}(z) \right\|_{L_{q_0}(\mathcal{M})}$$
$$\le \sup_{z \in \partial_0} \sum_{k=1}^m \lambda_k \|f_{1k}(z)\|_{L_\infty(\mathcal{N})} \|f_{2k}(z)\|_{L_{q_0}(\mathcal{M})} \|f_{3k}(z)\|_{L_\infty(\mathcal{N})}.$$
Then it follows from (2.8) and (2.9) that the expression above is bounded above by 1. On the other hand, since w_1 and w_3 are invertible, we can define the functions $h_{1k}, h_{3k} : \mathcal{S} \to \mathcal{N}$ by the relations
$$\sqrt{\lambda_k} f_{1k}(z) = w_1(z) h_{1k}(z) \quad \text{and} \quad \sqrt{\lambda_k} f_{3k}(z) = h_{3k}(z) w_3(z).$$
Then it is easy to check that for any $z \in \partial_1$
$$(2.10) \qquad \sum_{k=1}^m h_{1k}(z) h_{1k}(z)^* \le 1 \quad \text{and} \quad \sum_{k=1}^m h_{3k}(z)^* h_{3k}(z) \le 1.$$
Therefore, we obtain from (2.8) and (2.10) that
$$\sup_{z \in \partial_1} \|g(z)\|_{L_\infty(\mathcal{M})} = \sup_{z \in \partial_1} \left\| \sum_{k=1}^m h_{1k}(z) f_{2k}(z) h_{3k}(z) \right\|_{L_\infty(\mathcal{M})}$$
$$\le \sup_{z \in \partial_1} \left\| \left(\sum_{k=1}^m h_{1k}(z) h_{1k}(z)^* \right)^{1/2} \right\|_{L_\infty(\mathcal{N})}$$
$$\times \sup_{z \in \partial_1} \max_{1 \le k \le m} \|f_{2k}(z)\|_{L_\infty(\mathcal{M})}$$
$$\times \sup_{z \in \partial_1} \left\| \left(\sum_{k=1}^m h_{3k}(z)^* h_{3k}(z) \right)^{1/2} \right\|_{L_\infty(\mathcal{N})} < 1.$$
By Kosaki's interpolation we have $\|g(\theta)\|_{L_q(\mathcal{M})} \le 1$. This completes the proof. $\square$

Arguing as in the proof of Proposition 2.6 ii), we can see that any amalgamated space $L_u(\mathcal{N})L_q(\mathcal{M})L_v(\mathcal{N})$ is a γ-Banach space and that $j_{u,v}(\mathcal{N}_u L_q(\mathcal{M})\mathcal{N}_v)$ (with $j_{u,v}$ as defined in Proposition 2.6) is a dense subspace. Indeed, we just need to repeat the two given arguments (using geometric series) conveniently rewritten with exponents γ everywhere. We leave the details to the reader. Let us state this result for future reference.

PROPOSITION 2.8. *If $(1/u, 1/v, 1/q) \in \mathsf{K}$, there exists $0 < \gamma \leq 1$ such that*
  i) $L_u(\mathcal{N})L_q(\mathcal{M})L_v(\mathcal{N})$ *is a γ-Banach space.*
  ii) $j_{u,v}(\mathcal{N}_u L_q(\mathcal{M})\mathcal{N}_v)$ *is a dense subspace of $L_u(\mathcal{N})L_q(\mathcal{M})L_v(\mathcal{N})$.*

OBSERVATION 2.9. We do not claim at this moment that $j_{u,v}$ is an isometry.

CHAPTER 3

An interpolation theorem

In this chapter we prove that the solid K is an interpolation family on the indices (u, q, v). Of course, we need to know a priori that $L_u(\mathcal{N})L_q(\mathcal{M})L_v(\mathcal{N})$ is a Banach space for any $(1/u, 1/v, 1/q)$ in the solid K. In fact, as we shall see the proofs of both results depend on the proof of the other. Indeed, let us consider a parameter $0 \leq \tau \leq 1$. According to the terminology employed in Figure I, let us define K_τ to be the intersection of K with the plane P_τ which contains the point $(0, 0, \tau)$ and is parallel to the upper face ACDF of K. Roughly speaking, we first prove that
$$\mathcal{K}_\tau = \Big\{ L_u(\mathcal{N})L_q(\mathcal{M})L_v(\mathcal{N}) \,\Big|\, (1/u, 1/v, 1/q) \in \mathsf{K}_\tau \Big\}$$
is an interpolation family on the indices (u, q, v) for any $0 \leq \tau \leq 1$ and with ending points lying on $\partial_\infty \mathsf{K}$. Note that we already proved in Chapter 2 that the spaces associated to the points in $\partial_\infty \mathsf{K}$ are Banach spaces. Then, we use this result to prove that every point in K corresponds to a Banach space and derive our main interpolation theorem. More concretely, we first prove the following result.

LEMMA 3.1. *Let us assume that*
  i) $(1/u_j, 1/v_j, 1/q_j) \in \partial_\infty \mathsf{K}$ *for* $j = 0, 1$.
  ii) $1/u_0 + 1/q_0 + 1/v_0 = 1/u_1 + 1/q_1 + 1/v_1$.

Then $L_{u_\theta}(\mathcal{N})L_{q_\theta}(\mathcal{M})L_{v_\theta}(\mathcal{N})$ *is a Banach space isometrically isomorphic to*
$$\mathrm{X}_\theta(\mathcal{M}) = \Big[L_{u_0}(\mathcal{N})L_{q_0}(\mathcal{M})L_{v_0}(\mathcal{N}), L_{u_1}(\mathcal{N})L_{q_1}(\mathcal{M})L_{v_1}(\mathcal{N}) \Big]_\theta.$$

We know from Lemma 2.5 and Proposition 2.6 that the interpolation pairs considered in Lemma 3.1 are made of Banach spaces. After the proof of Lemma 3.1 we shall show that $L_u(\mathcal{N})L_q(\mathcal{M})L_v(\mathcal{N})$ is a Banach space for any $(1/u, 1/v, 1/q)$ in the solid K and we shall deduce our main result in this chapter.

THEOREM 3.2. *The amalgamated space* $L_u(\mathcal{N})L_q(\mathcal{M})L_v(\mathcal{N})$ *is a Banach space for any* $(1/u, 1/v, 1/q) \in \mathsf{K}$. *Moreover, if* $(1/u_j, 1/v_j, 1/q_j) \in \mathsf{K}$ *for* $j = 0, 1$, *the space* $L_{u_\theta}(\mathcal{N})L_{q_\theta}(\mathcal{M})L_{v_\theta}(\mathcal{N})$ *is isometrically isomorphic to*
$$\mathrm{X}_\theta(\mathcal{M}) = \Big[L_{u_0}(\mathcal{N})L_{q_0}(\mathcal{M})L_{v_0}(\mathcal{N}), L_{u_1}(\mathcal{N})L_{q_1}(\mathcal{M})L_{v_1}(\mathcal{N}) \Big]_\theta.$$

Note that the pairs of Banach spaces considered in Lemma 3.1 and Theorem 3.2 are compatible for complex interpolation. Indeed, since the amalgamated space $L_{u_j}(\mathcal{N})L_{q_j}(\mathcal{M})L_{v_j}(\mathcal{N})$ is continuously injected (by means of Hölder inequality) in $L_{p_j}(\mathcal{M})$ for $j = 0, 1$ and $1/p_j = 1/u_j + 1/q_j + 1/v_j$, any such pair lives in the sum $L_{p_0}(\mathcal{M}) + L_{p_1}(\mathcal{M})$. The main difficulty which appears to prove Lemma 3.1 lies in the fact that the intersection of amalgamated spaces is quite difficult to describe in the general case. Therefore, any attempt to work on a dense subspace meets this obstacle. We shall avoid this difficulty by proving this interpolation result in the particular case of finite von Neumann algebras whose density is bounded above and below. Under this assumption, we will be able to find a nice dense subspace to work with. Then we shall use the Haagerup construction [**12**] sketched in Section 2.1 (suitably modified to work in the present context) to get the result in the general case. After that, the proof of Theorem 3.2 follows easily by using similar techniques.

3.1. Finite von Neumann algebras

We begin by proving Lemma 3.1 for a finite von Neumann algebra $\mathcal{M}$ equipped with a n.f. state φ with respect to which, the corresponding density D satisfies the following property for some positive constants $0 < c_1 \leq c_2 < \infty$

$$(3.1) \qquad c_1 1 \leq \mathrm{D} \leq c_2 1.$$

LEMMA 3.3. *Assume that $(1/u_j, 1/v_j, 1/q_j) \in \partial_\infty \mathsf{K}$ for $j = 0, 1$. Then, given any $0 < \theta < 1$, the space $L_{q_0}(\mathcal{M}) \cap L_{q_1}(\mathcal{M})$ is dense in*

$$\mathrm{X}_\theta(\mathcal{M}) = \Big[L_{u_0}(\mathcal{N}) L_{q_0}(\mathcal{M}) L_{v_0}(\mathcal{N}), L_{u_1}(\mathcal{N}) L_{q_1}(\mathcal{M}) L_{v_1}(\mathcal{N}) \Big]_\theta.$$

PROOF. Let us write Δ for the intersection

$$L_{u_0}(\mathcal{N}) L_{q_0}(\mathcal{M}) L_{v_0}(\mathcal{N}) \cap L_{u_1}(\mathcal{N}) L_{q_1}(\mathcal{M}) L_{v_1}(\mathcal{N}).$$

According to the complex interpolation method, we know that Δ is dense in $\mathrm{X}_\theta(\mathcal{M})$. In particular, it suffices to show that we have density in Δ with respect to the norm of $\mathrm{X}_\theta(\mathcal{M})$. Let x be an element in Δ so that we have decompositions

$$\begin{array}{rcl} x & = & a_0 \tilde{y}_0 b_0 \\ x & = & a_1 \tilde{y}_1 b_1 \end{array} \quad \text{where} \quad a_j \in L_{u_j}(\mathcal{N}),\ \tilde{y}_j \in L_{q_j}(\mathcal{M}),\ b_j \in L_{v_j}(\mathcal{N}).$$

CASE I. Let us first assume that $\max\{u_0, u_1, v_0, v_1\} < \infty$. Then, we can define

$$a = \sum_{j=0,1} \left(a_j a_j^*\right)^{u_j/2} + \delta \mathrm{D},$$

$$b = \sum_{j=0,1} \left(b_j^* b_j\right)^{v_j/2} + \delta \mathrm{D}.$$

Note that, since $2/u_j$ and $2/v_j$ are in $(0,1]$, we have

$$\left.\begin{array}{l} \left(a_j a_j^*\right)^{u_j/2} \leq a \\ \left(b_j^* b_j\right)^{v_j/2} \leq b \end{array}\right\} \Rightarrow \left\{\begin{array}{l} a_j a_j^* \leq a^{2/u_j}, \\ b_j^* b_j \leq b^{2/v_j}. \end{array}\right.$$

Therefore, if we define α_j, β_j by $a_j = a^{1/u_j} \alpha_j$ and $b_j = \beta_j b^{1/v_j}$, we have $\alpha_j \alpha_j^* \leq 1$ and $\beta_j^* \beta_j \leq 1$. Thus, $\alpha_j, \beta_j \in \mathcal{N}$ for $j = 0, 1$ and we can set $y_j = \alpha_j \tilde{y}_j \beta_j$ in $L_{q_j}(\mathcal{M})$ so that

$$(3.2) \qquad x = a^{1/u_j} y_j b^{1/v_j}$$

for $j = 0, 1$. Now we consider the spectral projections

$$e_n = 1_{[\frac{1}{n}, n]}(a) \quad \text{and} \quad f_n = 1_{[\frac{1}{n}, n]}(b).$$

Then, since $e_n x f_n = (e_n a^{1/u_j}) y_j (b^{1/v_j} f_n)$ we find $e_n x f_n \in L_{q_0}(\mathcal{M}) \cap L_{q_1}(\mathcal{M})$. Thus, we have to show that $e_n x f_n$ tends to x in the norm of $\mathrm{X}_\theta(\mathcal{M})$. Let us write

$$x - e_n x f_n = (1 - e_n) x + e_n x (1 - f_n).$$

Then it is clear that

$$\begin{aligned}\|(1-e_n)x\|_\Delta &= \max\left\{\left\|(1-e_n)a^{\frac{1}{u_j}}y_j b^{\frac{1}{v_j}}\right\|_{u_j\cdot q_j\cdot v_j} \mid j=0,1\right\}\\ &\leq \max\left\{\left\|(1-e_n)a^{\frac{1}{u_j}}\right\|_{u_j}\|y_j\|_{q_j}\left\|b^{\frac{1}{v_j}}\right\|_{v_j} \mid j=0,1\right\}.\end{aligned}$$

The term on the right tends to 0 as $n \to \infty$ since $\max(u_0, u_1) < \infty$. Therefore, since the norm of $X_\theta(\mathcal{M})$ is controlled by the norm of Δ when $0 < \theta < 1$, we obtain that $(1-e_n)x$ tends to 0 in the norm of $X_\theta(\mathcal{M})$. The second term $e_n x(1-f_n)$ can be estimated in the same way. This concludes the proof of Case I.

CASE II. Now let us assume that there exists some indices among u_0, u_1, v_0, v_1 which are infinite. Then we can assume without lost of generality that a_j (resp. b_j) is 1 whenever u_j (resp. v_j) is infinite. According to this and assuming the convention $1^\infty = 1$, the previous definition of a and b still makes sense. Moreover, the relations obtained in (3.2) also hold in this case. Therefore, we need to prove again that $e_n x f_n$ tends to x with respect to the norm of $X_\theta(\mathcal{M})$. We only prove the convergence for the term $(1-e_n)x$ since again the second one can be estimated in the same way. By the three lines lemma, we have

$$\|(1-e_n)x\|_{X_\theta(\mathcal{M})} \leq \|(1-e_n)x\|_{u_0\cdot q_0\cdot v_0}^{1-\theta} \|(1-e_n)x\|_{u_1\cdot q_1\cdot v_1}^{\theta}.$$

Now, let us recall that both norms on the right hand side are uniformly bounded on $n \geq 1$. In particular, since $0 < \theta < 1$, it suffices to prove that $e_n x$ tends to x with respect to the norm of $L_{u_j}(\mathcal{N})L_{q_j}(\mathcal{M})L_{v_j}(\mathcal{N})$ for either $j = 0$ or $j = 1$. There are only three possible situations:

(a) Assume $\min(u_0, u_1) < \infty$. Let us suppose (w.l.o.g.) that $\min(u_0, u_1) = u_0$. Then, we apply the estimate

$$\|(1-e_n)x\|_{u_0\cdot q_0\cdot v_0} \leq \left\|(1-e_n)a^{\frac{1}{u_0}}\right\|_{L_{u_0}(\mathcal{N})} \|y_0\|_{L_{q_0}(\mathcal{M})} \left\|b^{\frac{1}{v_0}}\right\|_{L_{v_0}(\mathcal{N})}.$$

(b) Assume $\min(q_0, q_1) < \infty$. Let us suppose (w.l.o.g.) that $\min(q_0, q_1) = q_0$. Then, we apply the estimate

$$\|(1-e_n)x\|_{u_0\cdot q_0\cdot v_0} \leq \left\|a^{\frac{1}{u_0}}\right\|_{L_{u_0}(\mathcal{N})} \|(1-e_n)y_0\|_{L_{q_0}(\mathcal{M})} \left\|b^{\frac{1}{v_0}}\right\|_{L_{v_0}(\mathcal{N})}.$$

(c) Finally, assume that neither (a) nor (b) hold. Let us suppose (w.l.o.g.) that $\min(v_0, v_1) = v_0$. In this case we have $u_0 = q_0 = u_1 = q_1 = \infty$ and $\Delta = L_\infty(\mathcal{N})L_\infty(\mathcal{M})L_{v_1}(\mathcal{N})$. Therefore, $\mathcal{M} \,(= L_{q_0}(\mathcal{M}) \cap L_{q_1}(\mathcal{M}))$ is dense in Δ (by Proposition 2.6) and thereby in $X_\theta(\mathcal{M})$. $\square$

REMARK 3.4. Note that the finiteness of $\mathcal{M}$ is used in the proof of Lemma 3.3 to ensure that $e_n a^{1/u_j}$ and $b^{1/v_j}f_n$ are in $\mathcal{N}$. In particular, in the case of general von Neumann algebras we shall need to use a different approach.

PROOF OF LEMMA 3.1. The lower estimate is easy. Indeed, let

$$T_0 : L_{u_0}(\mathcal{N}) \times L_{q_0}(\mathcal{M}) \times L_{v_0}(\mathcal{N}) \to L_{u_0}(\mathcal{N})L_{q_0}(\mathcal{M})L_{v_0}(\mathcal{N}),$$

$$T_1 : L_{u_1}(\mathcal{N}) \times L_{q_1}(\mathcal{M}) \times L_{v_1}(\mathcal{N}) \to L_{u_1}(\mathcal{N})L_{q_1}(\mathcal{M})L_{v_1}(\mathcal{N}),$$

be the multilinear maps given by $T_0(a, y, b) = ayb = T_1(a, y, b)$. It is clear that both T_0 and T_1 are contractive. Hence, the lower estimate follows by multilinear interpolation. For the converse, we proceed in two steps.

STEP 1. Let x be of the form
$$x = \mathrm{D}^{1/u_\theta} \alpha y \beta \mathrm{D}^{1/v_\theta} \quad \text{with} \quad \alpha, \beta \in \mathcal{N},\ y \in L_{q_\theta}(\mathcal{M}).$$

Using the isometry j defined in Proposition 2.6, we have
$$x \in j_{u_\theta, v_\theta}\left(\mathcal{N}_{u_\theta} L_{q_\theta}(\mathcal{M}) \mathcal{N}_{v_\theta}\right).$$

We claim that it suffices to see that
$$\tag{3.3} |||j_{u_\theta, v_\theta}^{-1}(x)|||_{u_\theta \cdot q_\theta \cdot v_\theta} \leq \|x\|_{\mathrm{X}_\theta(\mathcal{M})}$$

for all x of the form considered above. Indeed, let us assume that inequality (3.3) holds. Then, combining this inequality with the lower estimate proved above, we obtain
$$\tag{3.4} |||j_{u_\theta, v_\theta}^{-1}(x)|||_{u_\theta \cdot q_\theta \cdot v_\theta} \leq \|x\|_{\mathrm{X}_\theta(\mathcal{M})} \leq \|x\|_{u_\theta \cdot q_\theta \cdot v_\theta} \leq |||j_{u_\theta, v_\theta}^{-1}(x)|||_{u_\theta \cdot q_\theta \cdot v_\theta}.$$

Note that, since $\mathrm{X}_\theta(\mathcal{M})$ is a Banach space, we deduce that $\mathcal{N}_{u_\theta} L_{q_\theta}(\mathcal{M}) \mathcal{N}_{v_\theta}$ satisfies the triangle inequality. Then, applying again Proposition 2.6 we deduce that the space $L_{u_\theta}(\mathcal{N}) L_{q_\theta}(\mathcal{M}) L_{v_\theta}(\mathcal{N})$ is the completion of $\mathcal{N}_{u_\theta} L_{q_\theta}(\mathcal{M}) \mathcal{N}_{v_\theta}$. Therefore, it follows that for any $x \in L_{u_\theta}(\mathcal{N}) L_{q_\theta}(\mathcal{M}) L_{v_\theta}(\mathcal{N})$, we have
$$\|x\|_{u_\theta \cdot q_\theta \cdot v_\theta} = \|x\|_{\mathrm{X}_\theta(\mathcal{M})}.$$

In particular, $L_{u_\theta}(\mathcal{N}) L_{q_\theta}(\mathcal{M}) L_{v_\theta}(\mathcal{N})$ embeds isometrically in $\mathrm{X}_\theta(\mathcal{M})$. To conclude, it remains to see that both spaces are the same. However, according to Lemma 3.3, we deduce that $L_{u_\theta}(\mathcal{N}) L_{q_\theta}(\mathcal{M}) L_{v_\theta}(\mathcal{N})$ is norm dense in $\mathrm{X}_\theta(\mathcal{M})$.

STEP 2. Now we prove inequality (3.3). We use one more time the interpolation trick based on Devinatz's factorization theorem [**8**]. Again, we refer the reader to Pisier's paper [**44**] for a precise statement of Devinatz's theorem adapted to our aims. Let x be an element of the form
$$x = \mathrm{D}^{1/u_\theta} \alpha y \beta \mathrm{D}^{1/v_\theta} \quad \text{with} \quad \alpha, \beta \in \mathcal{N},\ y \in L_{q_\theta}(\mathcal{M}).$$

Assume the norm of x in $\mathrm{X}_\theta(\mathcal{M})$ is less than one. Then, the complex interpolation method provides a bounded analytic function
$$f : \mathcal{S} \longrightarrow L_{u_0}(\mathcal{N}) L_{q_0}(\mathcal{M}) L_{v_0}(\mathcal{N}) + L_{u_1}(\mathcal{N}) L_{q_1}(\mathcal{M}) L_{v_1}(\mathcal{N})$$

satisfying $f(\theta) = x$ and
$$\tag{3.5} \begin{aligned} \sup_{z \in \partial_0} \|f(z)\|_{u_0 \cdot q_0 \cdot v_0} &< 1, \\ \sup_{z \in \partial_1} \|f(z)\|_{u_1 \cdot q_1 \cdot v_1} &< 1. \end{aligned}$$

On the other hand, we are assuming that the density D satisfies the boundedness condition (3.1). In particular, $z \in \mathcal{S} \mapsto \mathrm{D}^{\lambda z} \in \mathcal{M}$ is a bounded analytic function for any $\lambda \in \mathbb{C}$. Therefore (multiplying if necessary on the left and on the right by certain powers of D^z and its inverses), we may assume that f has the form
$$f(z) = \mathrm{D}^{\frac{1-z}{u_0}} \mathrm{D}^{\frac{z}{u_1}} f_1(z) \mathrm{D}^{\frac{z}{v_1}} \mathrm{D}^{\frac{1-z}{v_0}},$$

where $f_1 : \mathcal{S} \to L_{q_0}(\mathcal{M}) + L_{q_1}(\mathcal{M})$ is bounded analytic. Hence, we deduce from the boundary conditions (3.5) and Proposition 2.6 that f can be written on the boundary $\partial \mathcal{S}$ as follows
$$f(z) = \mathrm{D}^{\frac{1-z}{u_0}} \mathrm{D}^{\frac{z}{u_1}} g_1(z) g_2(z) g_3(z) \mathrm{D}^{\frac{z}{v_1}} \mathrm{D}^{\frac{1-z}{v_0}},$$

where $g_1, g_3 : \partial\mathcal{S} \to \mathcal{N}$ and $g_2 : \partial_j \to L_{q_j}(\mathcal{M})$ satisfy the following estimates

(3.6)
$$\sup_{z \in \partial_0} \max \left\{ \left\| \mathrm{D}^{\frac{1}{u_0}} g_1(z) \right\|_{u_0}, \|g_2(z)\|_{q_0}, \left\| g_3(z) \mathrm{D}^{\frac{1}{v_0}} \right\|_{v_0} \right\} < 1,$$
$$\sup_{z \in \partial_1} \max \left\{ \left\| \mathrm{D}^{\frac{1}{u_1}} g_1(z) \right\|_{u_1}, \|g_2(z)\|_{q_1}, \left\| g_3(z) \mathrm{D}^{\frac{1}{v_1}} \right\|_{v_1} \right\} < 1.$$

Given any $\delta > 0$, we define the following functions on the boundary
$$W_1 : z \in \partial\mathcal{S} \mapsto g_1(z)g_1(z)^* + \delta 1 \in \mathcal{N},$$
$$W_3 : z \in \partial\mathcal{S} \mapsto g_3(z)^* g_3(z) + \delta 1 \in \mathcal{N}.$$

According to Devinatz's factorization theorem [**44**], we can find invertible bounded analytic functions $w_1, w_3 : \mathcal{S} \to \mathcal{N}$ with bounded analytic inverse and satisfying the following identities on $\partial\mathcal{S}$

(3.7)
$$\begin{aligned} w_1(z) w_1(z)^* &= W_1(z), \\ w_3(z)^* w_3(z) &= W_3(z). \end{aligned}$$

Then, we consider the factorization
$$f(z) = h_1(z) h_2(z) h_3(z)$$
with $h_2(z) = h_1^{-1}(z) f(z) h_3^{-1}(z)$ and h_1, h_3 given by
$$h_1(z) = \mathrm{D}^{\frac{1-z}{u_0}} \mathrm{D}^{\frac{z}{u_1}} w_1(z) \mathrm{D}^{-\frac{z}{u_1}} \mathrm{D}^{\frac{z}{u_0}},$$
$$h_3(z) = \mathrm{D}^{\frac{z}{v_0}} \mathrm{D}^{-\frac{z}{v_1}} w_3(z) \mathrm{D}^{\frac{z}{v_1}} \mathrm{D}^{\frac{1-z}{v_0}}.$$

Note than our original hypothesis (3.1) implies the boundedness and analyticity of h_1, h_2, h_3. Then, recalling that D^ω is a unitary for any $\omega \in \mathbb{C}$ such that $\mathrm{Re}\,\omega = 0$ and that $2 \le u_j, v_j \le \infty$ (so that we have triangle inequality on $L_{u_j/2}(\mathcal{N})$ and $L_{v_j/2}(\mathcal{N})$), we obtain from (3.6) the following estimates for h_1 and h_3

$$\sup_{z \in \partial_0} \|h_1(z)\|^2_{L_{u_0}(\mathcal{N})} = \sup_{z \in \partial_0} \left\| \mathrm{D}^{\frac{1}{u_0}} w_1(z) w_1(z)^* \mathrm{D}^{\frac{1}{u_0}} \right\|_{L_{u_0/2}(\mathcal{N})}$$
$$\le \sup_{z \in \partial_0} \left\| \mathrm{D}^{\frac{1}{u_0}} g_1(z) \right\|^2_{L_{u_0}(\mathcal{N})} + \delta < 1 + \delta,$$

$$\sup_{z \in \partial_0} \|h_3(z)\|^2_{L_{v_0}(\mathcal{N})} = \sup_{z \in \partial_0} \left\| \mathrm{D}^{\frac{1}{v_0}} w_3(z)^* w_3(z) \mathrm{D}^{\frac{1}{v_0}} \right\|_{L_{v_0/2}(\mathcal{N})}$$
$$\le \sup_{z \in \partial_0} \left\| g_3(z) \mathrm{D}^{\frac{1}{v_0}} \right\|^2_{L_{v_0}(\mathcal{N})} + \delta < 1 + \delta,$$

$$\sup_{z \in \partial_1} \left\| h_1(z) \mathrm{D}^{\frac{1}{u_1} - \frac{1}{u_0}} \right\|^2_{L_{u_1}(\mathcal{N})} = \sup_{z \in \partial_1} \left\| \mathrm{D}^{\frac{1}{u_1}} w_1(z) w_1(z)^* \mathrm{D}^{\frac{1}{u_1}} \right\|_{L_{u_1/2}(\mathcal{N})}$$
$$\le \sup_{z \in \partial_1} \left\| \mathrm{D}^{\frac{1}{u_1}} g_1(z) \right\|^2_{L_{u_1}(\mathcal{N})} + \delta < 1 + \delta,$$

$$\sup_{z \in \partial_1} \left\| \mathrm{D}^{\frac{1}{v_1} - \frac{1}{v_0}} h_3(z) \right\|^2_{L_{v_1}(\mathcal{N})} = \sup_{z \in \partial_1} \left\| \mathrm{D}^{\frac{1}{v_1}} w_3(z)^* w_3(z) \mathrm{D}^{\frac{1}{v_1}} \right\|_{L_{v_1/2}(\mathcal{N})}$$
$$\le \sup_{z \in \partial_1} \left\| g_3(z) \mathrm{D}^{\frac{1}{v_1}} \right\|^2_{L_{v_1}(\mathcal{N})} + \delta < 1 + \delta.$$

By Kosaki's interpolation, we obtain
$$\left\| h_1(\theta) \mathrm{D}^{\frac{\theta}{u_1} - \frac{\theta}{u_0}} \right\|_{L_{u_\theta}(\mathcal{N})} < \sqrt{1 + \delta},$$
$$\left\| \mathrm{D}^{\frac{\theta}{v_1} - \frac{\theta}{v_0}} h_3(\theta) \right\|_{L_{v_\theta}(\mathcal{N})} < \sqrt{1 + \delta}.$$

On the other hand, using again the unitarity of D^ω when $\operatorname{Re}\omega = 0$, we have

$$\sup_{z \in \partial_0} \|h_2(z)\|^2_{L_{q_0}(\mathcal{M})}$$
$$= \sup_{z \in \partial_0} \|w_1^{-1}(z)g_1(z)g_2(z)g_3(z)w_3^{-1}(z)\|^2_{L_{q_0}(\mathcal{M})}$$
$$\leq \sup_{z \in \partial_0} \|w_1^{-1}(z)g_1(z)\|^2_{L_\infty(\mathcal{N})} \|g_2(z)\|^2_{L_{q_0}(\mathcal{M})} \|g_3(z)w_3^{-1}(z)\|^2_{L_\infty(\mathcal{N})}$$
$$< \sup_{z \in \partial_0} \|w_1^{-1}(z)g_1(z)g_1(z)^*w_1^{-1}(z)^*\|_{L_\infty(\mathcal{N})} \|w_3^{-1}(z)^*g_3(z)^*g_3(z)w_3^{-1}(z)\|_{L_\infty(\mathcal{N})},$$

$$\sup_{z \in \partial_1} \|D^{\frac{1}{u_0}-\frac{1}{u_1}} h_2(z) D^{\frac{1}{v_0}-\frac{1}{v_1}}\|^2_{L_{q_1}(\mathcal{M})}$$
$$= \sup_{z \in \partial_1} \|w_1^{-1}(z)g_1(z)g_2(z)g_3(z)w_3^{-1}(z)\|^2_{L_{q_1}(\mathcal{M})}$$
$$\leq \sup_{z \in \partial_1} \|w_1^{-1}(z)g_1(z)\|^2_{L_\infty(\mathcal{N})} \|g_2(z)\|^2_{L_{q_1}(\mathcal{M})} \|g_3(z)w_3^{-1}(z)\|^2_{L_\infty(\mathcal{N})}$$
$$< \sup_{z \in \partial_1} \|w_1^{-1}(z)g_1(z)g_1(z)^*w_1^{-1}(z)^*\|_{L_\infty(\mathcal{N})} \|w_3^{-1}(z)^*g_3(z)^*g_3(z)w_3^{-1}(z)\|_{L_\infty(\mathcal{N})}.$$

Then we combine (3.7) with the inequalities

$$g_1(z)g_1(z)^* \leq g_1(z)g_1(z)^* + \delta 1,$$

$$g_3(z)^*g_3(z) \leq g_3(z)^*g_3(z) + \delta 1,$$

to obtain by Kosaki's interpolation

$$\|D^{\frac{\theta}{u_0}-\frac{\theta}{u_1}} h_2(\theta) D^{\frac{\theta}{v_0}-\frac{\theta}{v_1}}\|_{L_{q_\theta}(\mathcal{M})} < 1.$$

In summary, we have obtained the following factorization

$$f(\theta) = \left(D^{\frac{1}{u_\theta}} w_1(\theta)\right) \left(D^{\frac{\theta}{u_0}-\frac{\theta}{u_1}} h_2(\theta) D^{\frac{\theta}{v_0}-\frac{\theta}{v_1}}\right) \left(w_3(\theta) D^{\frac{1}{v_\theta}}\right) = j_1(\theta) j_2(\theta) j_3(\theta),$$

with

$$\|j_1(\theta)\|_{L_{u_\theta}(\mathcal{N})} < \sqrt{1+\delta}, \quad \|j_2(\theta)\|_{L_{q_\theta}(\mathcal{M})} < 1, \quad \|j_3(\theta)\|_{L_{v_\theta}(\mathcal{N})} < \sqrt{1+\delta}.$$

Therefore, (3.3) follows by letting $\delta \to 0$ in

$$\left\|\left\|j^{-1}_{u_\theta, v_\theta}(f(\theta))\right\|\right\|_{u_\theta \cdot q_\theta \cdot v_\theta} < 1 + \delta.$$

This concludes the proof of Lemma 3.1 for finite von Neumann algebras. $\square$

REMARK 3.5. Note that condition ii) of Lemma 3.1 is not needed at any point in the proof given above for finite von Neumann algebras. This will be crucial in the proof of Theorem 3.2 below.

3.2. Conditional expectations on $\partial_\infty K$

Before the proof of Lemma 3.1 for general von Neumann algebras, we need some preliminary results in order to adapt Haagerup's construction to the present context. We begin with a technical lemma and some auxiliary interpolation results.

LEMMA 3.6. *Let $\mathcal{M}$ be a von Neumann algebra equipped with a n.f. state φ and let D be the associated density. Let us consider a bounded analytic function $f : \mathcal{S} \to \mathcal{M}$. Then, given $1 \leq p < \infty$, the following functions are also bounded analytic*

$$h_1 : z \in \mathcal{S} \mapsto D^{(1-z)/p} f(z) D^{z/p} \in L_p(\mathcal{M}),$$
$$h_2 : z \in \mathcal{S} \mapsto D^{z/p} f(z) D^{(1-z)/p} \in L_p(\mathcal{M}).$$

PROOF. The arguments to be used hold for both h_1 and h_2. Hence, we only prove the assertion for h_1. The continuity and boundedness of h_1 on the closure of $\mathcal{S}$ is trivial. To prove the analyticity of h_1 we may clearly assume that the function f is a finite power series

$$f(z) = \sum_{k=1}^{n} x_k z^k \quad \text{with} \quad x_k \in \mathcal{M}.$$

In particular, it suffices to see that for a fixed element $x_0 \in \mathcal{M}$, the function

$$h_0 : z \in \mathcal{S} \mapsto D^{(1-z)/p} x_0 D^{z/p} \in L_p(\mathcal{M})$$

is analytic. If $x_0 \in \mathcal{M}_a$ is an analytic element this is clear. Assume that x_0 is not an analytic element. According to Pedersen-Takesaki [42], the net $(x_\gamma)_{\gamma > 0} \subset \mathcal{M}_a$ given by

$$x_\gamma = \sqrt{\frac{\gamma}{\pi}} \int_\mathbb{R} \sigma_t(x_0) \exp(-\gamma t^2) \, dt$$

converges strongly to x_0 as $\gamma \to \infty$. Then Lemma 2.3 in [16] gives that

$$h_\gamma(z) = D^{(1-z)/p} x_\gamma D^{z/p}$$

converges pointwise to $h_0(z)$ in the norm of $L_p(\mathcal{M})$. Now, let us consider a linear functional $\varphi : L_p(\mathcal{M}) \to \mathbb{C}$ and a cycle Γ in $\mathcal{S}$ homologous to zero with respect to $\mathcal{S}$. Then, since $\|x_\gamma\|_\mathcal{M} \leq \|x_0\|_\mathcal{M}$ for all $\gamma > 0$, the dominated convergence theorem gives

$$\int_\Gamma \varphi(h_0(z)) dz = \lim_{\gamma \to \infty} \int_\Gamma \varphi(h_\gamma(z)) dz = 0.$$

Thus, $\varphi(h_0)$ is analytic for any linear functional $\varphi : L_p(\mathcal{M}) \to \mathbb{C}$ and so is h_0. □

LEMMA 3.7. *If $2 \leq u, v \leq \infty$, we have the following isometries*

$$\bigl[L_u(\mathcal{N}) L_\infty(\mathcal{M}), L_u(\mathcal{N}) L_2(\mathcal{M})\bigr]_\theta = L_u(\mathcal{N}) L_{2/\theta}(\mathcal{M}),$$
$$\bigl[L_\infty(\mathcal{M}) L_v(\mathcal{N}), L_2(\mathcal{M}) L_v(\mathcal{N})\bigr]_\theta = L_{2/\theta}(\mathcal{M}) L_v(\mathcal{N}).$$

PROOF. We shall only prove that

$$X_\theta(\mathcal{M}) = \bigl[L_u(\mathcal{N}) L_\infty(\mathcal{M}), L_u(\mathcal{N}) L_2(\mathcal{M})\bigr]_\theta = L_u(\mathcal{N}) L_{2/\theta}(\mathcal{M}).$$

The other isometry can be proved in the same way. The lower estimate follows by multilinear interpolation just as in the proof of Lemma 3.1 for finite von Neumann algebras. To prove the upper estimate we first note that, according to Proposition 2.6, we have a dense subset

$$D^{1/u} \mathcal{N} \bigl(L_2(\mathcal{M}) \cap L_\infty(\mathcal{M})\bigr) \subset L_u(\mathcal{N}) L_{2/\theta}(\mathcal{M}).$$

Since this subset is also dense in $X_\theta(\mathcal{M})$ (note that the intersection in this case is $L_u(\mathcal{N})L_\infty(\mathcal{M})$), it suffices to show that for any x of the form $D^{1/u}ay$ with $a \in \mathcal{N}$ and $y \in L_2(\mathcal{M}) \cap L_\infty(\mathcal{M})$, we have the following inequality

(3.8) $$\|x\|_{u \cdot \frac{2}{\theta}} \leq \|x\|_{X_\theta(\mathcal{M})}.$$

Assume that the norm of x in $X_\theta(\mathcal{M})$ is less than 1. Then we can find a bounded analytic function $f : \mathcal{S} \to L_u(\mathcal{N})L_\infty(\mathcal{M}) + L_u(\mathcal{N})L_2(\mathcal{M})$ satisfying $f(\theta) = x$ and the inequalities

(3.9) $$\begin{aligned} \sup_{z \in \partial_0} \|f(z)\|_{u \cdot \infty} &< 1, \\ \sup_{z \in \partial_1} \|f(z)\|_{u \cdot 2} &< 1. \end{aligned}$$

Moreover, by our previous considerations we can assume that f has the form
$$f(z) = D^{\frac{1}{u}} f_1(z),$$
where $f_1 : \mathcal{S} \to L_2(\mathcal{M}) \cap L_\infty(\mathcal{M})$ is a bounded analytic function. Then we deduce from the boundary conditions (3.9) and Proposition 2.6 that f can be written on $\partial \mathcal{S}$ as follows
$$f(z) = D^{\frac{1}{u}} g_1(z) g_2(z),$$
with $g_1 : \partial \mathcal{S} \to \mathcal{N}$ and $g_2 : \partial \mathcal{S} \to L_2(\mathcal{M}) \cap L_\infty(\mathcal{M})$ satisfying
$$\max \left\{ \sup_{z \in \partial \mathcal{S}} \|D^{\frac{1}{u}} g_1(z)\|_{L_u(\mathcal{N})}, \sup_{z \in \partial_0} \|g_2(z)\|_{L_\infty(\mathcal{M})}, \sup_{z \in \partial_1} \|g_2(z)\|_{L_2(\mathcal{M})} \right\} < 1.$$

Now we apply Devinatz's theorem [**44**] to $W : z \in \partial \mathcal{S} \to g_1(z)g_1(z)^* + \delta 1$, so that we find an invertible bounded analytic function $w : \mathcal{S} \to \mathcal{N}$ satisfying $ww^* = W$ on the boundary. Then we consider the factorization
$$f(z) = \left(D^{\frac{1}{u}} w(z) \right) \left(w^{-1}(z) f_1(z) \right).$$

Clearly both factors are bounded analytic and we have
$$\begin{aligned} \sup_{z \in \partial \mathcal{S}} \|D^{\frac{1}{u}} w(z)\|^2_{L_u(\mathcal{N})} &= \sup_{z \in \partial \mathcal{S}} \|D^{\frac{1}{u}} w(z) w(z)^* D^{\frac{1}{u}}\|_{L_{u/2}(\mathcal{N})} < 1 + \delta, \\ \sup_{z \in \partial_0} \|w^{-1}(z) f_1(z)\|_{L_\infty(\mathcal{M})} &\leq \sup_{z \in \partial_0} \|w^{-1}(z) g_1(z)\|_{L_\infty(\mathcal{N})} \|g_2(z)\|_{L_\infty(\mathcal{M})} < 1, \\ \sup_{z \in \partial_1} \|w^{-1}(z) f_1(z)\|_{L_2(\mathcal{M})} &\leq \sup_{z \in \partial_1} \|w^{-1}(z) g_1(z)\|_{L_\infty(\mathcal{N})} \|g_2(z)\|_{L_2(\mathcal{M})} < 1. \end{aligned}$$

Finally, by Kosaki's interpolation and letting $\delta \to 0$, we obtain inequality (3.8). $\square$

LEMMA 3.8. *Assume $1/u + 1/2 = 1/p = 1/2 + 1/v$ and $1/q_\theta = (1-\theta)/p + \theta/2$. Then, we have the following isometries*
$$\begin{aligned} \left[L_p(\mathcal{M}), L_u(\mathcal{N}) L_2(\mathcal{M}) \right]_\theta &= L_{u/\theta}(\mathcal{N}) L_{q_\theta}(\mathcal{M}), \\ \left[L_p(\mathcal{M}), L_2(\mathcal{M}) L_v(\mathcal{N}) \right]_\theta &= L_{q_\theta}(\mathcal{M}) L_{v/\theta}(\mathcal{N}). \end{aligned}$$

PROOF. One more time, the lower estimate follows by multilinear interpolation and we shall only prove the first isometry
$$X_\theta(\mathcal{M}) = \left[L_p(\mathcal{M}), L_u(\mathcal{N}) L_2(\mathcal{M}) \right]_\theta = L_{u/\theta}(\mathcal{N}) L_{q_\theta}(\mathcal{M}).$$

Let us point out that Hölder inequality gives
$$\Delta = L_p(\mathcal{M}) \cap L_u(\mathcal{N}) L_2(\mathcal{M}) = L_u(\mathcal{N}) L_2(\mathcal{M}).$$

3.2. CONDITIONAL EXPECTATIONS ON $\partial_\infty K$

On the other hand, it follows from Lemma 1.2 that
$$D^{1/p}\mathcal{M}_a = D^{1/u}\mathcal{N}_a\mathcal{M}_a D^{1/2} = D^{\theta/u}\mathcal{N}_a\mathcal{M}_a D^{1/q_\theta}.$$
Hence $D^{1/p}\mathcal{M}_a$ is dense in Δ and in $L_{u/\theta}(\mathcal{N})L_{q_\theta}(\mathcal{M})$ so that it suffices to see
$$\|x\|_{\frac{u}{\theta}\cdot q_\theta} \leq \|x\|_{X_\theta(\mathcal{M})}$$
for any element x of the form $D^{1/p}y$ for some $y \in \mathcal{M}_a$. Assume that the norm of x in $X_\theta(\mathcal{M})$ is less than 1. Then, according to the considerations above, we can find a bounded analytic function $f: \mathcal{S} \to L_p(\mathcal{M}) + L_u(\mathcal{N})L_2(\mathcal{M})$ of the form
$$f(z) = D^{1/p}f_1(z) \quad \text{with} \quad f_1: \mathcal{S} \to \mathcal{M}_a \quad \text{bounded analytic,}$$
satisfying $f(\theta) = x$ and such that

(3.10) $$\max\Big\{\sup_{z\in\partial_0}\|f(z)\|_p, \sup_{z\in\partial_1}\|f(z)\|_{u\cdot 2}\Big\} < 1.$$

Moreover, since f_1 takes values in $\mathcal{M}_a$ we can rewrite f as
$$f(z) = D^{\frac{z}{u}+\frac{1-z}{p}}f_2(z)D^{\frac{z}{2}} \quad \text{with} \quad f_2: \mathcal{S} \to \mathcal{M}_a \quad \text{bounded analytic.}$$
In particular, $f_{|\partial_1}$ has the form
$$f(1+it) = D^{\frac{1}{u}}\sigma_{-t/2}\big(f_2(1+it)\big)D^{\frac{1}{2}}.$$
According to Proposition 2.6 and the boundary estimate for f on ∂_1, we can write
$$f(1+it) = D^{\frac{1}{u}}g_1(1+it)g_2(1+it),$$
with $g_1: \partial_1 \to \mathcal{N}$ and $g_2: \partial_1 \to L_2(\mathcal{M})$ satisfying

(3.11) $$\max\Big\{\sup_{z\in\partial_1}\|D^{\frac{1}{u}}g_1(z)\|_{L_u(\mathcal{N})}, \sup_{z\in\partial_1}\|g_2(z)\|_{L_2(\mathcal{M})}\Big\} < 1.$$

By Devinatz's theorem, we can consider an invertible bounded analytic function $w: \mathcal{S} \to \mathcal{N}$ satisfying $w(z)w(z)^* = W(z)$ for all $z \in \partial\mathcal{S}$ where this time the function $W: \partial\mathcal{S} \to \mathcal{N}$ is given by
$$W(z) = \begin{cases} 1, & \text{if } z \in \partial_0, \\ \sigma_{-\text{Im}z/u}\big(g_1(z)g_1(z)^*\big) + \delta 1, & \text{if } z \in \partial_1, \end{cases}$$
with $z = \text{Re}z + i\text{Im}z$. Then we factorize f as $f(z) = h_1(z)D^{-\frac{1}{u}}h_2(z)$ where
$$\begin{aligned} h_1(z) &= D^{\frac{z}{u}}w(z)D^{\frac{1-z}{u}}, \\ h_2(z) &= D^{\frac{z}{u}}w^{-1}(z)D^{\frac{1-z}{u}}D^{\frac{1-z}{2}}f_2(z)D^{\frac{z}{2}}. \end{aligned}$$

Note that Lemma 3.6 provides the boundedness and analyticity of h_1 and h_2. As usual, we need to estimate the norms of h_1 and h_2 on the boundary. We begin with the estimates for h_1. On ∂_0 we have
$$\begin{aligned} \sup_{z\in\partial_0}\|h_1(z)D^{-\frac{1}{u}}\|_{L_\infty(\mathcal{N})} &= \sup_{t\in\mathbb{R}}\|\sigma_{t/u}(w(it))\|_{L_\infty(\mathcal{N})} \\ &= \sup_{z\in\partial_0}\|w(z)\|_{L_\infty(\mathcal{N})} \\ &= \sup_{z\in\partial_0}\|w(z)w(z)^*\|_{L_\infty(\mathcal{N})}^{1/2} = 1. \end{aligned}$$

On ∂_1 we apply the boundary condition (3.11)
$$\sup_{z\in\partial_1}\|h_1(z)\|_{L_u(\mathcal{N})}^2 = \sup_{t\in\mathbb{R}}\|\sigma_{t/u}(D^{\frac{1}{u}}w(1+it))\|_{L_u(\mathcal{N})}^2$$

$$\begin{aligned}
&= \sup_{z\in\partial_1} \left\|D^{\frac{1}{u}}w(z)\right\|^2_{L_u(\mathcal{N})} \\
&= \sup_{z\in\partial_1} \left\|D^{\frac{1}{u}}w(z)w(z)^*D^{\frac{1}{u}}\right\|_{L_{u/2}(\mathcal{N})} \\
&\leq \sup_{z\in\partial_1} \left\|\sigma_{-\mathrm{Im}z/u}\bigl(D^{\frac{1}{u}}g_1(z)g_1(z)^*D^{\frac{1}{u}}\bigr)\right\|_{L_{u/2}(\mathcal{N})} + \delta \\
&= \sup_{z\in\partial_1} \left\|D^{\frac{1}{u}}g_1(z)\right\|^2_{L_u(\mathcal{N})} + \delta < 1+\delta.
\end{aligned}$$

For the estimate of h_2 on ∂_0 we use (3.10)

$$\begin{aligned}
\sup_{z\in\partial_0}\left\|h_2(z)\right\|_{L_p(\mathcal{M})} &= \sup_{t\in\mathbb{R}} \left\|\sigma_{t/u}\bigl(w^{-1}(it)D^{\frac{1}{u}}\bigr)\sigma_{-t/2}\bigl(D^{\frac{1}{2}}f_2(it)\bigr)\right\|_{L_p(\mathcal{M})} \\
&= \sup_{t\in\mathbb{R}} \left\|\sigma_{t/u}\bigl(w^{-1}(it)\bigr)\sigma_{t/u}\bigl(\sigma_{-t/p}\bigl(D^{\frac{1}{p}}f_2(it)\bigr)\bigr)\right\|_{L_p(\mathcal{M})} \\
&= \sup_{t\in\mathbb{R}} \left\|w^{-1}(it)\sigma_{-t/p}\bigl(D^{\frac{1}{p}}f_2(it)\bigr)\right\|_{L_p(\mathcal{M})} \\
&\leq \sup_{t\in\mathbb{R}} \left\|\bigl(w(it)w(it)^*\bigr)^{-1}\right\|^{1/2}_{L_\infty(\mathcal{N})}\left\|D^{\frac{1}{p}}f_2(it)\right\|_{L_p(\mathcal{M})} \\
&= \sup_{t\in\mathbb{R}} \left\|\sigma_{-t/2}\bigl(D^{\frac{1}{p}}f_2(it)\bigr)\right\|_{L_p(\mathcal{M})} \\
&= \sup_{z\in\partial_0} \left\|f(z)\right\|_{L_p(\mathcal{M})} < 1.
\end{aligned}$$

Finally, we use again (3.11) to estimate h_2 on ∂_1

$$\begin{aligned}
&\sup_{z\in\partial_1}\left\|D^{-\frac{1}{u}}h_2(z)\right\|_{L_2(\mathcal{M})} \\
&= \sup_{t\in\mathbb{R}} \left\|\sigma_{t/u}\bigl(w^{-1}(1+it)\bigr)\sigma_{-t/2}\bigl(f_2(1+it)D^{\frac{1}{2}}\bigr)\right\|_{L_2(\mathcal{M})} \\
&= \sup_{t\in\mathbb{R}} \left\|\sigma_{t/u}\bigl(w^{-1}(1+it)\bigr)g_1(1+it)g_2(1+it)\right\|_{L_2(\mathcal{M})} \\
&\leq \sup_{t\in\mathbb{R}} \left\|\sigma_{t/u}\bigl(w^{-1}(1+it)\bigr)g_1(1+it)\right\|_{L_\infty(\mathcal{N})}\left\|g_2(1+it)\right\|_{L_2(\mathcal{M})} \\
&< \sup_{t\in\mathbb{R}} \left\|\sigma_{t/u}\bigl(w^{-1}(1+it)\bigr)g_1(1+it)g_1(1+it)^*\sigma_{t/u}\bigl(w^{-1}(1+it)\bigr)^*\right\|^{1/2}_{L_\infty(\mathcal{N})} \\
&= \sup_{t\in\mathbb{R}} \left\|w^{-1}(1+it)\sigma_{-t/u}\bigl(g_1(1+it)g_1(1+it)^*\bigr)w^{-1}(1+it)^*\right\|^{1/2}_{L_\infty(\mathcal{N})} \\
&\leq \sup_{z\in\partial_1} \left\|w^{-1}(z)w(z)w(z)^*w^{-1}(z)^*\right\|^{1/2}_{L_\infty(\mathcal{N})} = 1.
\end{aligned}$$

In summary, by Kosaki's interpolation we find

$$\left\|h_1(\theta)D^{-\frac{1-\theta}{u}}\right\|_{L_{u/\theta}(\mathcal{N})} < \sqrt{1+\delta} \quad \text{and} \quad \left\|D^{-\frac{\theta}{u}}h_2(\theta)\right\|_{L_{q_\theta}} < 1.$$

Therefore, since we have

$$f(\theta) = \bigl(h_1(\theta)D^{-\frac{1-\theta}{u}}\bigr)\bigl(D^{-\frac{\theta}{u}}h_2(\theta)\bigr),$$

the result follows by letting $\delta \to 0$. This completes the proof. $\square$

REMARK 3.9. For the proof of Lemmas 3.7 and 3.8 it has been essential to have an explicit description of the intersection Δ of the interpolation pair. The lack of this description in the general case is what forces us to use Haagerup's construction to extend the result from finite to general von Neumann algebras.

3.2. CONDITIONAL EXPECTATIONS ON $\partial_\infty K$

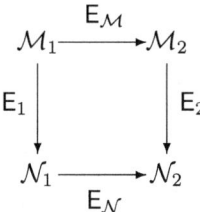

FIGURE II: A COMMUTATIVE SQUARE.

Let us consider a so-called commutative square of conditional expectations. That is, on one hand we have a von Neumann subalgebra $\mathcal{N}_1$ of $\mathcal{M}_1$. On the other, we consider a von Neumann subalgebra $\mathcal{N}_2$ of $\mathcal{M}_2$ such that $\mathcal{M}_2$ (resp. $\mathcal{N}_2$) is a von Neumann subalgebra of $\mathcal{M}_1$ (resp. $\mathcal{N}_1$) so that the diagram in Figure II below commutes. The interpolation results in Lemmas 3.7 and 3.8 will allow us to show that the conditional expectation $\mathsf{E}_\mathcal{M}$ extends to a contractive projection on the amalgamated spaces which correspond to points $(1/u, 1/v, 1/q)$ in the set $\partial_\infty K$.

OBSERVATION 3.10. Let us assume that the commutative square above satisfies that $\mathsf{E}_1(\mathcal{M}_2) \subset \mathcal{N}_2$. Then we claim that the restriction $\mathsf{E}_{1|\mathcal{M}_2}$ of E_1 to $\mathcal{M}_2$ coincides with E_2. Indeed, let us consider an element $x \in \mathcal{M}_2$. Then, since $\mathsf{E}_1(x) \in \mathcal{N}_2$, we have
$$\mathsf{E}_1(x) = \mathsf{E}_\mathcal{N} \circ \mathsf{E}_1(x) = \mathsf{E}_2 \circ \mathsf{E}_\mathcal{M}(x) = \mathsf{E}_2(x).$$
Similarly, we have $\mathsf{E}_{\mathcal{M}|\mathcal{N}_1} = \mathsf{E}_\mathcal{N}$ whenever $\mathsf{E}_\mathcal{M}(\mathcal{N}_1) \subset \mathcal{N}_2$. In what follows we shall assume these conditions on the commutative squares we are using. In fact, the commutative squares obtained from the Haagerup construction defined below will satisfy these assumptions.

PROPOSITION 3.11. *If $2 \le u, v \le \infty$, $\mathsf{E}_\mathcal{M}$ extends to contractions*
$$\mathsf{E}_\mathcal{M} : L_u(\mathcal{N}_1) L_2(\mathcal{M}_1) \to L_u(\mathcal{N}_2) L_2(\mathcal{M}_2),$$
$$\mathsf{E}_\mathcal{M} : L_2(\mathcal{M}_1) L_v(\mathcal{N}_1) \to L_2(\mathcal{M}_2) L_v(\mathcal{N}_2).$$

PROOF. Let us note that for any index $2 \le p \le \infty$, the natural inclusion
$$j : L_p^r(\mathcal{M}_2, \mathsf{E}_2) \to L_p^r(\mathcal{M}_1, \mathsf{E}_1)$$
is an isometry. Indeed, since the space $L_p(\mathcal{N}_2)$ embeds isometrically in $L_p(\mathcal{N}_1)$ and the conditional expectation E_2 is the restriction $\mathsf{E}_{1|\mathcal{M}_2}$ of E_1 to $\mathcal{M}_2$ (see Observation 3.10), the following identity holds for any $x \in L_p^r(\mathcal{M}_2, \mathsf{E}_2)$
$$\|j(x)\|_{L_p^r(\mathcal{M}_1, \mathsf{E}_1)} = \|\mathsf{E}_1(xx^*)^{1/2}\|_{L_p(\mathcal{N}_1)} = \|\mathsf{E}_2(xx^*)^{1/2}\|_{L_p(\mathcal{N}_2)} = \|x\|_{L_p^r(\mathcal{M}_2, \mathsf{E}_2)}.$$
Now let us consider the index $2 \le u \le \infty$ defined by $1/2 = 1/u + 1/p$. When $u > 2$ we have $p < \infty$ and Proposition 1.11 gives
$$L_u(\mathcal{N}_j) L_2(\mathcal{M}_j) = L_p^r(\mathcal{M}_j, \mathsf{E}_j)^* \quad \text{for} \quad j = 1, 2.$$
Therefore, our map $\mathsf{E}_\mathcal{M} : L_u(\mathcal{N}_1) L_2(\mathcal{M}_1) \to L_u(\mathcal{N}_2) L_2(\mathcal{M}_2)$ coincides with the adjoint of j so that we obtain a contraction. Finally, for $u = 2$ we just need to note that $L_u(\mathcal{N}_j) L_2(\mathcal{M}_j)$ embeds isometrically in $L_p^r(\mathcal{M}_j, \mathsf{E}_j)^*$ for $j = 0, 1$. Indeed, the first part of the proof of Proposition 1.11 holds even for $p = \infty$. Therefore, in this

case our map is a restriction of j^* to a closed subspace of $L_p^r(\mathcal{M}_1, \mathsf{E}_1)^*$ and hence contractive. The proof of the contractivity of the second mapping follows in the same way after replacing $L_p^r(\mathcal{M}, \mathsf{E})$ by $L_p^c(\mathcal{M}, \mathsf{E})$. □

PROPOSITION 3.12. *If $2 \leq u, v \leq \infty$, $\mathsf{E}_\mathcal{M}$ extends to a contraction*
$$\mathsf{E}_\mathcal{M} : L_u(\mathcal{N}_1) L_\infty(\mathcal{M}_1) L_v(\mathcal{N}_1) \to L_u(\mathcal{N}_2) L_\infty(\mathcal{M}_2) L_v(\mathcal{N}_2).$$

PROOF. According to Proposition 2.6, it suffices to prove the assertion on the dense subspace $j_{uv}(\mathcal{N}_{1u} L_\infty(\mathcal{M}_1) \mathcal{N}_{1v})$. Thus, let x be an element in such subspace so that it decomposes as $x = \mathrm{D}^{1/u} \alpha y_0 \beta \mathrm{D}^{1/v}$ with $\alpha, \beta \in \mathcal{N}_1$ and $y_0 \in \mathcal{M}_1$. Let us define the operators
$$a = (\alpha \alpha^* + \delta 1)^{1/2}, \quad b = (\beta^* \beta + \delta 1)^{1/2}, \quad y = a^{-1} \alpha y_0 \beta b^{-1},$$
so that $x = \mathrm{D}^{1/u} a y b \mathrm{D}^{1/v}$. Then, we proceed as in [**27**] by writing $\mathsf{E}_\mathcal{M}(x)$ as follows
$$\mathsf{E}_\mathcal{M}(x) = \mathrm{D}^{1/u} \mathsf{E}_\mathcal{M}(a^2)^{1/2} \Big[\mathsf{E}_\mathcal{M}(a^2)^{-1/2} \mathsf{E}_\mathcal{M}(ayb) \mathsf{E}_\mathcal{M}(b^2)^{-1/2} \Big] \mathsf{E}_\mathcal{M}(b^2)^{1/2} \mathrm{D}^{1/v}.$$
We clearly have
$$(3.12) \quad \begin{aligned} \left\| \mathrm{D}^{\frac{1}{u}} \mathsf{E}_\mathcal{M}(a^2)^{1/2} \right\|_{L_u(\mathcal{N}_2)} &\leq \left(\left\| \mathrm{D}^{\frac{1}{u}} \alpha \right\|_{L_u(\mathcal{N}_1)}^2 + \delta \right)^{1/2}, \\ \left\| \mathsf{E}_\mathcal{M}(b^2)^{1/2} \mathrm{D}^{\frac{1}{v}} \right\|_{L_v(\mathcal{N}_2)} &\leq \left(\left\| \beta \mathrm{D}^{\frac{1}{v}} \right\|_{L_v(\mathcal{N}_1)}^2 + \delta \right)^{1/2}. \end{aligned}$$

On the other hand, let us consider for a moment a von Neumann subalgebra $\mathcal{N}$ of a given von Neumann algebra $\mathcal{M}$ and let $\mathsf{E} : \mathcal{M} \to \mathcal{N}$ be the corresponding conditional expectation. Let us consider a positive element $\gamma \in \mathcal{M}_+$ satisfying $\gamma \geq \delta 1$. Then we define the map
$$\Lambda_\gamma : w \in \mathcal{M} \to \mathsf{E}(\gamma^2)^{-1/2} \mathsf{E}(\gamma w \gamma) \mathsf{E}(\gamma^2)^{-1/2} \in \mathcal{N}.$$

According to [**27**], this is a completely positive map so that $\|\Lambda_\gamma\| = \|\Lambda_\gamma(1)\| = 1$. Thus, Λ_γ is a contraction. Then we apply this result to
$$\gamma = \begin{pmatrix} a & 0 \\ 0 & b \end{pmatrix}.$$

More concretely, $\Lambda_\gamma : \mathrm{M}_2 \otimes \mathcal{M}_1 \to \mathrm{M}_2 \otimes \mathcal{M}_2$ is given by
$$\Lambda_\gamma(w) = \left[\widetilde{\mathsf{E}}_\mathcal{M} \begin{pmatrix} a^2 & 0 \\ 0 & b^2 \end{pmatrix} \right]^{-\frac{1}{2}} \widetilde{\mathsf{E}}_\mathcal{M}(\gamma w \gamma) \left[\widetilde{\mathsf{E}}_\mathcal{M} \begin{pmatrix} a^2 & 0 \\ 0 & b^2 \end{pmatrix} \right]^{-\frac{1}{2}} \quad \text{with } \widetilde{\mathsf{E}}_\mathcal{M} = id \otimes \mathsf{E}_\mathcal{M}.$$

Then we observe that
$$\Lambda_\gamma \begin{pmatrix} 0 & y \\ 0 & 0 \end{pmatrix} = \begin{pmatrix} 0 & \mathsf{E}_\mathcal{M}(a^2)^{-1/2} \mathsf{E}_\mathcal{M}(ayb) \mathsf{E}_\mathcal{M}(b^2)^{-1/2} \\ 0 & 0 \end{pmatrix}.$$

In particular, since $\|y\|_{\mathcal{M}_1} \leq \|a^{-1}\alpha\|_{\mathcal{N}_1} \|y_0\|_{\mathcal{M}_1} \|\beta b^{-1}\|_{\mathcal{N}_1} \leq \|y_0\|_{\mathcal{M}_1}$, we deduce
$$(3.13) \quad \left\| \mathsf{E}_\mathcal{M}(a^2)^{-1/2} \mathsf{E}_\mathcal{M}(ayb) \mathsf{E}_\mathcal{M}(b^2)^{-1/2} \right\|_{L_\infty(\mathcal{M}_2)} \leq \|y_0\|_{L_\infty(\mathcal{M}_1)}.$$

In summary, according to (3.12) and (3.13)
$$\left\| \mathrm{D}^{\frac{1}{u}} \mathsf{E}_\mathcal{M}(\alpha y_0 \beta) \mathrm{D}^{\frac{1}{v}} \right\|_{u \cdot \infty \cdot v} \leq \left(\left\| \mathrm{D}^{\frac{1}{u}} \alpha \right\|_{L_u(\mathcal{N}_1)}^2 + \delta \right)^{\frac{1}{2}} \|y_0\|_{L_\infty(\mathcal{M}_1)} \left(\left\| \beta \mathrm{D}^{\frac{1}{v}} \right\|_{L_v(\mathcal{N}_1)}^2 + \delta \right)^{\frac{1}{2}}.$$

The proof is completed by letting $\delta \to 0$ and taking the infimum on the right. □

COROLLARY 3.13. *If* $(1/u, 1/v, 1/q) \in \partial_\infty \mathsf{K}$, $\mathsf{E}_\mathcal{M}$ *extends to a contraction*
$$\mathsf{E}_\mathcal{M} : L_u(\mathcal{N}_1) L_q(\mathcal{M}_1) L_v(\mathcal{N}_1) \to L_u(\mathcal{N}_2) L_q(\mathcal{M}_2) L_v(\mathcal{N}_2).$$

PROOF. The case $1/q = 0$ has already been considered in Proposition 3.12. Thus, assume (without lost of generality) that $1/v = 0$. Then, we know from Propositions 3.11 and 3.12 that
$$\begin{aligned} \mathsf{E}_\mathcal{M} : & L_u(\mathcal{N}_1) L_2(\mathcal{M}_1) & \to & L_u(\mathcal{N}_2) L_2(\mathcal{M}_2), \\ \mathsf{E}_\mathcal{M} : & L_u(\mathcal{N}_1) L_\infty(\mathcal{M}_1) & \to & L_u(\mathcal{N}_2) L_\infty(\mathcal{M}_2), \end{aligned}$$
are contractions. Therefore, it follows from Lemma 3.7 that the same holds for
$$\mathsf{E}_\mathcal{M} : L_u(\mathcal{N}_1) L_q(\mathcal{M}_1) \to L_u(\mathcal{N}_2) L_q(\mathcal{M}_2)$$
when $2 \le q \le \infty$. Finally, it remains to consider the case $1 \le q \le 2$. Given $1 \le q \le 2$ and $2 \le u \le \infty$ such that $1/u + 1/q \le 1$, we consider the index defined by $1/p = 1/u + 1/q$. Then, according to Lemma 3.8 we have
$$L_u(\mathcal{N}_1) L_q(\mathcal{M}_1) = \bigl[L_p(\mathcal{M}_1), L_{u_1}(\mathcal{N}_1) L_2(\mathcal{M}_1) \bigr]_\theta,$$
for some $0 \le \theta \le 1$ and some $2 \le u_1 \le \infty$. Note that the fact that u_1 can be chosen so that $2 \le u_1 \le \infty$ follows by a quick look at Figure I. By Proposition 3.11, $\mathsf{E}_\mathcal{M}$ is contractive on $L_{u_1}(\mathcal{N}_1) L_2(\mathcal{M}_1)$ and of course on $L_p(\mathcal{M}_1)$. Therefore, the result follows by complex interpolation. □

3.3. General von Neumann algebras I

In this section we prove Lemma 3.1 under the assumption $\min(q_0, q_1) < \infty$. However, most of the arguments we are giving here will be useful for the remaining case. Before starting the proof we fix some notation. As in Lemma 3.1, we are given a von Neumann subalgebra $\mathcal{N}$ of $\mathcal{M}$ and we denote by $\mathsf{E} : \mathcal{M} \to \mathcal{N}$ the corresponding conditional expectation. According to the Haagerup construction sketched in Section 2.1, we have
$$\mathcal{R}_\mathcal{M} = \mathcal{M} \rtimes_\sigma \mathrm{G} = \overline{\bigcup_{k \ge 1} \mathcal{M}_k}.$$

As above, we consider the conditional expectation
$$\mathsf{E}_\mathcal{M} : \sum_{g \in \mathrm{G}} x_g \lambda(g) \in \mathcal{R}_\mathcal{M} \mapsto x_0 \in \mathcal{M}$$
and the state $\widehat{\varphi} = \varphi \circ \mathsf{E}_\mathcal{M}$ on $\mathcal{R}_\mathcal{M}$. Moreover, we have conditional expectations $\mathcal{E}_{\mathcal{M}_k} : \mathcal{R}_\mathcal{M} \to \mathcal{M}_k$ for each $k \ge 1$ so that (if φ_k denotes the restriction of $\widehat{\varphi}$ to $\mathcal{M}_k$ and D_{φ_k} stands for the corresponding density) the family of finite von Neumann subalgebras $(\mathcal{M}_k)_{k \ge 1}$ satisfy

(3.14) $$c_1(k) 1_{\mathcal{M}_k} \le \mathrm{D}_{\varphi_k} \le c_2(k) 1_{\mathcal{M}_k}.$$

Moreover, given $0 \le \eta \le 1$ and $1 \le p < \infty$, we have

(3.15) $$\lim_{k \to \infty} \left\| \mathrm{D}_{\widehat{\varphi}}^{(1-\eta)/p} \bigl(\hat{x} - \mathcal{E}_k(\hat{x}) \bigr) \mathrm{D}_{\widehat{\varphi}}^{\eta/p} \right\|_{L_p(\mathcal{R}_\mathcal{M})} = 0$$

for any $\hat{x} \in \mathcal{R}_\mathcal{M}$. Clearly, we can consider another Haagerup construction for the von Neumann subalgebra $\mathcal{N}$ so that the analogous properties hold. In summary, we sketch the situation in Figure III below. Now let us consider two points

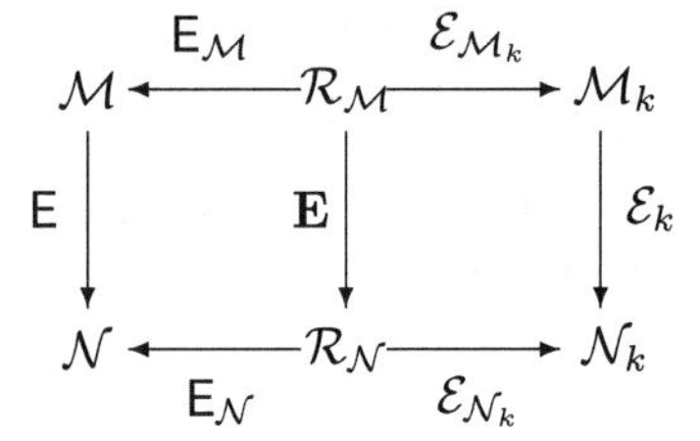

$$\begin{array}{ccc}
\mathcal{M} & \xleftarrow{\mathsf{E}_\mathcal{M}} \mathcal{R}_\mathcal{M} \xrightarrow{\mathcal{E}_{\mathcal{M}_k}} & \mathcal{M}_k \\
\mathsf{E} \downarrow & \mathbf{E} \downarrow & \downarrow \mathcal{E}_k \\
\mathcal{N} & \xleftarrow[\mathsf{E}_\mathcal{N}]{} \mathcal{R}_\mathcal{N} \xrightarrow[\mathcal{E}_{\mathcal{N}_k}]{} & \mathcal{N}_k
\end{array}$$

FIGURE III: HAAGERUP'S CONSTRUCTION.

$(1/u_0, 1/v_0, 1/q_0)$ and $(1/u_1, 1/v_1, 1/q_1)$ in $\partial_\infty \mathsf{K}$. Then it is clear that the natural inclusion mappings

$$id : L_{u_0}(\mathcal{N})L_{q_0}(\mathcal{M})L_{v_0}(\mathcal{N}) \to L_{u_0}(\mathcal{R}_\mathcal{N})L_{q_0}(\mathcal{R}_\mathcal{M})L_{v_0}(\mathcal{R}_\mathcal{N}),$$
$$id : L_{u_1}(\mathcal{N})L_{q_1}(\mathcal{M})L_{v_1}(\mathcal{N}) \to L_{u_1}(\mathcal{R}_\mathcal{N})L_{q_1}(\mathcal{R}_\mathcal{M})L_{v_1}(\mathcal{R}_\mathcal{N}),$$

are contractions. In particular, defining for $0 \leq \theta \leq 1$

$$\mathrm{X}_\theta(\mathcal{M}) = \Big[L_{u_0}(\mathcal{N})L_{q_0}(\mathcal{M})L_{v_0}(\mathcal{N}), L_{u_1}(\mathcal{N})L_{q_1}(\mathcal{M})L_{v_1}(\mathcal{N})\Big]_\theta,$$
$$\mathrm{X}_\theta(\mathcal{R}_\mathcal{M}) = \Big[L_{u_0}(\mathcal{R}_\mathcal{N})L_{q_0}(\mathcal{R}_\mathcal{M})L_{v_0}(\mathcal{R}_\mathcal{N}), L_{u_1}(\mathcal{R}_\mathcal{N})L_{q_1}(\mathcal{R}_\mathcal{M})L_{v_1}(\mathcal{R}_\mathcal{N})\Big]_\theta,$$

we obtain by interpolation that $id : \mathrm{X}_\theta(\mathcal{M}) \to \mathrm{X}_\theta(\mathcal{R}_\mathcal{M})$ is also a contraction. Now, considering the commutative square on the right of Figure III, it follows from Corollary 3.13 that for each $k \geq 1$ we have contractions

$$\mathcal{E}_{\mathcal{M}_k} : L_{u_0}(\mathcal{R}_\mathcal{N})L_{q_0}(\mathcal{R}_\mathcal{M})L_{v_0}(\mathcal{R}_\mathcal{N}) \to L_{u_0}(\mathcal{N}_k)L_{q_0}(\mathcal{M}_k)L_{v_0}(\mathcal{N}_k),$$
$$\mathcal{E}_{\mathcal{M}_k} : L_{u_1}(\mathcal{R}_\mathcal{N})L_{q_1}(\mathcal{R}_\mathcal{M})L_{v_1}(\mathcal{R}_\mathcal{N}) \to L_{u_1}(\mathcal{N}_k)L_{q_1}(\mathcal{M}_k)L_{v_1}(\mathcal{N}_k).$$

Moreover, $\mathcal{M}_k$ satisfies (3.14) for each $k \geq 1$. In particular, since we have already proved that Lemma 3.1 holds in this case, we obtain by complex interpolation a contraction $\mathcal{E}_{\mathcal{M}_k} : \mathrm{X}_\theta(\mathcal{R}_\mathcal{M}) \to L_{u_\theta}(\mathcal{N}_k)L_{q_\theta}(\mathcal{M}_k)L_{v_\theta}(\mathcal{N}_k)$. In summary, writing

$$\mathcal{E}_{\mathcal{M}_k} = \mathcal{E}_{\mathcal{M}_k} \circ id,$$

we have found a contraction

(3.16) $$\mathcal{E}_{\mathcal{M}_k} : \mathrm{X}_\theta(\mathcal{M}) \to L_{u_\theta}(\mathcal{N}_k)L_{q_\theta}(\mathcal{M}_k)L_{v_\theta}(\mathcal{N}_k).$$

On the other hand, regarding the space $L_{u_\theta}(\mathcal{N}_k)L_{q_\theta}(\mathcal{M}_k)L_{v_\theta}(\mathcal{N}_k)$ as a subspace of $L_{u_\theta}(\mathcal{R}_\mathcal{N})L_{q_\theta}(\mathcal{R}_\mathcal{M})L_{v_\theta}(\mathcal{R}_\mathcal{N})$, we can consider the restriction of the conditional expectation $\mathsf{E}_\mathcal{M}$ to $\mathcal{M}_k$ and define the following map

$$\mathsf{E}_\mathcal{M} : L_{u_\theta}(\mathcal{N}_k)L_{q_\theta}(\mathcal{M}_k)L_{v_\theta}(\mathcal{N}_k) \to L_{u_\theta}(\mathcal{N})L_{q_\theta}(\mathcal{M})L_{v_\theta}(\mathcal{N}).$$

The following technical result will be the key to prove Lemma 3.1.

PROPOSITION 3.14. *If $k \geq 1$ and $0 \leq \theta \leq 1$, $\mathsf{E}_\mathcal{M}$ extends to a contraction*

$$\mathsf{E}_\mathcal{M} : L_{u_\theta}(\mathcal{N}_k)L_{q_\theta}(\mathcal{M}_k)L_{v_\theta}(\mathcal{N}_k) \to L_{u_\theta}(\mathcal{N})L_{q_\theta}(\mathcal{M})L_{v_\theta}(\mathcal{N}).$$

PROOF. It clearly suffices to see our assertion on the dense subspace

$$\mathcal{A}_k = \mathrm{D}_{\varphi_k}^{1/u_\theta} \mathcal{N}_{k,a} \mathrm{D}_{\varphi_k}^{1/2q_\theta} \mathcal{M}_{k,a} \mathrm{D}_{\varphi_k}^{1/2q_\theta} \mathcal{N}_{k,a} \mathrm{D}^{1/v_\theta},$$

where $\mathcal{M}_{k,a}$ and $\mathcal{N}_{k,a}$ denote the $*$-algebras of analytic elements in $\mathcal{M}_k$ and $\mathcal{N}_k$ respectively. Thus, let us consider an element x in $\mathcal{A}_k$ and assume that the norm

3.3. GENERAL VON NEUMANN ALGEBRAS I

of x in $L_{u_\theta}(\mathcal{N}_k)L_{q_\theta}(\mathcal{M}_k)L_{v_\theta}(\mathcal{N}_k)$ is less that 1. Then, since we know that Lemma 3.1 holds for finite von Neumann algebras satisfying (3.14), we can find a bounded analytic function $f : \mathcal{S} \to \mathcal{A}_k$ satisfying $f(\theta) = x$ and the boundary estimate

$$\max\Big\{\sup_{z \in \partial_0} \|f(z)\|_{u_0 \cdot q_0 \cdot v_0}, \sup_{z \in \partial_1} \|f(z)\|_{u_1 \cdot q_1 \cdot v_1}\Big\} < 1.$$

Then we use the hypothesis $1/u_\theta + 1/q_\theta + 1/v_\theta = 1/u_0 + 1/v_0 + 1/q_0$ and Lemma 1.2 to rewrite $\mathcal{A}_k$ as follows

$$\mathcal{A}_k = \mathrm{D}_{\varphi_k}^{1/u_0 + 1/2q_0} \mathcal{M}_{k,a} \mathrm{D}_{\varphi_k}^{1/2q_0 + 1/v_0}.$$

Therefore, multiplying if necessary on the left and on the right by certain powers of $\mathrm{D}_{\varphi_k}^z$ and its inverses, we may assume that f has the following form (recall that we are using here the property (3.14))

$$(3.17) \qquad f(z) = \mathrm{D}_{\varphi_k}^{\frac{1-z}{u_0}} \mathrm{D}_{\varphi_k}^{\frac{z}{u_1}} \mathrm{D}_{\varphi_k}^{\frac{1-z}{2q_0}} \mathrm{D}_{\varphi_k}^{\frac{z}{2q_1}} f_1(z) \mathrm{D}_{\varphi_k}^{\frac{z}{2q_1}} \mathrm{D}_{\varphi_k}^{\frac{1-z}{2q_0}} \mathrm{D}_{\varphi_k}^{\frac{z}{v_1}} \mathrm{D}_{\varphi_k}^{\frac{1-z}{v_0}},$$

with $f_1 : \mathcal{S} \to \mathcal{M}_{k,a}$ bounded analytic. In particular, we have

$$\sup_{z \in \partial_0} \left\|\mathrm{D}_{\varphi_k}^{\frac{1}{u_0} + \frac{1}{2q_0}} f_1(z) \mathrm{D}_{\varphi_k}^{\frac{1}{2q_0} + \frac{1}{v_0}}\right\|_{u_0 \cdot q_0 \cdot v_0} < 1,$$

$$\sup_{z \in \partial_1} \left\|\mathrm{D}_{\varphi_k}^{\frac{1}{u_1} + \frac{1}{2q_1}} f_1(z) \mathrm{D}_{\varphi_k}^{\frac{1}{2q_1} + \frac{1}{v_1}}\right\|_{u_1 \cdot q_1 \cdot v_1} < 1.$$

Note that here we have used the fact that $\mathrm{D}_{\varphi_k}^w$ is a unitary for any $w \in \mathbb{C}$ such that $\mathrm{Re}\, w = 0$. Now we apply Corollary 3.13 to the commutative square on the left of Figure III. In other words, we have contractions

$$\mathsf{E}_{\mathcal{M}} : L_{u_j}(\mathcal{N}_k)L_{q_j}(\mathcal{M}_k)L_{v_j}(\mathcal{N}_k) \to L_{u_j}(\mathcal{N})L_{q_j}(\mathcal{M})L_{v_j}(\mathcal{N})$$

since $(1/u_j, 1/v_j, 1/q_j) \in \partial_\infty \mathsf{K}$ for $j = 0, 1$. Therefore,

$$\sup_{z \in \partial_0} \left\|\mathrm{D}_{\varphi}^{\frac{1}{u_0} + \frac{1}{2q_0}} \mathsf{E}_{\mathcal{M}}(f_1(z)) \mathrm{D}_{\varphi}^{\frac{1}{2q_0} + \frac{1}{v_0}}\right\|_{u_0 \cdot q_0 \cdot v_0} < 1,$$

$$\sup_{z \in \partial_1} \left\|\mathrm{D}_{\varphi}^{\frac{1}{u_1} + \frac{1}{2q_1}} \mathsf{E}_{\mathcal{M}}(f_1(z)) \mathrm{D}_{\varphi}^{\frac{1}{2q_1} + \frac{1}{v_1}}\right\|_{u_1 \cdot q_1 \cdot v_1} < 1.$$

Then, according to Proposition 2.6, we can find functions $g_1, g_3 : \partial \mathcal{S} \to \mathcal{N}$ and $g_2 : \partial_j \to L_{q_j}(\mathcal{M})$ such that

$$\mathrm{D}_{\varphi}^{1/2q_j} \mathsf{E}_{\mathcal{M}}(f_1(z)) \mathrm{D}_{\varphi}^{1/2q_j} = g_1(z)g_2(z)g_3(z) \quad \text{for all} \quad z \in \partial_j$$

and satisfying the following boundary estimates

$$\sup_{z \in \partial_0} \Big\{\big\|\mathrm{D}_{\varphi}^{\frac{1}{u_0}} g_1(z)\big\|_{L_{u_0}(\mathcal{N})}, \big\|g_2(z)\big\|_{L_{q_0}(\mathcal{M})}, \big\|g_3(z)\mathrm{D}_{\varphi}^{\frac{1}{v_0}}\big\|_{L_{v_0}(\mathcal{N})}\Big\} < 1,$$

$$\sup_{z \in \partial_1} \Big\{\big\|\mathrm{D}_{\varphi}^{\frac{1}{u_1}} g_1(z)\big\|_{L_{u_1}(\mathcal{N})}, \big\|g_2(z)\big\|_{L_{q_1}(\mathcal{M})}, \big\|g_3(z)\mathrm{D}_{\varphi}^{\frac{1}{v_1}}\big\|_{L_{v_1}(\mathcal{N})}\Big\} < 1.$$

Then we can proceed as in the proof of Lemma 3.1 for finite von Neumann algebras. Namely, we apply Devinatz's factorization theorem to the functions

$$\mathrm{W}_1 : z \in \partial \mathcal{S} \mapsto \sigma_{\mathrm{Im} z/2q_1 - \mathrm{Im} z/2q_0}\big(g_1(z)g_1(z)^*\big) + \delta 1 \in \mathcal{N},$$

$$\mathrm{W}_3 : z \in \partial \mathcal{S} \mapsto \sigma_{\mathrm{Im} z/2q_0 - \mathrm{Im} z/2q_1}\big(g_3(z)^* g_3(z)\big) + \delta 1 \in \mathcal{N},$$

so that we can find invertible bounded analytic functions $\mathrm{w}_1, \mathrm{w}_3 : \mathcal{S} \to \mathcal{N}$ satisfying

$$\mathrm{w}_1(z)\mathrm{w}_1(z)^* = \mathrm{W}_1(z),$$

$$w_3(z)^* w_3(z) = W_3(z),$$

for all $z \in \partial \mathcal{S}$. Then we consider the factorization

$$\mathsf{E}_{\mathcal{M}}(f(z)) = h_1(z) h_2(z) h_3(z)$$

with $h_2(z) = h_1^{-1}(z) \mathsf{E}_{\mathcal{M}}(f(z)) h_3^{-1}(z)$ and h_1, h_3 given by

$$h_1(z) = \mathrm{D}_\varphi^{\frac{1-z}{u_0}} \mathrm{D}_\varphi^{\frac{z}{u_1}} \mathrm{w}_1(z) \mathrm{D}_\varphi^{-\frac{z}{u_1}} \mathrm{D}_\varphi^{\frac{z}{u_0}},$$

$$h_3(z) = \mathrm{D}_\varphi^{\frac{z}{v_0}} \mathrm{D}_\varphi^{-\frac{z}{v_1}} \mathrm{w}_3(z) \mathrm{D}_\varphi^{\frac{z}{v_1}} \mathrm{D}_\varphi^{\frac{1-z}{v_0}}.$$

To estimate the norm of $h_j(\theta)$ for $j = 1, 2, 3$ in the corresponding L_p space we proceed as in the proof of Lemma 3.1 for finite von Neumann algebras. That is, we first show the boundedness and analyticity of h_1, h_2, h_3. This enables us to estimate the norm on the boundary $\partial \mathcal{S}$ and apply Kosaki's interpolation. Let us start with the function h_1. To that aim we note that

(a) If $u_0 \leq u_1$, we can write

$$h_1(z) = \mathrm{D}_\varphi^{\frac{1}{u_1}} \mathrm{D}_\varphi^{(\frac{1}{u_0} - \frac{1}{u_1})(1-z)} \mathrm{w}_1(z) \mathrm{D}_\varphi^{(\frac{1}{u_0} - \frac{1}{u_1})z}.$$

(b) If $u_0 \geq u_1$, we can write

$$h_1(z) = \left(\mathrm{D}_\varphi^{\frac{1}{u_0}} \mathrm{D}_\varphi^{(\frac{1}{u_1} - \frac{1}{u_0})z} \mathrm{w}_1(z) \mathrm{D}_\varphi^{(\frac{1}{u_1} - \frac{1}{u_0})(1-z)} \right) \mathrm{D}_\varphi^{(\frac{1}{u_0} - \frac{1}{u_1})}.$$

By Lemma 3.6, h_1 is either bounded analytic or can be written as

$$h_1(z) = j_1(z) \mathrm{D}_\varphi^{(\frac{1}{u_0} - \frac{1}{u_1})},$$

with j_1 bounded analytic. In any case, Kosaki's interpolation gives

$$\left\| h_1(\theta) \mathrm{D}_\varphi^{\frac{\theta}{u_1} - \frac{\theta}{u_0}} \right\|_{L_{u_\theta}(\mathcal{N})} \leq \max \left\{ \sup_{z \in \partial_0} \left\| h_1(z) \right\|_{L_{u_0}(\mathcal{N})}, \sup_{z \in \partial_1} \left\| h_1(z) \mathrm{D}_\varphi^{\frac{1}{u_1} - \frac{1}{u_0}} \right\|_{L_{u_1}(\mathcal{N})} \right\}.$$

Moreover, arguing as in the proof of Lemma 3.1 for finite von Neumann algebras

$$\sup_{z \in \partial_0} \left\| h_1(z) \right\|_{L_{u_0}(\mathcal{N})} < \sqrt{1+\delta},$$

$$\sup_{z \in \partial_1} \left\| h_1(z) \mathrm{D}_\varphi^{\frac{1}{u_1} - \frac{1}{u_0}} \right\|_{L_{u_1}(\mathcal{N})} < \sqrt{1+\delta}.$$

Indeed, the only significant difference (with respect to the proof for finite von Neumann algebras) is that we can not assume any longer that D_φ^w is a unitary for a purely imaginary complex number w. However, this difficulty is easily avoided by using the norm-invariance property of the one-parameter modular automorphism group. In fact, that is the reason why we used the modular automorphism group in the definition of W_1 and W_3, see also the proof of Lemma 3.8. Our previous estimates give rise to

(3.18) $$\left\| h_1(\theta) \mathrm{D}_\varphi^{\frac{\theta}{u_1} - \frac{\theta}{u_0}} \right\|_{L_{u_\theta}(\mathcal{N})} < \sqrt{1+\delta}.$$

Similarly, we have

(3.19) $$\left\| \mathrm{D}_\varphi^{\frac{\theta}{v_1} - \frac{\theta}{v_0}} h_3(\theta) \right\|_{L_{v_\theta}(\mathcal{N})} < \sqrt{1+\delta}.$$

3.3. GENERAL VON NEUMANN ALGEBRAS I

Finally, we consider the function h_2. To study the boundedness and analyticity of h_2, we recall the expression for f obtained in (3.17). Then, we can rewrite h_2 in the following way

$$h_2(z) = D_\varphi^{\frac{z}{u_1}} D_\varphi^{-\frac{z}{u_0}} w_1^{-1}(z) D_\varphi^{\frac{1-z}{2q_0}} D_\varphi^{\frac{z}{2q_1}} \mathsf{E}_\mathcal{M}(f_1(z)) D_\varphi^{\frac{z}{2q_1}} D_\varphi^{\frac{1-z}{2q_0}} w_3^{-1}(z) D_\varphi^{-\frac{z}{v_0}} D_\varphi^{\frac{z}{v_1}}.$$

Using one more time that $1/u_0 + 1/q_0 + 1/v_0 = 1/u_1 + 1/q_1 + 1/v_1$, we can write

(a) If $\max(u_0, u_1) = u_0$ and $\max(v_0, v_1) = v_0$, we have

$$\begin{aligned} h_2(z) &= D_\varphi^{\frac{z}{u_1} - \frac{z}{u_0}} w_1^{-1}(z) D_\varphi^{\frac{1-z}{u_1} - \frac{1-z}{u_0}} \\ &\times \mathsf{E}_\mathcal{M}\left(D_{\varphi_k}^{\frac{1-z}{u_0} - \frac{1-z}{u_1} + \frac{1-z}{2q_0} + \frac{z}{2q_1}} f_1(z) D_{\varphi_k}^{\frac{z}{2q_1} + \frac{1-z}{2q_0} - \frac{1-z}{v_1} + \frac{1-z}{v_0}} \right) \\ &\times D_\varphi^{\frac{1-z}{v_1} - \frac{1-z}{v_0}} w_3^{-1}(z) D_\varphi^{\frac{z}{v_1} - \frac{z}{v_0}}. \end{aligned}$$

The first and third terms on the right hand side are bounded analytic by Lemma 3.6. The middle term is clearly bounded analytic in $L_{q_1}(\mathcal{M})$. Indeed, it follows easily from the fact that D_{φ_k} satisfies (3.14).

(b) If $\max(u_0, u_1) = u_1$ and $\max(v_0, v_1) = v_1$, we proceed as above with

$$D_\varphi^{\frac{1}{u_0} - \frac{1}{u_1}} h_2(z) D_\varphi^{\frac{1}{v_0} - \frac{1}{v_1}}.$$

(c) If $\max(u_0, u_1) = u_0$ and $\max(v_0, v_1) = v_1$, we proceed as above with

$$h_2(z) D_\varphi^{\frac{1}{v_0} - \frac{1}{v_1}}.$$

(d) If $\max(u_0, u_1) = u_1$ and $\max(v_0, v_1) = v_0$, we proceed as above with

$$D_\varphi^{\frac{1}{u_0} - \frac{1}{u_1}} h_2(z).$$

In any of the possible situations considered, Kosaki's interpolation provides the following estimate

$$\left\| D_\varphi^{\frac{\theta}{u_0} - \frac{\theta}{u_1}} h_2(\theta) D_\varphi^{\frac{\theta}{v_0} - \frac{\theta}{v_1}} \right\|_{L_{q_\theta}(\mathcal{M})}$$
$$\leq \max\left\{ \sup_{z \in \partial_0} \|h_2(z)\|_{L_{q_0}(\mathcal{M})}, \sup_{z \in \partial_1} \left\| D_\varphi^{\frac{1}{u_0} - \frac{1}{u_1}} h_2(z) D_\varphi^{\frac{1}{v_0} - \frac{1}{v_1}} \right\|_{L_{q_1}(\mathcal{M})} \right\}.$$

On the other hand, arguing one more time as in the proof of Lemma 3.1,

$$\sup_{z \in \partial_0} \|h_2(z)\|_{L_{q_0}(\mathcal{M})} < 1,$$
$$\sup_{z \in \partial_1} \left\| D_\varphi^{\frac{1}{u_0} - \frac{1}{u_1}} h_2(z) D_\varphi^{\frac{1}{v_0} - \frac{1}{v_1}} \right\|_{L_{q_1}(\mathcal{M})} < 1.$$

In particular,

(3.20) $$\left\| D_\varphi^{\frac{\theta}{u_0} - \frac{\theta}{u_1}} h_2(\theta) D_\varphi^{\frac{\theta}{v_0} - \frac{\theta}{v_1}} \right\|_{L_{q_\theta}(\mathcal{M})} < 1.$$

In summary, we have found a factorization of $\mathsf{E}_\mathcal{M}(x)$

$$\mathsf{E}_\mathcal{M}(f(\theta)) = \left(h_1(\theta) D_\varphi^{\frac{\theta}{u_1} - \frac{\theta}{u_0}} \right) \left(D_\varphi^{\frac{\theta}{u_0} - \frac{\theta}{u_1}} h_2(\theta) D_\varphi^{\frac{\theta}{v_0} - \frac{\theta}{v_1}} \right) \left(D_\varphi^{\frac{\theta}{v_1} - \frac{\theta}{v_0}} h_3(\theta) \right)$$

which, according to (3.18, 3.19, 3.20), provides the estimate

$$\|\mathsf{E}_\mathcal{M}(x)\|_{u_\theta \cdot q_\theta \cdot v_\theta} < 1 + \delta.$$

Therefore, the proof is concluded by letting $\delta \to 0$ in the inequality above. □

COROLLARY 3.15. *If $k \geq 1$ and $0 \leq \theta \leq 1$, we have*
$$\left\|\mathsf{E}_\mathcal{M}\mathcal{E}_{\mathcal{M}_k}(x)\right\|_{u_\theta \cdot q_\theta \cdot v_\theta} \leq \|x\|_{\mathrm{X}_\theta(\mathcal{M})} \quad \text{for all} \quad x \in \mathrm{X}_\theta(\mathcal{M}).$$

PROOF. The result follows automatically from (3.16) and Proposition 3.14. □

PROOF OF LEMMA 3.1. As in Lemmas 3.7 and 3.8 and also as in the proof for finite von Neumann algebras, the lower estimate follows by using multilinear interpolation. This means that we have a contraction
$$(3.21) \qquad id : L_{u_\theta}(\mathcal{N}) L_{q_\theta}(\mathcal{M}) L_{v_\theta}(\mathcal{N}) \to \mathrm{X}_\theta(\mathcal{M}).$$
To prove the converse, let us consider the space
$$\mathcal{A} = \mathrm{D}_\varphi^{1/u_\theta} \mathcal{N} \mathrm{D}_\varphi^{1/2q_\theta} \mathcal{M} \mathrm{D}_\varphi^{1/2q_\theta} \mathcal{N} \mathrm{D}_\varphi^{1/v_\theta}.$$
According to Proposition 2.8 and Corollary 3.15, given $x \in \mathcal{A}$ we have
$$\|x\|_{u_\theta \cdot q_\theta \cdot v_\theta}^\gamma \leq \|x\|_{\mathrm{X}_\theta(\mathcal{M})}^\gamma + \lim_{k \to \infty} \left\|x - \mathsf{E}_\mathcal{M}\mathcal{E}_{\mathcal{M}_k}(x)\right\|_{u_\theta \cdot q_\theta \cdot v_\theta}^\gamma.$$
Therefore, we need to see that the limit above is 0. To that aim, we use that $x \in \mathcal{A}$ so that we can write
$$x = \mathrm{D}_\varphi^{1/u_\theta} y \mathrm{D}_\varphi^{1/v_\theta} \quad \text{with} \quad y \in L_{q_\theta}(\mathcal{M}).$$
Then we use the contractivity of $\mathsf{E}_\mathcal{M}$ on $L_{q_\theta}(\mathcal{R}_\mathcal{M})$ to obtain
$$\begin{aligned}
\lim_{k\to\infty} \left\|x - \mathsf{E}_\mathcal{M}\mathcal{E}_{\mathcal{M}_k}(x)\right\|_{u_\theta \cdot q_\theta \cdot v_\theta} &= \lim_{k\to\infty} \left\|\mathrm{D}_\varphi^{1/u_\theta}(y - \mathsf{E}_\mathcal{M}\mathcal{E}_{\mathcal{M}_k}(y))\mathrm{D}_\varphi^{1/v_\theta}\right\|_{u_\theta \cdot q_\theta \cdot v_\theta} \\
&\leq \lim_{k\to\infty} \left\|y - \mathsf{E}_\mathcal{M}\mathcal{E}_{\mathcal{M}_k}(y)\right\|_{L_{q_\theta}(\mathcal{M})} \\
&= \lim_{k\to\infty} \left\|\mathsf{E}_\mathcal{M}(y - \mathcal{E}_{\mathcal{M}_k}(y))\right\|_{L_{q_\theta}(\mathcal{M})} \\
&\leq \lim_{k\to\infty} \left\|y - \mathcal{E}_{\mathcal{M}_k}(y)\right\|_{L_{q_\theta}(\mathcal{R}_\mathcal{M})}.
\end{aligned}$$
Then, recalling our hypothesis $\min(q_0, q_1) < \infty$ assumed at the beginning of this paragraph, we deduce that $q_\theta < \infty$ for any $0 < \theta < 1$. Therefore, according to (3.15) we conclude that the limit above is 0. In particular, we have
$$\|x\|_{u_\theta \cdot q_\theta \cdot v_\theta} = \|x\|_{\mathrm{X}_\theta(\mathcal{M})} \quad \text{for all} \quad x \in \mathcal{A}.$$
Now, using the density of $\mathcal{A}$ in $L_{u_\theta}(\mathcal{N}) L_{q_\theta}(\mathcal{M}) L_{v_\theta}(\mathcal{N})$ and (3.21) we deduce that the same holds for any $x \in L_{u_\theta}(\mathcal{N}) L_{q_\theta}(\mathcal{M}) L_{v_\theta}(\mathcal{N})$. Hence, it remains to see that $L_{u_\theta}(\mathcal{N}) L_{q_\theta}(\mathcal{M}) L_{v_\theta}(\mathcal{N})$ is dense in $\mathrm{X}_\theta(\mathcal{M})$. To that aim, it suffices to prove that the subspace
$$\mathsf{E}_\mathcal{M}\mathcal{E}_{\mathcal{M}_k}(\mathrm{X}_\theta(\mathcal{M})) \subset L_{u_\theta}(\mathcal{N}) L_{q_\theta}(\mathcal{M}) L_{v_\theta}(\mathcal{N})$$
is dense in $\mathrm{X}_\theta(\mathcal{M})$. Moreover, since the intersection space
$$\Delta = L_{u_0}(\mathcal{N}) L_{q_0}(\mathcal{M}) L_{v_0}(\mathcal{N}) \cap L_{u_1}(\mathcal{N}) L_{q_1}(\mathcal{M}) L_{v_1}(\mathcal{N})$$
is dense in $\mathrm{X}_\theta(\mathcal{M})$ for any $0 < \theta < 1$, we just need to approximate any element x in Δ by an element in $\mathsf{E}_\mathcal{M}\mathcal{E}_{\mathcal{M}_k}(\mathrm{X}_\theta(\mathcal{M}))$ with respect to the norm of $\mathrm{X}_\theta(\mathcal{M})$. Using one more time that $\min(q_0, q_1) < \infty$ we assume (without lost of generality) that $q_0 < \infty$. Then, applying the three lines lemma
$$\left\|x - \mathsf{E}_\mathcal{M}\mathcal{E}_{\mathcal{M}_k}(x)\right\|_{\mathrm{X}_\theta(\mathcal{M})} \leq \left\|x - \mathsf{E}_\mathcal{M}\mathcal{E}_{\mathcal{M}_k}(x)\right\|_{u_0 \cdot q_0 \cdot v_0}^{1-\theta} \left\|x - \mathsf{E}_\mathcal{M}\mathcal{E}_{\mathcal{M}_k}(x)\right\|_{u_1 \cdot q_1 \cdot v_1}^{\theta},$$

we just need to show that the first term on the right tends to 0 as $k \to \infty$ while the second term is uniformly bounded on k. The uniform boundedness follows from Corollary 3.15 since
$$\left\| x - \mathsf{E}_{\mathcal{M}} \mathcal{E}_{\mathcal{M}_k}(x) \right\|_{u_1 \cdot q_1 \cdot v_1} \leq \|x\|_{u_1 \cdot q_1 \cdot v_1} + \left\| \mathsf{E}_{\mathcal{M}} \mathcal{E}_{\mathcal{M}_k}(x) \right\|_{u_1 \cdot q_1 \cdot v_1} \leq 2\|x\|_{u_1 \cdot q_1 \cdot v_1}.$$
For the first term, we pick $y \in L_{q_0}(\mathcal{M})$ so that
$$\left\| x - \mathrm{D}_\varphi^{1/u_0} y \mathrm{D}_\varphi^{1/v_0} \right\|_{u_0 \cdot q_0 \cdot v_0} \leq \delta.$$
Then we have
$$\begin{aligned}
\left\| x - \mathsf{E}_{\mathcal{M}} \mathcal{E}_{\mathcal{M}_k}(x) \right\|_{u_0 \cdot q_0 \cdot v_0} &\leq \left\| x - \mathrm{D}_\varphi^{1/u_0} y \mathrm{D}_\varphi^{1/v_0} \right\|_{u_0 \cdot q_0 \cdot v_0} \\
&+ \left\| \mathrm{D}_\varphi^{1/u_0} \big(y - \mathsf{E}_{\mathcal{M}} \mathcal{E}_{\mathcal{M}_k}(y)\big) \mathrm{D}_\varphi^{1/v_0} \right\|_{u_0 \cdot q_0 \cdot v_0} \\
&+ \left\| \mathrm{D}_\varphi^{1/u_0} \mathsf{E}_{\mathcal{M}} \mathcal{E}_{\mathcal{M}_k}(y) \mathrm{D}_\varphi^{1/v_0} - \mathsf{E}_{\mathcal{M}} \mathcal{E}_{\mathcal{M}_k}(x) \right\|_{u_0 \cdot q_0 \cdot v_0}.
\end{aligned}$$
In particular, according to Corollary 3.15
$$\begin{aligned}
\left\| x - \mathsf{E}_{\mathcal{M}} \mathcal{E}_{\mathcal{M}_k}(x) \right\|_{u_0 \cdot q_0 \cdot v_0} &\leq 2\delta + \left\| \mathrm{D}_\varphi^{1/u_0} \big(y - \mathsf{E}_{\mathcal{M}} \mathcal{E}_{\mathcal{M}_k}(y)\big) \mathrm{D}_\varphi^{1/v_0} \right\|_{u_0 \cdot q_0 \cdot v_0} \\
&\leq 2\delta + \left\| y - \mathsf{E}_{\mathcal{M}} \mathcal{E}_{\mathcal{M}_k}(y) \right\|_{L_{q_0}(\mathcal{M})} \\
&= 2\delta + \left\| \mathsf{E}_{\mathcal{M}} \big(y - \mathcal{E}_{\mathcal{M}_k}(y)\big) \right\|_{L_{q_0}(\mathcal{M})} \\
&\leq 2\delta + \left\| y - \mathcal{E}_{\mathcal{M}_k}(y) \right\|_{L_{q_0}(\mathcal{R}_{\mathcal{M}})}.
\end{aligned}$$
Finally, since $q_0 < \infty$ by hypothesis, we know that the second term on the right tends to 0 as $k \to \infty$. Then, we let $\delta \to 0$. This completes the proof of Lemma 3.1 for general von Neumann algebras with $\min(q_0, q_1) < \infty$. $\square$

3.4. General von Neumann algebras II

To complete the proof of Lemma 3.1 we have to study the case $\min(q_0, q_1) = \infty$. Note that the proof above fails in this case since (3.15) does not hold for $p = \infty$. However, Corollary 3.15 is still valid in this case so that it suffices to see that

$$(3.22) \qquad \|x\|_{u_\theta \cdot \infty \cdot v_\theta} \leq \sup_{k \geq 1} \left\| \mathsf{E}_{\mathcal{M}} \mathcal{E}_{\mathcal{M}_k}(x) \right\|_{u_\theta \cdot \infty \cdot v_\theta} \quad \text{for all} \quad x \in \mathrm{X}_\theta(\mathcal{M}).$$

Indeed, since the lower estimate follows one more time by multilinear interpolation, inequality (3.22) and Corollary 3.15 are enough to conclude the proof of Lemma 3.1. In order to prove (3.22) we shall need to consider the spaces
$$L_p^r(\mathcal{M}, \mathsf{E}) \otimes_{\mathcal{M}} L_q^c(\mathcal{M}, \mathsf{E}) = \Big\{ \sum_k w_{1k} w_{2k} \,\Big|\, w_{1k} \in L_p^r(\mathcal{M}, \mathsf{E}),\ w_{2k} \in L_q^c(\mathcal{M}, \mathsf{E}) \Big\}$$
for $2 \leq p, q \leq \infty$ and equipped with
$$\|y\|_{r_p \cdot c_q} = \inf \left\{ \left\| \Big(\sum_k \mathsf{E}(w_{1k} w_{1k}^*)\Big)^{1/2} \right\|_{L_p(\mathcal{N})} \left\| \Big(\sum_k \mathsf{E}(w_{2k}^* w_{2k})\Big)^{1/2} \right\|_{L_q(\mathcal{N})} \right\}$$
where the infimum runs over all possible decompositions
$$y = \sum_k w_{1k} w_{2k}.$$
It is not hard to check that $L_p^r(\mathcal{M}, \mathsf{E}) \otimes_{\mathcal{M}} L_q^c(\mathcal{M}, \mathsf{E})$ is a normed space, see e.g. Lemma 3.5 in [**16**] for a similar result. The notation $\otimes_{\mathcal{M}}$ is motivated by the fact that the norm given above comes from an amalgamated Haagerup tensor product, we refer the reader to Chapter 6 below for a more detailed explanation. The following result is the key to conclude the proof of Lemma 3.1. We use the well

known Grothendieck-Pietsch version of the Hahn-Banach theorem, see [**43**] for more on this topic.

THEOREM 3.16. *Let $2 < p, q, u, v < \infty$ related by*
$$1/u + 1/p = 1/2 = 1/v + 1/q.$$
Then, we have the following isometry via the anti-linear bracket $\langle x, y \rangle = \mathrm{tr}(x^*y)$
$$L_u(\mathcal{N})L_\infty(\mathcal{M})L_v(\mathcal{N}) = \big(L_p^r(\mathcal{M}, \mathsf{E}) \otimes_\mathcal{M} L_q^c(\mathcal{M}, \mathsf{E})\big)^*.$$

PROOF. Given $x = ayb$ in $L_u(\mathcal{N})L_\infty(\mathcal{M})L_v(\mathcal{N})$, Hölder inequality gives

$$\begin{aligned}
&\left| \mathrm{tr}\Big(x^* \sum_k w_{1k}w_{2k}\Big)\right| \\
&= \left| \mathrm{tr}\Big(y^*\Big[\sum_k a^* w_{1k} \otimes e_{1k}\Big]\Big[\sum_k w_{2k} b^* \otimes e_{k1}\Big]\Big)\right| \\
&\leq \left\| a^*\Big(\sum_k w_{1k} \otimes e_{1k}\Big)\right\|_2 \|y\|_\infty \left\|\Big(\sum_k w_{2k} \otimes e_{k1}\Big)b^*\right\|_2 \\
&= \mathrm{tr}\Big(aa^*\Big[\sum_k \mathsf{E}(w_{1k}w_{1k}^*)\Big]\Big)^{1/2} \|y\|_\infty \mathrm{tr}\Big(\Big[\sum_k \mathsf{E}(w_{2k}^* w_{2k})\Big] b^* b\Big)^{1/2} \\
&\leq \|a\|_u \|y\|_\infty \|b\|_v \left\|\Big(\sum_k \mathsf{E}(w_{1k}w_{1k}^*)\Big)^{1/2}\right\|_p \left\|\Big(\sum_k \mathsf{E}(w_{2k}^* w_{2k})\Big)^{1/2}\right\|_q.
\end{aligned}$$

Thus, taking the infimum on the right, we have a contraction
$$x \in L_u(\mathcal{N})L_\infty(\mathcal{M})L_v(\mathcal{N}) \mapsto \mathrm{tr}(x^* \cdot) \in \big(L_p^r(\mathcal{M}, \mathsf{E}) \otimes_\mathcal{M} L_q^c(\mathcal{M}, \mathsf{E})\big)^*.$$
To prove the converse, we take a norm one functional φ on $L_p^r(\mathcal{M}, \mathsf{E}) \otimes_\mathcal{M} L_q^c(\mathcal{M}, \mathsf{E})$. If $1/s = 1/p + 1/q$ it is clear that (see Remark 1.7)
$$L_s(\mathcal{M}) = L_p(\mathcal{M})L_q(\mathcal{M}) \to L_p^r(\mathcal{M}, \mathsf{E}) \otimes_\mathcal{M} L_q^c(\mathcal{M}, \mathsf{E})$$
is a dense contractive inclusion. In particular, we can assume that there exists
$$x \in L_{s'}(\mathcal{M}) = L_u(\mathcal{M})L_v(\mathcal{M})$$
satisfying
$$\varphi(y) = \varphi_x(y) = \mathrm{tr}(x^*y). \tag{3.23}$$
To conclude, it suffices to see that $\|x\|_{u \cdot \infty \cdot v} \leq 1$. Let us consider a finite family $y_1, y_2, \ldots, y_m$ in the dense subspace $L_p(\mathcal{M})L_q(\mathcal{M})$ with decompositions
$$y_k = w_{1k}w_{2k}.$$
Since φ_x has norm one
$$\left|\sum_k \varphi_x(w_{1k}w_{2k})\right| \leq \left\|\sum_k \mathsf{E}(w_{1k}w_{1k}^*)\right\|_{L_{p/2}(\mathcal{N})}^{1/2} \left\|\sum_k \mathsf{E}(w_{2k}^* w_{2k})\right\|_{L_{q/2}(\mathcal{N})}^{1/2}.$$
Moreover, since the right hand side remains unchanged under multiplication with unimodular complex numbers $z_k \in \mathbb{T}$, we have the following inequality
$$\sum_k |\varphi_x(w_{1k}w_{2k})| \leq \left\|\sum_k \mathsf{E}(w_{1k}w_{1k}^*)\right\|_{L_{p/2}(\mathcal{N})}^{1/2} \left\|\sum_k \mathsf{E}(w_{2k}^* w_{2k})\right\|_{L_{q/2}(\mathcal{N})}^{1/2}.$$
Now we consider the unit balls in $L_u(\mathcal{N})$ and $L_v(\mathcal{N})$
$$\begin{aligned}
\mathsf{B}_1 &= \big\{\alpha \in L_u(\mathcal{N}) \mid \|\alpha\|_{L_u(\mathcal{N})} \leq 1\big\}, \\
\mathsf{B}_2 &= \big\{\beta \in L_v(\mathcal{N}) \mid \|\beta\|_{L_v(\mathcal{N})} \leq 1\big\}.
\end{aligned}$$

3.4. GENERAL VON NEUMANN ALGEBRAS II

By the arithmetic-geometric mean inequality

$$
\begin{aligned}
(3.24) \quad \sum_k & |\varphi_x(w_{1k}w_{2k})| \\
& \leq \frac{1}{2}\Big(\sup_{\alpha \in \mathsf{B}_1} \sum_k \operatorname{tr}\big(\alpha \mathsf{E}(w_{1k}w_{1k}^*)\alpha^*\big) + \sup_{\beta \in \mathsf{B}_2} \sum_k \operatorname{tr}\big(\beta^*\mathsf{E}(w_{2k}^*w_{2k})\beta\big)\Big) \\
& = \frac{1}{2}\Big(\sup_{\alpha \in \mathsf{B}_1} \sum_k \|\alpha w_{1k}\|_{L_2(\mathcal{M})}^2 + \sup_{\beta \in \mathsf{B}_2} \sum_k \|w_{2k}\beta\|_{L_2(\mathcal{M})}^2\Big).
\end{aligned}
$$

Note that B_1 and B_2 are compact when equipped with the $\sigma(L_u(\mathcal{N}), L_{u'}(\mathcal{N}))$ and the $\sigma(L_v(\mathcal{N}), L_{v'}(\mathcal{N}))$ topologies respectively. Now, labelling $(w_{1k})_{k\geq 1}$ and $(w_{2k})_{k\geq 1}$ by w_1 and w_2, we consider $f_{\mathsf{w}_1\mathsf{w}_2} : \mathsf{B}_1 \times \mathsf{B}_2 \to \mathbb{R}$ defined by

$$ f_{\mathsf{w}_1\mathsf{w}_2}(\alpha, \beta) = \sum_k \|\alpha w_{1k}\|_{L_2(\mathcal{M})}^2 + \sum_k \|w_{2k}\beta\|_{L_2(\mathcal{M})}^2 - 2\sum_k |\varphi_x(w_{1k}w_{2k})|. $$

This gives rise to the cone

$$ \mathsf{C}_+ = \Big\{ f_{\mathsf{w}_1\mathsf{w}_2} \in \mathcal{C}(\mathsf{B}_1 \times \mathsf{B}_2) \,\Big|\, w_{1k}w_{2k} \in L_p(\mathcal{M})L_q(\mathcal{M}) \Big\}. $$

Then we consider the open cone

$$ \mathsf{C}_- = \Big\{ f \in \mathcal{C}(\mathsf{B}_1 \times \mathsf{B}_2) \,\Big|\, \sup f < 0 \Big\}. $$

According to (3.24), the cones C_+ and C_- are disjoint. Therefore, the geometric Hahn-Banach theorem provides a norm one functional $\xi : \mathcal{C}(\mathsf{B}_1 \times \mathsf{B}_2) \to \mathbb{R}$ satisfying

$$ \xi(f_-) < \rho \leq \xi(f_+) $$

for some $\rho \in \mathbb{R}$ and all $(f_+, f_-) \in \mathsf{C}_+ \times \mathsf{C}_-$. Moreover, since we are dealing with cones, it turns out that $\rho = 0$ and ξ is a positive functional. Then, according to Riesz representation theorem, there exists a unique (positive) Radon measure μ_ξ on $\mathsf{B}_1 \times \mathsf{B}_2$ satisfying

$$ (3.25) \qquad \xi(f) = \int_{\mathsf{B}_1 \times \mathsf{B}_2} f \, d\mu_\xi \quad \text{for all} \quad f \in \mathcal{C}(\mathsf{B}_1 \times \mathsf{B}_2). $$

In fact, since ξ is a norm one positive functional, μ_ξ is a probability measure. Now we use that $\xi_{|\mathsf{C}_+}$ takes values in $\mathbb{R}_+$ and (3.25) to obtain the following inequality

$$
\begin{aligned}
2\sum_k |\varphi_x(w_{1k}w_{2k})| & \leq \sum_k \int_{\mathsf{B}_1 \times \mathsf{B}_2} \operatorname{tr}(w_{1k}w_{1k}^*\alpha^*\alpha) \, d\mu_\xi(\alpha, \beta) \\
& \quad + \sum_k \int_{\mathsf{B}_1 \times \mathsf{B}_2} \operatorname{tr}(w_{2k}^*w_{2k}\beta\beta^*) \, d\mu_\xi(\alpha, \beta) \\
& = \sum_k \|\alpha_0 w_{1k}\|_{L_2(\mathcal{M})}^2 + \sum_k \|w_{2k}\beta_0\|_{L_2(\mathcal{M})}^2,
\end{aligned}
$$

where $(\alpha_0, \beta_0) \in \mathsf{B}_1 \times \mathsf{B}_2$ are given by

$$
\begin{aligned}
\alpha_0 & = \Big(\int_{\mathsf{B}_1 \times \mathsf{B}_2} \alpha^*\alpha \, d\mu_\xi(\alpha, \beta)\Big)^{1/2} \in \mathsf{B}_1, \\
\beta_0 & = \Big(\int_{\mathsf{B}_1 \times \mathsf{B}_2} \beta\beta^* \, d\mu_\xi(\alpha, \beta)\Big)^{1/2} \in \mathsf{B}_2.
\end{aligned}
$$

Then, using the identity $2rs = \inf_{\gamma > 0} (\gamma r)^2 + (s/\gamma)^2$, we conclude

$$ \sum_k |\varphi_x(w_{1k}w_{2k})| \leq \Big(\sum_k \|\alpha_0 w_{1k}\|_{L_2(\mathcal{M})}^2\Big)^{1/2} \Big(\sum_k \|w_{2k}\beta_0\|_{L_2(\mathcal{M})}^2\Big)^{1/2}. $$

In particular, given any pair $(w_1, w_2) \in L_p(\mathcal{M}) \times L_q(\mathcal{M})$, we have
$$(3.26) \qquad |\varphi_x(w_1 w_2)| \leq \|\alpha_0 w_1\|_{L_2(\mathcal{M})} \|w_2 \beta_0\|_{L_2(\mathcal{M})}.$$

Let us write q_{α_0} and q_{β_0} for the support projections of α_0 and β_0. Then we define
$$\begin{aligned} d_{\alpha_0} &= \alpha_0^u + (1 - q_{\alpha_0}) \mathrm{D}(1 - q_{\alpha_0}), \\ d_{\beta_0} &= \beta_0^v + (1 - q_{\beta_0}) \mathrm{D}(1 - q_{\beta_0}). \end{aligned}$$

Note that $\phi_{\alpha_0} = \mathrm{tr}(d_{\alpha_0} \cdot)$ and $\phi_{\beta_0} = \mathrm{tr}(d_{\beta_0} \cdot)$ are n.f. finite weights on $\mathcal{M}$. In particular, by Theorem 1.3 we know that
$$d_{\alpha_0}^{1/2} \mathcal{M} \to L_2(\mathcal{M}) \quad \text{and} \quad \mathcal{M} d_{\beta_0}^{1/2} \to L_2(\mathcal{M})$$
are dense inclusions. Therefore, since
$$\begin{aligned} q_{\alpha_0} d_{\alpha_0}^{1/2} \mathcal{M} &= \alpha_0^{u/2} \mathcal{M} = \alpha_0 \alpha_0^{u/p} \mathcal{M} \subset \alpha_0 L_p(\mathcal{M}), \\ \mathcal{M} d_{\beta_0}^{1/2} q_{\beta_0} &= \mathcal{M} \beta_0^{v/2} = \mathcal{M} \beta_0^{v/q} \beta_0 \subset L_q(\mathcal{M}) \beta_0, \end{aligned}$$
it follows that $\alpha_0 L_p(\mathcal{M})$ (resp. $L_q(\mathcal{M}) \beta_0$) is dense in $q_{\alpha_0} L_2(\mathcal{M})$ (resp. $L_2(\mathcal{M}) q_{\beta_0}$). Hence we can consider the linear map
$$T_x : q_{\alpha_0} L_2(\mathcal{M}) \to q_{\beta_0} L_2(\mathcal{M})$$
determined by the relation
$$\langle \beta_0 w_2^*, T_x(\alpha_0 w_1) \rangle = \mathrm{tr}(w_2 \beta_0 T_x(\alpha_0 w_1)) = \varphi_x(w_1 w_2).$$

According to (3.26), T_x is contractive. Moreover, T_x is clearly a right $\mathcal{M}$ module map so that it commutes with the right action on $\mathcal{M}$. This means that there exists a contraction $m \in \mathcal{M}$ satisfying $T_x(\alpha_0 w_1) = m \alpha_0 w_1$. Finally, applying (3.23) we deduce the following identity
$$\mathrm{tr}(x^* w_1 w_2) = \varphi_x(w_1 w_2) = \mathrm{tr}(T_x(\alpha_0 w_1) w_2 \beta_0) = \mathrm{tr}(\beta_0 m \alpha_0 w_1 w_2),$$
which holds for any pair $(w_1, w_2) \in L_p(\mathcal{M}) \times L_q(\mathcal{M})$. Therefore, by the density of $L_p(\mathcal{M}) L_q(\mathcal{M})$ in $L_p^r(\mathcal{M}, \mathsf{E}) \otimes_{\mathcal{M}} L_q^c(\mathcal{M}, \mathsf{E})$ we have
$$x = \alpha_0^* m^* \beta_0^*.$$
Then, since $(\alpha_0, \beta_0) \in \mathsf{B}_1 \times \mathsf{B}_2$ and m is contractive, we have $\|x\|_{u \cdot \infty \cdot v} \leq 1$. $\square$

OBSERVATION 3.17. With a slight change in the arguments used, we can see that Theorem 3.16 holds for any $(u, v) \in [2, \infty] \times [2, \infty]$ such that $\max(u, v) > 2$. Indeed, by symmetry it suffices to see that

(a) $L_u(\mathcal{N}) L_\infty(\mathcal{M}) L_2(\mathcal{N}) = \big(L_p^r(\mathcal{M}, \mathsf{E}) \otimes_{\mathcal{M}} L_\infty^c(\mathcal{M}, \mathsf{E}) \big)^*$ for any $2 < u \leq \infty$.

(b) $L_u(\mathcal{N}) L_\infty(\mathcal{M}) L_\infty(\mathcal{N}) = \big(L_p^r(\mathcal{M}, \mathsf{E}) \otimes_{\mathcal{M}} L_2^c(\mathcal{M}, \mathsf{E}) \big)^*$ for any $2 < u \leq \infty$.

Since the proofs are similar, we only prove (a). Recalling that $L_p(\mathcal{M}) L_\infty(\mathcal{M})$ is norm dense in $L_p^r(\mathcal{M}, \mathsf{E}) \otimes_{\mathcal{M}} L_\infty^c(\mathcal{M}, \mathsf{E})$, we deduce that every norm one functional $\varphi : L_p^r(\mathcal{M}, \mathsf{E}) \otimes_{\mathcal{M}} L_\infty^c(\mathcal{M}, \mathsf{E}) \to \mathbb{C}$ is given by $\varphi(y) = \varphi_x(y) = \mathrm{tr}(x^* y)$ for some $x \in L_{p'}(\mathcal{M})$. Using one more time the Grothendieck-Pietsch separation trick we get
$$|\mathrm{tr}(x^* w_1 w_2)| \leq \|\alpha_0 w_1\|_2 \psi\big(\mathsf{E}(w_2^* w_2) \big)^{1/2}$$
for some α_0 in the unit ball of $L_u(\mathcal{N})$ and $\psi \in \mathcal{N}^*$. Let (e_α) be a net such that $e_\alpha \to 1$ strongly so that $\lim_\alpha \varphi(e_\alpha y) = \varphi_n(y)$ gives the normal part. Now replace y by $y e_\alpha$ and we get in the limit
$$|\mathrm{tr}(x^* w_1 w_2)| \leq \|\alpha_0 w_1\|_2 \varphi_n\big(\mathsf{E}(w_2^* w_2) \big)^{1/2}.$$

Thus $|\mathrm{tr}(x^* w_1 w_2)| \le \|\alpha_0 w_1\|_2 \|w_2 \beta_0\|_2$ and we may continue as in Theorem 3.16.

PROOF OF LEMMA 3.1. As we already pointed out at the beginning of this section, we have to prove inequality (3.22). Before doing it we recall that the indices (u_θ, v_θ) satisfy

(3.27) $\qquad 2 < \max(u_\theta, v_\theta) \quad \text{and} \quad \min(u_\theta, v_\theta) < \infty.$

for any $0 < \theta < 1$. Indeed, otherwise we would have $u_\theta = v_\theta = 2$ or $u_\theta = v_\theta = \infty$. However, $(1/2, 1/2, 0)$ and $(0, 0, 0)$ are extreme points of $\mathsf{K} \cap \{z = 0\}$ and this is not possible for $0 < \theta < 1$. The first inequality in (3.27) allows us to apply Theorem 3.16 after Observation 3.17, while the second inequality will be used below. Now, let x be an element of $L_{u_\theta}(\mathcal{N}) L_\infty(\mathcal{M}) L_{v_\theta}(\mathcal{N})$. According to Theorem 3.16, we know that we can find a finite family $(w_{1k}, w_{2k}) \in L_{p_\theta}(\mathcal{M}) \times L_{q_\theta}(\mathcal{M})$ with

$$1/u_\theta + 1/p_\theta = 1/2 = 1/v_\theta + 1/q_\theta$$

so that
(3.28)
$$\max\left\{ \left\| \left(\sum_k \mathsf{E}(w_{1k} w_{1k}^*)\right)^{1/2} \right\|_{L_{p_\theta}(\mathcal{N})}, \left\| \left(\sum_k \mathsf{E}(w_{2k}^* w_{2k})\right)^{1/2} \right\|_{L_{q_\theta}(\mathcal{N})} \right\} \le 1,$$

and
$$\|x\|_{u_\theta \cdot \infty \cdot v_\theta} \le \left|\mathrm{tr}\left(x^* \sum_k w_{1k} w_{2k}\right)\right| + \delta.$$

Moreover, given $1/s = 1/u_\theta + 1/v_\theta$, we have

$$\left|\mathrm{tr}\left([x - \mathsf{E}_\mathcal{M} \mathcal{E}_{\mathcal{M}_k}(x)]^* \sum_j w_{1j} w_{2j}\right)\right| \le \left\|x - \mathsf{E}_\mathcal{M} \mathcal{E}_{\mathcal{M}_k}(x)\right\|_s \left\|\sum_j w_{1j} w_{2j}\right\|_{s'}.$$

According to (3.27) we have $1 < s < \infty$. Then, it follows from (3.15) that the first factor on the right hand side tends to 0 as $k \to \infty$ while the second factor belongs to $L_{s'}(\mathcal{M})$. In conclusion, we obtain the following estimate

$$\|x\|_{u_\theta \cdot \infty \cdot v_\theta} \le \lim_{k \to \infty} \left|\mathrm{tr}\left(\mathsf{E}_\mathcal{M} \mathcal{E}_{\mathcal{M}_k}(x)^* \sum_k w_{1k} w_{2k}\right)\right| + \delta.$$

Using (3.28) and applying Theorem 3.16 one more time

$$\|x\|_{u_\theta \cdot \infty \cdot v_\theta} \le \sup_{k \ge 1} \left\|\mathsf{E}_\mathcal{M} \mathcal{E}_{\mathcal{M}_k}(x)\right\|_{u_\theta \cdot \infty \cdot v_\theta} + \delta.$$

Thus, (3.22) follows for $x \in L_{u_\theta}(\mathcal{N}) L_\infty(\mathcal{M}) L_{v_\theta}(\mathcal{N})$ by letting $\delta \to 0$. Finally, as in the case $\min(q_0, q_1) < \infty$, it remains to see that $L_{u_\theta}(\mathcal{N}) L_\infty(\mathcal{M}) L_{v_\theta}(\mathcal{N})$ is dense in $\mathsf{X}_\theta(\mathcal{M})$. Here we also need a different argument. Let us keep the notation $1/u_0 + 1/v_0 = 1/s = 1/u_1 + 1/v_1$. Then, we may assume that $(u_0, v_0) = (s, \infty)$ and $(u_1, v_1) = (\infty, s)$. Indeed, if we conclude the proof in this particular case, the general case follows from the reiteration theorem for complex interpolation, see e.g. [2]. This can be justified by means of Figure I, since the segment joining the points $(1/u_0, 1/v_0, 0)$ and $(1/u_1, 1/v_1, 0)$ is always contained in the segment joining $(1/s, 0, 0)$ and $(0, 1/s, 0)$. Thus, we assume in what follows that

$$\mathsf{X}_\theta(\mathcal{M}) = \Big[L_s(\mathcal{N}) L_\infty(\mathcal{M}) L_\infty(\mathcal{N}), L_\infty(\mathcal{N}) L_\infty(\mathcal{M}) L_s(\mathcal{N})\Big]_\theta$$

so that $1/u_\theta = (1 - \theta)/s$ and $1/v_\theta = \theta/s$. By the density of

$$\Delta = L_s(\mathcal{N}) L_\infty(\mathcal{M}) L_\infty(\mathcal{N}) \cap L_\infty(\mathcal{N}) L_\infty(\mathcal{M}) L_s(\mathcal{N})$$

in $X_\theta(\mathcal{M})$, it suffices to approximate any element $x \in \Delta$. In particular, we can write $x = a_0 b_0$ and $x = b_1 a_1$ where $a_0, a_1 \in L_s(\mathcal{N})$ and $b_0, b_1 \in \mathcal{M}$. Moreover, we can assume that $a_0 = a_1$. Indeed, taking

$$a = \left(a_0 a_0^* + a_1^* a_1 + \delta D_\varphi^{2/s}\right)^{1/2}$$

we have $x = aa^{-1} a_0 b_0 = a c_0$ and $x = b_1 a_1 a^{-1} a = c_1 a$ with

$$\|c_j\|_\mathcal{M} \leq \|b_j\|_\mathcal{M} \quad \text{for} \quad j = 0, 1.$$

Then $c_1 = a c_0 a^{-1}$ and we claim that $a^\theta c_0 a^{-\theta}$ is in $\mathcal{M}$ for all $0 < \theta < 1$. Indeed, let ϕ be the n.f. finite weight $\phi(\cdot) = \text{tr}(a^s \cdot)$ and let $\mathcal{M} \rtimes_{\sigma^\phi} \mathbb{R}$ be the crossed product with respect to the modular automorphism group associated to ϕ. Let us consider the spectral projection $p_n = 1_{[1/n, n]}(a)$. Then, for a fixed integer n the function

$$f_n(z) = p_n a^z c_0 a^{-z} p_n = p_n a^z p_n c_0 p_n a^{-z} p_n$$

is analytic so that

$$\|f_n(\theta)\| \leq \sup_{t \in \mathbb{R}} \left\|\sigma_{t/s}^\phi(c_0)\right\|^{1-\theta} \left\|\sigma_{t/s}^\phi(a c_0 a^{-1})\right\|^\theta = \|c_0\|^{1-\theta} \|a c_0 a^{-1}\|^\theta = \|c_0\|^{1-\theta} \|c_1\|^\theta.$$

Sending n to infinity, we deduce that $a^\theta c_0 a^{-\theta}$ is a bounded element of $\mathcal{M} \rtimes_{\sigma^\phi} \mathbb{R}$. Moreover, using the dual action with respect to ϕ, we find that $a^\theta c_0 a^{-\theta}$ belongs to $\mathcal{M}$. Therefore, we obtain

$$x = a^{1-\theta} a^\theta c_0 a^{-\theta} a^\theta \in L_{u_\theta}(\mathcal{N}) L_\infty(\mathcal{M}) L_{v_\theta}(\mathcal{N}).$$

We have seen that the intersection space Δ is included in $L_{u_\theta}(\mathcal{N}) L_\infty(\mathcal{M}) L_{v_\theta}(\mathcal{N})$. Hence, the result follows since Δ is dense in $X_\theta(\mathcal{M})$. The proof of Lemma 3.1 (for any von Neumann algebra) is therefore completed. $\square$

3.5. Proof of the main interpolation theorem

To prove Theorem 3.2, we need to know a priori that $L_u(\mathcal{N}) L_q(\mathcal{M}) L_v(\mathcal{N})$ is a Banach space for any indices (u, q, v) associated to a point $(1/u, 1/v, 1/q) \in \mathsf{K}$. This is a simple consequence of Lemma 3.1. Indeed, according to Lemma 2.5 and Proposition 2.6, we know that our assertion is true for any $(1/u, 1/v, 1/q) \in \partial_\infty \mathsf{K}$. In particular, it follows from Lemma 3.1 that

$$L_{u_\theta}(\mathcal{N}) L_{q_\theta}(\mathcal{M}) L_{v_\theta}(\mathcal{N})$$

is a Banach space for any $0 \leq \theta \leq 1$ whenever $(u_j, q_j, v_j) \in \partial_\infty \mathsf{K}$ for $j = 0, 1$ and $1/u_0 + 1/q_0 + 1/v_0 = 1/u_1 + 1/q_1 + 1/v_1$. In other words, according to the notation introduced at the beginning of this chapter, this condition holds whenever $(1/u_j, 1/v_j, 1/q_j) \in \mathsf{K}_\tau \cap \partial_\infty \mathsf{K}$ for $j = 0, 1$ and some $0 \leq \tau \leq 1$. Therefore, it suffices to see that K_τ is the convex hull of $\mathsf{K}_\tau \cap \partial_\infty \mathsf{K}$ for any $0 \leq \tau \leq 1$. However, this follows easily from Figure I. Note that K_τ is either a point ($\tau = 0$), a triangle ($0 < \tau \leq 1/2$), a pentagon ($1/2 < \tau < 1$) or a parallelogram ($\tau = 1$).

OBSERVATION 3.18. In fact, in the case of finite von Neumann algebras, Lemma 3.1 provides more information. Namely, according to (3.4) and the fact that K_τ is the convex hull of $\mathsf{K}_\tau \cap \partial_\infty \mathsf{K}$, we deduce that $\mathcal{N}_u L_q(\mathcal{M}) \mathcal{N}_v$ is a normed space when equipped with $\||\ \||_{u \cdot q \cdot v}$ for any $(1/u, 1/v, 1/q) \in \mathsf{K}$. Moreover, $\mathcal{N}_u L_q(\mathcal{M}) \mathcal{N}_v$ embeds isometrically in $L_u(\mathcal{N}) L_q(\mathcal{M}) L_v(\mathcal{N})$ as a dense subspace, something we did not know up to now (see Observation 2.9). This means that Lemma 2.5 and Proposition 2.6 hold for any point in K in the case of finite von Neumann algebras.

PROOF OF THEOREM 3.2. Now we are ready to prove Theorem 3.2. The arguments to be used follow the same strategy used for the proof of Lemma 3.1. In particular, we only need to point out how to proceed. In first place, as usual, the lower estimate follows by multilinear interpolation.

STEP 1. Let us show the validity of Theorem 3.2 for finite von Neumann algebras satisfying the boundedness condition (3.1). First we note that, once we know that $L_u(\mathcal{N})L_q(\mathcal{M})L_v(\mathcal{N})$ is always a Banach space, the proof of Lemma 3.3 is still valid for ending points $(1/u_j, 1/v_j, 1/q_j)$ lying on $\mathsf{K} \setminus \partial_\infty \mathsf{K}$. Then we follow the proof of Lemma 3.1 for finite von Neumann algebras verbatim to deduce Theorem 3.2 in this case. Here is essential to observe (as we did in Remark 3.5) that the proof of Lemma 3.1 for finite von Neumann algebras does not use at any point the restriction $1/u_0 + 1/q_0 + 1/v_0 = 1/u_1 + 1/q_1 + 1/v_1$. Note also that, in order to obtain the boundary estimates (3.6), Proposition 2.6 is needed. Here is where we apply Observation 3.18. This proves Theorem 3.2 for finite von Neumann algebras.

STEP 2. The next goal is to show that the corresponding conditional expectations are contractive. First we observe that, according to Lemma 3.1, we can extend the validity of Corollary 3.13 to any point $(1/u, 1/v, 1/q) \in \mathsf{K}$ by complex interpolation. Here we use again that $\mathsf{K}_\tau = \mathrm{conv}(\mathsf{K}_\tau \cap \partial_\infty \mathsf{K})$. Then, it is straightforward to see that Corollary 3.15 also holds in this case. Indeed, first we apply complex interpolation to obtain a contraction
$$\mathcal{E}_{\mathcal{M}_k} : X_\theta(\mathcal{M}) \to X_\theta(\mathcal{R}_\mathcal{M}) \to L_{u_\theta}(\mathcal{N}_k)L_{q_\theta}(\mathcal{M}_k)L_{v_\theta}(\mathcal{N}_k).$$
Second, the contractivity of
$$\mathsf{E}_\mathcal{M} : L_{u_\theta}(\mathcal{N}_k)L_{q_\theta}(\mathcal{M}_k)L_{v_\theta}(\mathcal{N}_k) \to L_{u_\theta}(\mathcal{N})L_{q_\theta}(\mathcal{M})L_{v_\theta}(\mathcal{N})$$
follows since, as we have seen, Corollary 3.13 holds for any point $(1/u, 1/v, 1/q) \in \mathsf{K}$.

STEP 3. We now prove Theorem 3.2 in the case $\min(q_0, q_1) < \infty$. It follows easily from Step 2. Indeed, recalling again that the restriction $1/u_0 + 1/q_0 + 1/v_0 = 1/u_1 + 1/q_1 + 1/v_1$ is not used in the proof of Lemma 3.1 (once we know the validity of Corollary 3.15), the proof follows verbatim.

STEP 4. Finally, we consider the case $\min(q_0, q_1) = \infty$. First we observe that the first half of the proof of Lemma 3.1 for this case holds for any two ending points $\mathsf{p}_j = (1/u_j, 1/v_j, 0)$ in the square $\mathsf{K} \cap \{z = 0\}$. Thus it only remains to check that $L_{u_\theta}(\mathcal{N})L_\infty(\mathcal{M})L_{v_\theta}(\mathcal{N})$ is dense in $X_\theta(\mathcal{M})$. Applying the reiteration theorem as we did in the proof of Lemma 3.1, we may assume that p_0 and p_1 are in the boundary of $\mathsf{K} \cap \{z = 0\}$. We have three possible situations. First we assume that p_0 and p_1 *live in the same edge of* $\mathsf{K} \cap \{z = 0\}$. Let Δ be the intersection of the interpolation pair. In this case, we have $\Delta = L_{u_0}(\mathcal{N})L_\infty(\mathcal{M})L_{v_0}(\mathcal{N})$ or $\Delta = L_{u_1}(\mathcal{N})L_\infty(\mathcal{M})L_{v_1}(\mathcal{N})$ since the points of any edge of $\mathsf{K} \cap \{z = 0\}$ are directed by inclusion. In particular, we deduce $\Delta \subset L_{u_\theta}(\mathcal{N})L_\infty(\mathcal{M})L_{v_\theta}(\mathcal{N})$ from which the result follows. If p_0 and p_1 *live in consecutive edges of* $\mathsf{K} \cap \{z = 0\}$, we have four choices according to the common vertex v of the corresponding (consecutive) edges. Following Figure I we may have $\mathsf{v} = 0, \mathsf{E}, \mathsf{F}, \mathsf{G}$. When $\mathsf{v} = \mathsf{E}, \mathsf{G}$ we are back to the situation above (one endpoint is contained in the other) and there is nothing to prove. When $\mathsf{v} = 0$, we may assume w.l.o.g. that
$$\begin{aligned} L_{u_0}(\mathcal{N})L_\infty(\mathcal{M})L_{v_0}(\mathcal{N}) &= L_{s_0}(\mathcal{N})L_\infty(\mathcal{M}), \\ L_{u_1}(\mathcal{N})L_\infty(\mathcal{M})L_{v_1}(\mathcal{N}) &= L_\infty(\mathcal{M})L_{s_1}(\mathcal{N}), \end{aligned}$$

for some $2 \leq s_0, s_1 \leq \infty$. Moreover, we may also assume w.l.o.g. that $s_0 \leq s_1$. This allows us to write $1/s_0 = 1/s_1 + 1/r$ for some index $2 \leq r \leq \infty$. In particular, $L_{s_0}(\mathcal{N})L_\infty(\mathcal{M}) = (L_{s_1}(\mathcal{N})L_r(\mathcal{N}))L_\infty(\mathcal{M})$. Let $x \in \Delta$ so that
$$x = \alpha\gamma m_0 = m_1\beta$$
with $\alpha, \beta \in L_{s_1}(\mathcal{N})$, $\gamma \in L_r(\mathcal{N})$ and $m_0, m_1 \in \mathcal{M}$. Taking
$$a = \left(\alpha\alpha^* + \beta^*\beta + \delta \mathrm{D}^{2/s_1}\right)^{1/2},$$
we may write $x = ac_0 = c_1 a$ (so that $c_1 = ac_0 a^{-1}$) with
$$c_0 = (a^{-1}\alpha\gamma)m_0 \in L_r(\mathcal{N})L_\infty(\mathcal{M}) \quad \text{and} \quad c_1 = m_1\beta a^{-1} \in L_\infty(\mathcal{M}).$$
On the other hand, $(1/r, 0, 0)$ and $(0, 0, 0)$ are in the same edge of $\mathsf{K} \cap \{z = 0\}$. Thus $[L_r(\mathcal{N})L_\infty(\mathcal{M}), L_\infty(\mathcal{M})]_\theta = L_{r_\theta}(\mathcal{N})L_\infty(\mathcal{M})$ with $1/r_\theta = (1-\theta)/r$. Using this interpolation result and arguing as in the proof of Lemma 3.1 for $\min(q_0, q_1) = \infty$, we easily obtain that $a^\theta c_0 a^{-\theta} \in L_{r_\theta}(\mathcal{N})L_\infty(\mathcal{M})$. Thus we deduce
$$x = a^{1-\theta}a^\theta c_0 a^{-\theta}a^\theta \in (L_{s_1/(1-\theta)}(\mathcal{N})L_{r_\theta}(\mathcal{N}))L_\infty(\mathcal{M})L_{s_1/\theta}(\mathcal{N}).$$
Then, since the latter space is $L_{u_\theta}(\mathcal{N})L_\infty(\mathcal{M})L_{v_\theta}(\mathcal{N})$, we have seen that Δ is included in this space. This completes the proof for $\mathsf{v} = 0$. When $\mathsf{v} = \mathsf{F}$, we may assume w.l.o.g. that
$$\begin{aligned}L_{u_0}(\mathcal{N})L_\infty(\mathcal{M})L_{v_0}(\mathcal{N}) &= L_{s_0}(\mathcal{N})L_\infty(\mathcal{M})L_2(\mathcal{N}),\\ L_{u_1}(\mathcal{N})L_\infty(\mathcal{M})L_{v_1}(\mathcal{N}) &= L_2(\mathcal{N})L_\infty(\mathcal{M})L_{s_1}(\mathcal{N}).\end{aligned}$$
Writing $1/2 = 1/s_0 + 1/r_0 = 1/s_1 + 1/r_1$ for some $2 \leq r_0, r_1 \leq \infty$ we have
$$\begin{aligned}L_{s_0}(\mathcal{N})L_\infty(\mathcal{M})L_2(\mathcal{N}) &= L_{s_0}(\mathcal{N})L_\infty(\mathcal{M})L_{r_1}(\mathcal{N})L_{s_1}(\mathcal{N}),\\ L_2(\mathcal{N})L_\infty(\mathcal{M})L_{s_1}(\mathcal{N}) &= L_{s_0}(\mathcal{N})L_{r_0}(\mathcal{N})L_\infty(\mathcal{M})L_{s_1}(\mathcal{N}).\end{aligned}$$
Thus, using our result for $\mathsf{v} = 0$ we find
$$\begin{aligned}\Delta &= L_{s_0}(\mathcal{N})\big(L_\infty(\mathcal{M})L_{r_1}(\mathcal{N}) \cap L_{r_0}(\mathcal{N})L_\infty(\mathcal{M})\big)L_{s_1}(\mathcal{N})\\ &\subset L_{s_0}(\mathcal{N})\big(L_{r_0/\theta}(\mathcal{N})L_\infty(\mathcal{M})L_{r_1/(1-\theta)}(\mathcal{N})\big)L_{s_1}(\mathcal{N})\\ &\subset L_{s_0(\theta)}(\mathcal{N})L_\infty(\mathcal{M})L_{s_1(\theta)}(\mathcal{N}),\end{aligned}$$
with $1/s_0(\theta) = (1-\theta)/s_0 + \theta/2$ and $1/s_1(\theta) = (1-\theta)/2 + \theta/s_1$. This completes the proof for consecutive edges. Finally, we assume that p_0 and p_1 live in opposite edges of $\mathsf{K} \cap \{z = 0\}$. Since the two possible situations are symmetric, we only consider the case
$$\begin{aligned}L_{u_0}(\mathcal{N})L_\infty(\mathcal{M})L_{v_0}(\mathcal{N}) &= L_{s_0}(\mathcal{N})L_\infty(\mathcal{M}),\\ L_{u_1}(\mathcal{N})L_\infty(\mathcal{M})L_{v_1}(\mathcal{N}) &= L_{s_1}(\mathcal{N})L_\infty(\mathcal{M})L_2(\mathcal{N}).\end{aligned}$$
If $s_0 \geq s_1$ we clearly have
$$\Delta = L_{u_0}(\mathcal{N})L_\infty(\mathcal{M})L_{v_0}(\mathcal{N}) \subset L_{u_\theta}(\mathcal{N})L_\infty(\mathcal{M})L_{v_\theta}(\mathcal{N})$$
and there is nothing to prove. If $s_0 < s_1$ we have $1/s_0 = 1/s_1 + 1/r$ so that
$$\begin{aligned}\Delta &= L_{s_1}(\mathcal{N})\big(L_r(\mathcal{N})L_\infty(\mathcal{M}) \cap L_\infty(\mathcal{M})L_2(\mathcal{N})\big)\\ &\subset L_{s_1}(\mathcal{N})\big(L_{r/(1-\theta)}(\mathcal{N})L_\infty(\mathcal{M})L_{2/\theta}(\mathcal{N})\big)\\ &= L_{u_\theta}(\mathcal{N})L_\infty(\mathcal{M})L_{v_\theta}(\mathcal{N}).\end{aligned}$$
This proves the assertion for opposite edges and so the space $L_{u_\theta}(\mathcal{N})L_\infty(\mathcal{M})L_{v_\theta}(\mathcal{N})$ is always dense in $\mathsf{X}_\theta(\mathcal{M})$. The proof of Theorem 3.2 is completed. □

3.5. PROOF OF THE MAIN INTERPOLATION THEOREM

REMARK 3.19. The key points to see that the proof of Lemma 3.1 applies whenever we start with any two ending points $(1/u_j, 1/v_j, 1/q_j)$ lying on K are the following:
  (a) $L_{u_j}(\mathcal{N})L_{q_j}(\mathcal{M})L_{v_j}(\mathcal{N})$ is a Banach space.
  (b) Lemma 2.5 holds on K for finite von Neumann algebras.
  (c) Corollary 3.15 also holds with ending points in $\mathsf{K} \setminus \partial_\infty \mathsf{K}$.

Thus, it suffices to see that Lemma 3.1 gives (a), (b), (c) by complex interpolation. On the other hand, it is worthy to explain with some more details why restriction $1/u_0 + 1/q_0 + 1/v_0 = 1/u_1 + 1/q_1 + 1/v_1$ can be dropped. The only two points where this restriction is needed (apart from the case $\min(q_0, q_1) = \infty$ which has been discussed in Step 4 above) are in the proofs of Lemma 3.8 and Proposition 3.14. However, Lemma 3.8 is only needed to obtain Corollary 3.13 (which we have *auto-improved* in Step 2 above by using Lemma 3.1). Moreover, as we also pointed out in Step 2, Proposition 3.14 now follows from our improvement of Corollary 3.13. Therefore, restriction $1/u_0 + 1/q_0 + 1/v_0 = 1/u_1 + 1/q_1 + 1/v_1$ can be ignored.

CHAPTER 4

Conditional L_p spaces

We conclude the first part of this paper by studying the duals of amalgamated L_p spaces and the subsequent applications of Theorem 3.2. Let us consider a von Neumann algebra $\mathcal{M}$ equipped with a n.f. state φ and a von Neumann subalgebra $\mathcal{N}$ of $\mathcal{M}$. Let $\mathsf{E}: \mathcal{M} \to \mathcal{N}$ denote the corresponding conditional expectation. We consider any three indices (u, p, v) such that $(1/u, 1/v, 1/p)$ belongs to K and we define $1 \leq s \leq \infty$ by $1/s = 1/u + 1/p + 1/v$. Then, the *conditional L_p space*

$$L^p_{(u,v)}(\mathcal{M}, \mathsf{E})$$

is defined as the completion of $L_p(\mathcal{M})$ with respect to the norm

$$\|x\|_{L^p_{(u,v)}(\mathcal{M},\mathsf{E})} = \sup \big\{ \|\alpha x \beta\|_{L_s(\mathcal{M})} \,\big|\, \|\alpha\|_{L_u(\mathcal{N})}, \|\beta\|_{L_v(\mathcal{N})} \leq 1 \big\}.$$

We shall show below that amalgamated and conditional L_p spaces are related by duality. According to our main result in Chapter 3, this immediately provides interpolation isometries of the form

$$(4.1) \qquad \big[L^{p_0}_{(u_0,v_0)}(\mathcal{M}, \mathsf{E}), L^{p_1}_{(u_1,v_1)}(\mathcal{M}, \mathsf{E})\big]_\theta = L^{p_\theta}_{(u_\theta,v_\theta)}(\mathcal{M}, \mathsf{E}).$$

Our aim now is to explore these identities, since they will be useful in the sequel.

EXAMPLE 4.1. As in Chapter 2, several noncommutative function spaces arise as particular cases of our notion of conditional L_p space. Let us mention four particularly relevant examples:

(a) The noncommutative L_p spaces arise as

$$L_p(\mathcal{M}) = L^p_{(\infty,\infty)}(\mathcal{M}, \mathsf{E}).$$

(b) If $p \geq q$ and $1/r = 1/q - 1/p$, the spaces $L_p(\mathcal{N}_1; L_q(\mathcal{N}_2))$ arise as

$$L^p_{(2r,2r)}(\mathcal{N}_1 \bar\otimes \mathcal{N}_2, \mathsf{E}),$$

where the conditional expectation $\mathsf{E}: \mathcal{N}_1 \bar\otimes \mathcal{N}_2 \to \mathcal{N}_1$ is $\mathsf{E} = 1_{\mathcal{N}_1} \otimes \varphi_{\mathcal{N}_2}$.

(c) If $2 \leq p \leq \infty$ and $1/p + 1/q = 1/2$, Lemma 1.8 gives

$$\begin{aligned} L^r_p(\mathcal{M}, \mathsf{E}) &= L^p_{(q,\infty)}(\mathcal{M}, \mathsf{E}), \\ L^c_p(\mathcal{M}, \mathsf{E}) &= L^p_{(\infty,q)}(\mathcal{M}, \mathsf{E}). \end{aligned}$$

As we shall see below, $L_p(\mathcal{M}; R^n_p)$ and $L_p(\mathcal{M}; C^n_p)$ are particular cases.

(d) In Chapter 7 we will also identify certain asymmetric noncommutative L_p spaces as particular cases of conditional L_p spaces. We prefer in this case to leave the details for Chapter 7.

4.1. Duality

Note that given $(1/u, 1/v, 1/q) \in \mathsf{K}$, we usually take $1/p = 1/u + 1/q + 1/v$. In the following it will be more convenient to replace p by p', the index conjugate to p. Let us consider the following restriction of (2.1)

$$(4.2) \qquad 1 < q < \infty \quad \text{and} \quad 2 < u, v \leq \infty \quad \text{and} \quad 0 < \frac{1}{u} + \frac{1}{q} + \frac{1}{v} = \frac{1}{p'} < 1.$$

THEOREM 4.2. *Let* $\mathsf{E} : \mathcal{M} \to \mathcal{N}$ *denote the conditional expectation of* $\mathcal{M}$ *onto* $\mathcal{N}$ *and let* $1 < p < \infty$ *given by* $1/q' = 1/u + 1/p + 1/v$, *where the indices* (u, q, v) *satisfy* (4.2) *and* q' *is conjugate to* q. *Then, the following isometric isomorphisms hold via the anti-linear duality bracket* $\langle x, y \rangle = \operatorname{tr}(x^* y)$

$$\bigl(L_u(\mathcal{N}) L_q(\mathcal{M}) L_v(\mathcal{N})\bigr)^* = L^p_{(u,v)}(\mathcal{M}, \mathsf{E}),$$

$$\bigl(L^p_{(u,v)}(\mathcal{M}, \mathsf{E})\bigr)^* = L_u(\mathcal{N}) L_q(\mathcal{M}) L_v(\mathcal{N}).$$

PROOF. Let us consider the map

$$\Lambda_p : x \in L^p_{(u,v)}(\mathcal{M}, \mathsf{E}) \mapsto \operatorname{tr}(x^* \,\cdot\,) \in \bigl(L_u(\mathcal{N}) L_q(\mathcal{M}) L_v(\mathcal{N})\bigr)^*.$$

We first show that Λ_p is an isometry

$$\begin{aligned}
\|\Lambda_p(x)\|_{(u \cdot q \cdot v)^*} &= \sup \Bigl\{ |\operatorname{tr}(x^* y)| \,\Big|\, \inf_{y = \alpha z \beta} \|\alpha\|_{L_u(\mathcal{N})} \|z\|_{L_q(\mathcal{M})} \|\beta\|_{L_v(\mathcal{N})} \leq 1 \Bigr\} \\
&= \sup \Bigl\{ |\operatorname{tr}(\beta x^* \alpha z)| \,\Big|\, \|\alpha\|_{L_u(\mathcal{N})}, \|z\|_{L_q(\mathcal{M})}, \|\beta\|_{L_v(\mathcal{N})} \leq 1 \Bigr\} \\
&= \sup \Bigl\{ \|\beta x^* \alpha\|_{L_{q'}(\mathcal{M})} \,\Big|\, \|\alpha\|_{L_u(\mathcal{N})} \leq 1, \|\beta\|_{L_v(\mathcal{N})} \leq 1 \Bigr\} \\
&= \sup \Bigl\{ \|\alpha^* x \beta^*\|_{L_{q'}(\mathcal{M})} \,\Big|\, \|\alpha\|_{L_u(\mathcal{N})} \leq 1, \|\beta\|_{L_v(\mathcal{N})} \leq 1 \Bigr\} \\
&= \|x\|_{L^p_{(u,v)}(\mathcal{M}, \mathsf{E})}.
\end{aligned}$$

It remains to see that Λ_p is surjective. To that aim we use again the solid K in Figure I. Note that, since the case $u = v = \infty$ is clear, we may assume that $\min(u, v) < \infty$. In that case any point $(1/u, 1/v, 1/q)$ with (u, q, v) satisfying (4.2) lies in the interior of a segment S contained in K and satisfying

(a) One end point of S lies in the open interval $(\mathsf{0}, \mathsf{A})$.
(b) The segment S belongs to a plane parallel to ACDF.

According to Theorem 3.2, this means that

$$(4.3) \qquad L_u(\mathcal{N}) L_q(\mathcal{M}) L_v(\mathcal{N}) = \bigl[L_{p'}(\mathcal{M}), L_{u_1}(\mathcal{N}) L_{q_1}(\mathcal{M}) L_{v_1}(\mathcal{N})\bigr]_\theta,$$

for some $0 < \theta < 1$ and some $(1/u_1, 1/v_1, 1/q_1) \in \mathsf{K}$. Recalling that $1 < p < \infty$, we know that $L_{p'}(\mathcal{M})$ is reflexive. In particular, the same holds for the interpolation space in (4.3) and we obtain the following isometric isomorphism

$$\bigl(L_u(\mathcal{N}) L_q(\mathcal{M}) L_v(\mathcal{N})\bigr)^* = \bigl[L_p(\mathcal{M}), \bigl(L_{u_1}(\mathcal{N}) L_{q_1}(\mathcal{M}) L_{v_1}(\mathcal{N})\bigr)^*\bigr]_\theta.$$

Moreover, since $0 < \theta < 1$, we know that the intersection

$$L_p(\mathcal{M}) \cap \bigl(L_{u_1}(\mathcal{N}) L_{q_1}(\mathcal{M}) L_{v_1}(\mathcal{N})\bigr)^*$$

is norm dense in the space $\big(L_u(\mathcal{N})L_q(\mathcal{M})L_v(\mathcal{N})\big)^*$. On the other hand, recalling that $1/u_1 + 1/q_1 + 1/v_1 = 1/p'$, we know from the definition of amalgamated spaces that
$$L_{p'}(\mathcal{M}) = L_{p'}(\mathcal{M}) + L_{u_1}(\mathcal{N})L_{q_1}(\mathcal{M})L_{v_1}(\mathcal{N}).$$
Hence, $L_p(\mathcal{M}) = L_p(\mathcal{M}) \cap \big(L_{u_1}(\mathcal{N})L_{q_1}(\mathcal{M})L_{v_1}(\mathcal{N})\big)^*$ is norm dense in
$$\big(L_u(\mathcal{N})L_q(\mathcal{M})L_v(\mathcal{N})\big)^*.$$
Therefore, Λ_p has dense range since
$$L_p(\mathcal{M}) \subset L_{(u,v)}^p(\mathcal{M}, \mathsf{E}).$$
For the second part, we use from (4.3) that $L_u(\mathcal{N})L_q(\mathcal{M})L_v(\mathcal{N})$ is reflexive. $\square$

REMARK 4.3. The first part of the proof of Theorem 4.2 holds for any point $(1/u, 1/v, 1/q)$ in the solid K. In particular, we always have an isometric embedding
$$L_{(u,v)}^p(\mathcal{M}, \mathsf{E}) \longrightarrow \big(L_u(\mathcal{N})L_q(\mathcal{M})L_v(\mathcal{N})\big)^*.$$

REMARK 4.4. Note that the indices excluded in Theorem 4.2 by the restriction imposed by property (4.2) are the natural ones. For instance, the last restriction $0 < 1/u + 1/q + 1/v < 1$ only affects conditional/amalgamated L_1 and L_∞ spaces, which are not expected to be reflexive. Moreover, the spaces $L_u(\mathcal{N})L_\infty(\mathcal{M})L_v(\mathcal{N})$ are not reflexive in general. Indeed, let us consider the particular case in which $\mathcal{N}$ is the complex field. These spaces collapse into $L_\infty(\mathcal{M})$ which is not reflexive.

4.2. Conditional L_∞ spaces

Among the non-reflexive conditional spaces, we concentrate on some properties of conditional L_∞ spaces that will be needed in the second half of this paper. Note that, given indices (u, q, v) satisfying (2.1) with $1/u + 1/q + 1/v = 1$, we have defined the space
$$L_{(u,v)}^\infty(\mathcal{M}, \mathsf{E})$$
as the completion of $L_\infty(\mathcal{M})$ with respect to the norm
$$\|x\|_{L_{(u,v)}^\infty(\mathcal{M},\mathsf{E})} = \sup\Big\{\|\alpha x \beta\|_{L_{q'}(\mathcal{M})} \,\Big|\, \|\alpha\|_{L_u(\mathcal{N})}, \|\beta\|_{L_v(\mathcal{N})} \leq 1\Big\}.$$
According to Remark 4.3, we know that this space embeds isometrically in
$$\mathcal{L}_{(u,v)}^\infty(\mathcal{M}, \mathsf{E}) = \big(L_u(\mathcal{N})L_q(\mathcal{M})L_v(\mathcal{N})\big)^*.$$

PROPOSITION 4.5. *The following properties hold:*
  i) $\mathcal{L}_{(u,v)}^\infty(\mathcal{M}, \mathsf{E})$ *is contractively included in* $L_{q'}(\mathcal{M})$.
  ii) $L_\infty(\mathcal{M})$ *and* $L_{(u,v)}^\infty(\mathcal{M}, \mathsf{E})$ *are weak* dense subspaces of* $\mathcal{L}_{(u,v)}^\infty(\mathcal{M}, \mathsf{E})$.

PROOF. Let us consider the map
$$j : \varphi \in \mathcal{L}_{(u,v)}^\infty(\mathcal{M}, \mathsf{E}) \to \varphi\big(\mathrm{D}^{\frac{1}{u}} \cdot \mathrm{D}^{\frac{1}{v}}\big) \in L_{q'}(\mathcal{M}).$$
By Proposition 2.6, the map j is clearly injective. On the other hand,
$$\big\|\varphi\big(\mathrm{D}^{\frac{1}{u}} \cdot \mathrm{D}^{\frac{1}{v}}\big)\big\|_{L_{q'}(\mathcal{M})}$$
$$= \sup\Big\{\big|\varphi\big(\mathrm{D}^{\frac{1}{u}} y \mathrm{D}^{\frac{1}{v}}\big)\big| \,\big|\, \|y\|_q \leq 1\Big\}$$
$$\leq \sup\Big\{\big|\varphi\big(\mathrm{D}^{\frac{1}{u}} \alpha y \beta \mathrm{D}^{\frac{1}{v}}\big)\big| \,\big|\, \|\mathrm{D}^{\frac{1}{u}}\alpha\|_{L_u(\mathcal{N})}, \|y\|_q, \|\beta \mathrm{D}^{\frac{1}{v}}\|_{L_v(\mathcal{N})} \leq 1\Big\}$$

$$= \|\varphi\|_{\mathcal{L}^\infty_{(u,v)}(\mathcal{M},\mathsf{E})}.$$

The last identity follows from Proposition 2.6. This shows that j is a contraction. For the second part, it suffices to see that $L_\infty(\mathcal{M})$ is weak* dense. However, it is clear from the definition of amalgamated spaces that the inclusion map

$$L_u(\mathcal{N})L_q(\mathcal{M})L_v(\mathcal{N}) \longrightarrow L_1(\mathcal{M})$$

is injective. In particular, taking adjoints we get the announced weak* density. □

4.3. Interpolation results and applications

In this last section we consider some interesting particular cases of the dual version (4.1) of Theorem 3.2. One of the applications we shall consider generalizes Pisier's interpolation result [44] and Xu's recent extension [69].

THEOREM 4.6. *Let $\mathcal{N}$ be a von Neumann subalgebra of $\mathcal{M}$ and let $\mathsf{E}: \mathcal{M} \to \mathcal{N}$ be the corresponding conditional expectation. Assume that (u_j, q_j, v_j) satisfy (4.2) for $j = 0, 1$ and that $1/u_j + 1/q_j + 1/v_j = 1/p'_j$. Then, if p_j denotes the conjugate index to p'_j, the following isometric isomorphism holds*

$$\left[L^{p_0}_{(u_0,v_0)}(\mathcal{M},\mathsf{E}), L^{p_1}_{(u_1,v_1)}(\mathcal{M},\mathsf{E})\right]_\theta = L^{p_\theta}_{(u_\theta,v_\theta)}(\mathcal{M},\mathsf{E}).$$

Moreover, if $2 \le u_j, v_j \le \infty$ for $j = 0, 1$, we also have

$$\left[\mathcal{L}^\infty_{(u_0,v_0)}(\mathcal{M},\mathsf{E}), \mathcal{L}^\infty_{(u_1,v_1)}(\mathcal{M},\mathsf{E})\right]^\theta = \mathcal{L}^\infty_{(u_\theta,v_\theta)}(\mathcal{M},\mathsf{E}).$$

PROOF. The first part follows automatically from Theorem 3.2 and Theorem 4.2. The second part follows from Theorem 3.2 and the duality properties which link the complex interpolation brackets $[\ ,\ ]_\theta$ and $[\ ,\ ]^\theta$, see e.g. [2]. □

Now we study some consequences of Theorem 4.6. We shall content ourselves by exploring only the case $p_0 = p_1$. This restriction is motivated by the applications we are using in the successive chapters. The last part of the following result requires to introduce some notation. As usual, we shall write R_p^n (resp. C_p^n) to denote the interpolation space $[R_n, C_n]_{1/p}$ (resp. $[C_n, R_n]_{1/p}$), where R_n and C_n denote the n-dimensional row and column Hilbert spaces. Alternatively, we may define R_p^n and C_p^n as the first row and column subspaces of the Schatten class S_p^n. On the other hand, given an element x_0 in a von Neumann algebra $\mathcal{M}$, we shall consider the mappings L_{x_0} and R_{x_0} on $\mathcal{M}$ defined respectively as follows

$$L_{x_0}(x) = x_0 x \quad \text{and} \quad R_{x_0}(x) = x x_0.$$

COROLLARY 4.7. *Let $\mathcal{N}$ be a von Neumann subalgebra of $\mathcal{M}$ and let $\mathsf{E}: \mathcal{M} \to \mathcal{N}$ be the corresponding conditional expectation of $\mathcal{M}$ onto $\mathcal{N}$. Then, we have the following isometric isomorphisms:*

i) *If $2 \le p < \infty$ and $2 < q \le \infty$ are such that $1/2 = 1/p + 1/q$, we have*

$$\begin{aligned}\left[L_p(\mathcal{M}), L_p^r(\mathcal{M},\mathsf{E})\right]_\theta &= L^p_{(s,\infty)}(\mathcal{M},\mathsf{E}), \\ \left[L_p(\mathcal{M}), L_p^c(\mathcal{M},\mathsf{E})\right]_\theta &= L^p_{(\infty,s)}(\mathcal{M},\mathsf{E}),\end{aligned} \quad \text{with} \quad \frac{1}{s} = \frac{\theta}{q}.$$

In the case $p = \infty$, we obtain

$$\begin{aligned}\left[L_\infty(\mathcal{M}), L_\infty^r(\mathcal{M},\mathsf{E})\right]_\theta &= L^\infty_{(2/\theta,\infty)}(\mathcal{M},\mathsf{E}), \\ \left[L_\infty(\mathcal{M}), L_\infty^c(\mathcal{M},\mathsf{E})\right]_\theta &= L^\infty_{(\infty,2/\theta)}(\mathcal{M},\mathsf{E}).\end{aligned}$$

ii) *If $2 \leq p < \infty$ and $2 < q \leq \infty$ are such that $1/2 = 1/p + 1/q$, we have*
$$\left[L_p^c(\mathcal{M}, \mathsf{E}), L_p^r(\mathcal{M}, \mathsf{E})\right]_\theta = L_{(u,v)}^p(\mathcal{M}, \mathsf{E})$$
with
$$(1/u, 1/v) = (\theta/q, (1-\theta)/q).$$

iii) *Let us define for $2 \leq p < \infty$*
$$\mathrm{X}_\theta(\mathcal{M}) = \left[L_p(\mathcal{M}; C_p^n), L_p(\mathcal{M}; R_p^n)\right]_\theta.$$
Then, if $1/w = 1/p + 1/v$ with $1/v = (1-\theta)/q$, we deduce
$$\Big\| \sum_{k=1}^n x_k \otimes \delta_k \Big\|_{\mathrm{X}_\theta(\mathcal{M})}^2 = \Big\| \sum_{k=1}^n L_{x_k} R_{x_k^*} : L_{v/2}(\mathcal{M}) \to L_{w/2}(\mathcal{M}) \Big\|.$$

PROOF. The assertions in the first part of i) and ii) follow from Theorem 4.6 after the obvious identifications, see Example 4.1. For the last part of i) we only prove the first identity since the second one follows in the same way. According to Remark 4.3, $L_\infty^r(\mathcal{M}, \mathsf{E})$ is the closure of $L_\infty(\mathcal{M})$ in $\mathcal{L}_{(2,\infty)}^\infty(\mathcal{M}, \mathsf{E})$. Then, applying a well-known property of the complex method (see e.g. [**2**, Theorem 4.2.2]) we find
$$\left[L_\infty(\mathcal{M}), L_\infty^r(\mathcal{M}, \mathsf{E})\right]_\theta = \left[L_\infty(\mathcal{M}), \mathcal{L}_{(2,\infty)}^\infty(\mathcal{M}, \mathsf{E})\right]_\theta \quad \text{for} \quad 0 < \theta < 1.$$
By Berg's theorem, we have an isometric inclusion
$$\left[L_\infty(\mathcal{M}), \mathcal{L}_{(2,\infty)}^\infty(\mathcal{M}, \mathsf{E})\right]_\theta \subset \left[L_\infty(\mathcal{M}), \mathcal{L}_{(2,\infty)}^\infty(\mathcal{M}, \mathsf{E})\right]^\theta.$$
Therefore given $x \in L_\infty(\mathcal{M})$, Theorem 4.6 gives
$$\begin{aligned} \|x\|_{[L_\infty(\mathcal{M}), L_\infty^r(\mathcal{M},\mathsf{E})]_\theta} &= \|x\|_{[L_\infty(\mathcal{M}), \mathcal{L}_{(2,\infty)}^\infty(\mathcal{M},\mathsf{E})]_\theta} \\ &= \|x\|_{[L_\infty(\mathcal{M}), \mathcal{L}_{(2,\infty)}^\infty(\mathcal{M},\mathsf{E})]^\theta} \\ &= \|x\|_{L_{(2/\theta,\infty)}^\infty(\mathcal{M},\mathsf{E})}.\end{aligned}$$
The assertion then follows by a simple density argument
$$\left[L_\infty(\mathcal{M}), L_\infty^r(\mathcal{M}, \mathsf{E})\right]_\theta = L_{(2/\theta,\infty)}^\infty(\mathcal{M}, \mathsf{E}).$$
Finally, for part iii) we consider the direct sum $\mathcal{M}_{\oplus n} = \mathcal{M} \oplus \mathcal{M} \oplus \cdots \oplus \mathcal{M}$ with n terms and equipped with the *n.f.* state $\varphi_n(x_1, x_2, \cdots, x_n) = \frac{1}{n} \sum_k \varphi(x_k)$. The natural conditional expectation is given by
$$\mathsf{E}_n : \sum_{k=1}^n x_k \otimes \delta_k \in \mathcal{M}_{\oplus n} \mapsto \frac{1}{n} \sum_{k=1}^n x_k \in \mathcal{M}.$$
It is clear that we have the isometries
$$\begin{aligned} L_p(\mathcal{M}; R_p^n) &= \sqrt{n}\, L_p^r(\mathcal{M}_{\oplus n}, \mathsf{E}_n), \\ L_p(\mathcal{M}; C_p^n) &= \sqrt{n}\, L_p^c(\mathcal{M}_{\oplus n}, \mathsf{E}_n).\end{aligned}$$
According to ii) and the definition of the norm in $L_{(u,v)}^p(\mathcal{M}_{\oplus n}, \mathsf{E}_n)$, we have
$$\Big\| \sum_{k=1}^n x_k \otimes \delta_k \Big\|_{\mathrm{X}_\theta(\mathcal{M})}^2$$
$$= \sup\left\{ n \Big\| \sum_{k=1}^n \alpha x_k \beta \otimes \delta_k \Big\|_{L_2(\mathcal{M}_{\oplus n})}^2 \;\Big|\; \|\alpha\|_{L_u(\mathcal{M})}, \|\beta\|_{L_v(\mathcal{M})} \leq 1 \right\}$$

$$= \sup\left\{\sum_{k=1}^n \operatorname{tr}(\alpha x_k \beta\beta^* x_k^* \alpha^*) \mid \|\alpha\|_{L_u(\mathcal{M})}, \|\beta\|_{L_v(\mathcal{M})} \le 1\right\}$$

$$= \sup\left\{\left\|\sum_{k=1}^n L_{x_k} R_{x_k^*}(\beta\beta^*)\right\|_{L_{(u/2)'}(\mathcal{M})} \mid \|\beta\|_{L_v(\mathcal{M})} \le 1\right\}$$

$$= \sup\left\{\left\|\sum_{k=1}^n L_{x_k} R_{x_k^*}(\gamma)\right\|_{L_{(u/2)'}(\mathcal{M})} \mid \gamma \ge 0,\ \|\gamma\|_{L_{v/2}(\mathcal{M})} \le 1\right\}.$$

Recalling that $\sum_k L_{x_k} R_{x_k^*}$ is positive and that $(u/2)' = w/2$, we conclude. $\square$

OBSERVATION 4.8. Arguing as in Corollary 4.7, we easily obtain
$$\bigl[L_\infty(\mathcal{M}), L^\infty_{(u,v)}(\mathcal{M},\mathsf{E})\bigr]_\theta = L^\infty_{(u/\theta, v/\theta)}(\mathcal{M},\mathsf{E}).$$

These results shall be frequently used in the successive chapters. Moreover, let us also mention that Corollary 4.7 i) is needed in [**21**] to study the noncommutative John-Nirenberg theorem.

At the time of this writing we do not know whether Corollary 4.7 ii) extends to $p = \infty$ in full generality. We will now show that the equality holds when restricted to elements in $\mathcal{M}$. This result will play a very important role in the sequel.

LEMMA 4.9. *If $1 < p, q < \infty$ and $z \in \mathcal{M}$, we have*
$$\inf\left\{\|x\|_{L^\infty_{(2p,\infty)}(\mathcal{M},\mathsf{E})} \|y\|_{L^\infty_{(\infty,2q)}(\mathcal{M},\mathsf{E})} \mid z = xy,\ x, y \in \mathcal{M}\right\} \le \|z\|_{L^\infty_{(2p,2q)}(\mathcal{M},\mathsf{E})}.$$

PROOF. On $\mathcal{M}$ we define the norm
$$\|z\|_h = \inf_{z=xy} \|x\|_{L^\infty_{(2p,\infty)}(\mathcal{M},\mathsf{E})} \|y\|_{L^\infty_{(\infty,2q)}(\mathcal{M},\mathsf{E})}$$

where the infimum is taken over $x, y \in L_\infty(\mathcal{M})$. It is easy to check that we do not need to consider sums here. Let us assume that $\|z\|_h = 1$. By the Hahn-Banach theorem there exists a linear functional $\phi : \mathcal{M} \to \mathbb{C}$ such that $\phi(z) = 1$ and

$$(4.4) \quad \left|\sum_k \phi(x_k y_k)\right| \le \sup_{\|a\|_{2p}\le 1}\left\|\sum_k a x_k x_k^* a^*\right\|_p^{1/2} \sup_{\|b\|_{2q}\le 1}\left\|\sum_k b^* y_k^* y_k b\right\|_q^{1/2}.$$

Note that $\|z\|_h \le \|z\|_\infty$ and thus ϕ is continuous. It follows immediately that we may move the absolute values in (4.4) inside. Thus, we get

$$\sum_k |\phi(x_k y_k)| \le \sup_{a,c} \operatorname{tr}\left(\sum_k a x_k x_k^* a^* c\right)^{1/2} \sup_{b,d} \operatorname{tr}\left(\sum_k b^* y_k^* y_k b d\right)^{1/2}.$$

Here we take the supremum over (a, c, b, d) in
$$\mathsf{B}_{L_{2p}(\mathcal{N})} \times \mathsf{B}^+_{L_{p'}(\mathcal{M})} \times \mathsf{B}_{L_{2q}(\mathcal{N})} \times \mathsf{B}^+_{L_{q'}(\mathcal{M})},$$

all equipped with the weak* topology. Using the standard Grothendieck-Pietsch separation argument as in Theorem 3.16 we obtain two probability measures μ_1 and μ_2 such that
$$|\phi(xy)| \le \left(\int \operatorname{tr}(a x x^* a^* c)\, d\mu_1(a,c)\right)^{1/2}\left(\int \operatorname{tr}(b^* y^* y b d)\, d\mu_2(b,d)\right)^{1/2}.$$

Since $L_{2p}(\mathcal{N}) L_{p'}(\mathcal{M}) L_{2p}(\mathcal{N})$ and $L_{2q}(\mathcal{N}) L_{q'}(\mathcal{M}) L_{2q}(\mathcal{N})$ are Banach spaces,
$$\alpha = \int a^* c a\, d\mu_1(a,c) \quad \text{and} \quad \beta = \int b d b^*\, d\mu_2(b,d)$$

are positive elements respectively in the unit balls of
$$L_{2p}(\mathcal{N})L_{p'}(\mathcal{M})L_{2p}(\mathcal{N}) \quad \text{and} \quad L_{2q}(\mathcal{N})L_{q'}(\mathcal{M})L_{2q}(\mathcal{N}).$$
Therefore we find $a_1, a_2 \in \mathsf{B}_{L_{2p}(\mathcal{N})}$ and $c_1, c_2 \in \mathsf{B}_{L_{2p'}(\mathcal{M})}$ such that $\alpha = a_1 c_1 c_2 a_2$. We deduce from the Cauchy-Schwartz inequality and the arithmetic-geometric mean inequality that
$$\begin{aligned} \operatorname{tr}(xx^*\alpha) &= \left|\operatorname{tr}(xx^*a_1c_1c_2a_2)\right| \\ &= \left|\operatorname{tr}(c_2a_2xx^*a_1c_1)\right| \\ &\leq \operatorname{tr}\left(c_2a_2xx^*a_2^*c_2^*\right)^{1/2}\operatorname{tr}\left(c_1^*a_1^*xx^*a_1c_1\right)^{1/2} \\ &\leq \operatorname{tr}\left(xx^*\frac{a_1c_1c_1^*a_1^* + a_2^*c_2^*c_2a_2}{2}\right). \end{aligned}$$
We could consider $a = (a_1^*a_1 + a_2^*a_2)^{1/2}$ to deduce that
$$a^{-1}\frac{a_1c_1c_1^*a_1^* + a_2^*c_2^*c_2a_2}{2}a^{-1}$$
is a positive element in $L_{p'}(\mathcal{M})$ of norm ≤ 1. This is not enough for our purposes. However, following the proof of the triangle inequality in Lemma 2.5, we may apply Devinatz's theorem one more time to find an operator $a \in (1+\varepsilon)\mathsf{B}_{L_{2p}(\mathcal{N})}$ with full support and $c \in \mathsf{B}_{L_{p'}(\mathcal{M})}$ such that
$$\frac{a_1c_1c_1^*a_1^* + a_2^*c_2^*c_2a_2}{2} = a^*ca.$$
We leave the details to the interested reader. This implies that c is positive and
$$\operatorname{tr}(xx^*\alpha) \leq \operatorname{tr}(xx^*a^*ca) = \|c^{1/2}ax\|_2^2.$$
The same argument for β gives $b \in (1+\varepsilon)\mathsf{B}_{L_{2q}(\mathcal{N})}$ and $d \in \mathsf{B}_{L_{q'}(\mathcal{M})}^+$ such that
$$\operatorname{tr}(y^*y\beta) \leq \operatorname{tr}(y^*ybdb^*) = \|ybd^{1/2}\|_2^2.$$
This yields
$$|\phi(xy)| \leq \|c^{1/2}ax\|_2 \|ybd^{1/2}\|_2.$$
From this it is easy to find a contraction $u \in \mathcal{M}$ such that
$$\phi(xy) = \operatorname{tr}(uc^{\frac{1}{2}}axybd^{\frac{1}{2}}).$$
If $1/r = 1/2p + 1/2q$, we deduce from Hölder's inequality that
$$\|z\|_h = |\phi(z)| = \left|\operatorname{tr}(uc^{\frac{1}{2}}azbd^{\frac{1}{2}})\right| \leq \|azb\|_{L_r(\mathcal{M})} \leq (1+\varepsilon)^2 \|z\|_{L^\infty_{(2p,2q)}(\mathcal{M},\mathsf{E})}.$$
Finally, recalling that $\varepsilon > 0$ is arbitrary, the assertion follows by taking $\varepsilon \to 0$. $\square$

COROLLARY 4.10. *Assume that $1 \leq p_0, p_1, q_0, q_1 \leq \infty$ satisfy*
$$\max(p_0, p_1), \max(q_0, q_1) > 1 \quad \text{and} \quad \min(p_0, p_1), \min(q_0, q_1) < \infty.$$
Then, given $x \in \mathcal{M}$ we have
$$\|x\|_{L^\infty_{(2p_\theta, 2q_\theta)}(\mathcal{M}, \mathsf{E})} = \|x\|_{[L^\infty_{(2p_0, 2q_0)}(\mathcal{M}, \mathsf{E}), L^\infty_{(2p_1, 2q_1)}(\mathcal{M}, \mathsf{E})]_\theta}.$$
In particular, for $\theta = 1/q$ we obtain
$$\|x\|_{[L^c_\infty(\mathcal{M}, \mathsf{E}), L^r_\infty(\mathcal{M}, \mathsf{E})]_\theta} = \sup\left\{\|axb\|_{L_2(\mathcal{M})} \mid \|a\|_{L_{2q}(\mathcal{N})}, \|b\|_{L_{2q'}(\mathcal{N})} \leq 1\right\}.$$

PROOF. The upper estimate is an easy application of trilinear interpolation. For the converse we apply Lemma 4.9 so that for any $\varepsilon > 0$ we can always find a factorization $x = x_1 x_2$ satisfying
$$\|x_1\|_{L^\infty_{(2p_\theta,\infty)}(\mathcal{M},\mathsf{E})} \|x_2\|_{L^\infty_{(\infty,2q_\theta)}(\mathcal{M},\mathsf{E})} \leq (1+\varepsilon) \|x\|_{L^\infty_{(2p_\theta,2q_\theta)}(\mathcal{M},\mathsf{E})}.$$
According to Corollary 4.7 i) we know that
$$L^\infty_{(2p_\theta,\infty)}(\mathcal{M},\mathsf{E}) = \big[L_\infty(\mathcal{M}), L^r_\infty(\mathcal{M},\mathsf{E})\big]_{1/p_\theta}.$$
Taking $\theta_0 = 1/p_0$ and $\theta_1 = 1/p_1$, the reiteration theorem implies that
$$\big[L_\infty(\mathcal{M}), L^r_\infty(\mathcal{M},\mathsf{E})\big]_{1/p_\theta} = \big[[L_\infty(\mathcal{M}), L^r_\infty(\mathcal{M},\mathsf{E})]_{\theta_0}, [L_\infty(\mathcal{M}), L^r_\infty(\mathcal{M},\mathsf{E})]_{\theta_1}\big]_\theta.$$
In particular,
$$\big[L_\infty(\mathcal{M}), L^r_\infty(\mathcal{M},\mathsf{E})\big]_{1/p_\theta} = \big[L^\infty_{(2p_0,\infty)}(\mathcal{M},\mathsf{E}), L^\infty_{(2p_1,\infty)}(\mathcal{M},\mathsf{E})\big]_\theta.$$
Therefore, we get
$$\|x_1\|_{[L^\infty_{(2p_0,\infty)}(\mathcal{M},\mathsf{E}), L^\infty_{(2p_1,\infty)}(\mathcal{M},\mathsf{E})]_\theta} = \|x_1\|_{L^\infty_{(2p_\theta,\infty)}(\mathcal{M},\mathsf{E})}.$$
Similarly, we have
$$\|x_2\|_{[L^\infty_{(\infty,2q_0)}(\mathcal{M},\mathsf{E}), L^\infty_{(\infty,2q_1)}(\mathcal{M},\mathsf{E})]_\theta} = \|x_2\|_{L^\infty_{(\infty,2q_\theta)}(\mathcal{M},\mathsf{E})}.$$
On the other hand the inequality
$$\|xy\|_{L^\infty_{(2p,2q)}(\mathcal{M},\mathsf{E})} \leq \|x\|_{L^\infty_{(2p,\infty)}(\mathcal{M},\mathsf{E})} \|y\|_{L^\infty_{(\infty,2q)}(\mathcal{M},\mathsf{E})}$$
holds for all $1 \leq p, q \leq \infty$. Thus, by bilinear interpolation we deduce
$$\|x\|_{[L^\infty_{(2p_0,2q_0)}(\mathcal{M},\mathsf{E}), L^\infty_{(2p_1,2q_1)}(\mathcal{M},\mathsf{E})]_\theta}$$
$$\leq \quad \|x_1\|_{[L^\infty_{(2p_0,\infty)}(\mathcal{M},\mathsf{E}), L^\infty_{(2p_1,\infty)}(\mathcal{M},\mathsf{E})]_\theta} \|x_2\|_{[L^\infty_{(\infty,2q_0)}(\mathcal{M},\mathsf{E}), L^\infty_{(\infty,2q_1)}(\mathcal{M},\mathsf{E})]_\theta}.$$
Combining the previous estimates and taking $\varepsilon \to 0$ we obtain the assertion. $\square$

REMARK 4.11. If $\mathcal{M}$ is dense in the intersection $L^r_\infty(\mathcal{M},\mathsf{E}) \cap L^c_\infty(\mathcal{M},\mathsf{E})$, then Corollary 4.7 ii) extends to $p = \infty$. This holds for instance when the inclusion $\mathcal{N} \subset \mathcal{M}$ has finite index or when $\mathcal{M} = \mathcal{A} \bar\otimes \mathcal{N}$ with $\mathcal{A}$ finite-dimensional. However, at the time of this writing it is not clear whether the density assumption is satisfied for general conditional expectations. On the other hand, we note that the validity of Corollary 4.7 ii) for $p = \infty$ can be understood as a *conditional version* of Pisier's interpolation result [44]. Indeed, taking $\theta = 1/q$ we easily find that
$$\|x\|_{[L^c_\infty(\mathcal{M},\mathsf{E}), L^r_\infty(\mathcal{M},\mathsf{E})]_\theta} = \sup\Big\{\|x^*ax\|_{L_q(\mathcal{M})} \,\Big|\, \|a\|_{L_q(\mathcal{N})} \leq 1\Big\}$$
$$= \sup\Big\{\|xbx^*\|_{L_{q'}(\mathcal{M})} \,\Big|\, \|b\|_{L_{q'}(\mathcal{N})} \leq 1\Big\}.$$
Furthermore, according to Theorem 4.6, this also applies for $2 \leq p \leq \infty$ (we leave the details to the reader). In particular, we also find a conditional version of Xu's interpolation result [69]. Moreover, when $1 \leq p \leq 2$ this result follows from Theorem 3.2 instead of Theorem 4.6.

CHAPTER 5

Intersections of L_p spaces

In a second part of this paper, we study intersections of certain generalized L_p spaces. Our main goal is to prove a noncommutative version of (Σ_{pq}), see the Introduction. In this chapter we begin by proving certain interpolation results for intersections. As usual we consider a von Neumann subalgebra $\mathcal{N}$ of $\mathcal{M}$ with corresponding conditional expectation $\mathsf{E} : \mathcal{M} \to \mathcal{N}$. Then given a positive integer n, $1 \le q \le p \le \infty$ and $1/r = 1/q - 1/p$, we define the following intersection spaces

$$\mathcal{R}_{2p,q}^n(\mathcal{M}, \mathsf{E}) = n^{\frac{1}{2p}} L_{2p}(\mathcal{M}) \cap n^{\frac{1}{2q}} L_{(2r,\infty)}^{2p}(\mathcal{M}, \mathsf{E}),$$
$$\mathcal{C}_{2p,q}^n(\mathcal{M}, \mathsf{E}) = n^{\frac{1}{2p}} L_{2p}(\mathcal{M}) \cap n^{\frac{1}{2q}} L_{(\infty,2r)}^{2p}(\mathcal{M}, \mathsf{E}).$$

Our main result in this chapter shows that the two families of intersection spaces considered above are interpolation scales in the index q. That is, the intersections commute with the complex interpolation functor. Indeed, we obtain the following isomorphisms with relevant constants independent of n

(5.1) $$\begin{aligned}\left[\mathcal{R}_{2p,1}^n(\mathcal{M}, \mathsf{E}), \mathcal{R}_{2p,p}^n(\mathcal{M}, \mathsf{E})\right]_\theta &\simeq \mathcal{R}_{2p,q}^n(\mathcal{M}, \mathsf{E}), \\ \left[\mathcal{C}_{2p,1}^n(\mathcal{M}, \mathsf{E}), \mathcal{C}_{2p,p}^n(\mathcal{M}, \mathsf{E})\right]_\theta &\simeq \mathcal{C}_{2p,q}^n(\mathcal{M}, \mathsf{E}),\end{aligned}$$

and with $1/q = 1 - \theta + \theta/p$. Moreover, we shall also prove that

(5.2) $$\left[\mathcal{R}_{2p,1}^n(\mathcal{M}, \mathsf{E}), \mathcal{C}_{2p,1}^n(\mathcal{M}, \mathsf{E})\right]_\theta \simeq \bigcap_{u,v \in \{2p',\infty\}} n^{\frac{1-\theta}{u} + \frac{1}{2p} + \frac{\theta}{v}} L_{(\frac{u}{1-\theta}, \frac{v}{\theta})}^{2p}(\mathcal{M}, \mathsf{E}).$$

5.1. Free Rosenthal inequalities

Our aim in this section is to present the free analogue given in [26] of Rosenthal inequalities [56], where mean-zero independent random variables are replaced by free random variables. This will be one of the key tools needed for the proof of the isomorphisms (5.1) and (5.2). For the sake of completeness we first recall the construction of reduced amalgamated free products.

5.1.1. Amalgamated free products. The basics of free products without amalgamation can be found in [67]. Let $\mathcal{A}$ be a von Neumann algebra equipped with a *n.f.* state ϕ and let $\mathcal{N}$ be a von Neumann subalgebra of $\mathcal{A}$. Let $\mathsf{E}_\mathcal{N} : \mathcal{A} \to \mathcal{N}$ be the corresponding conditional expectation onto $\mathcal{N}$. A family $\mathsf{A}_1, \mathsf{A}_2, \ldots, \mathsf{A}_n$ of von Neumann subalgebras of $\mathcal{A}$, having $\mathcal{N}$ as a common subalgebra, is called *freely independent* over $\mathsf{E}_\mathcal{N}$ if

$$\mathsf{E}_\mathcal{N}(a_1 a_2 \cdots a_m) = 0$$

whenever $\mathsf{E}_\mathcal{N}(a_k) = 0$ for all $1 \le k \le m$ and $a_k \in \mathsf{A}_{j_k}$ with $j_1 \ne j_2 \ne \cdots \ne j_m$. As in the scalar-valued case, operator-valued freeness admits a Fock space representation.

We first assume that $A_1, A_2, \ldots, A_n$ are C*-algebras having $\mathcal{N}$ as a common C*-subalgebra and that $\mathsf{E}_k = \mathsf{E}_{\mathcal{N}|_{A_k}}$ are faithful conditional expectations. Let us consider the mean-zero subspaces

$$\overset{\circ}{\mathsf{A}}_k = \Big\{ a_k \in \mathsf{A}_k \,\Big|\, \mathsf{E}_k(a_k) = 0 \Big\}.$$

Following [**9, 65**], we consider the Hilbert $\mathcal{N}$-module $\overset{\circ}{\mathsf{A}}_{j_1} \otimes \overset{\circ}{\mathsf{A}}_{j_2} \otimes \cdots \overset{\circ}{\mathsf{A}}_{j_m}$ (where the tensor products are amalgamated over the von Neumann subalgebra $\mathcal{N}$) equipped with the $\mathcal{N}$-valued inner product

$$\langle a_1 \otimes \cdots \otimes a_m, b_1 \otimes \cdots \otimes b_m \rangle = \mathsf{E}_{j_m}\big(a_m^* \cdots \mathsf{E}_{j_2}(a_2^* \,\mathsf{E}_{j_1}(a_1^* b_1)\, b_2) \cdots b_m\big).$$

Then, the usual Fock space is replaced by the Hilbert $\mathcal{N}$-module

$$\mathcal{H}_{\mathcal{N}} = \mathcal{N} \oplus \bigoplus_{m \geq 1} \bigoplus_{j_1 \neq j_2 \neq \cdots \neq j_m} \overset{\circ}{\mathsf{A}}_{j_1} \otimes \overset{\circ}{\mathsf{A}}_{j_2} \otimes \cdots \otimes \overset{\circ}{\mathsf{A}}_{j_m}.$$

The direct sums above are assumed to be $\mathcal{N}$-orthogonal. Let $\mathcal{L}(\mathcal{H}_{\mathcal{N}})$ stand for the algebra of adjointable maps on $\mathcal{H}_{\mathcal{N}}$. A linear right $\mathcal{N}$-module map $\mathrm{T}: \mathcal{H}_{\mathcal{N}} \to \mathcal{H}_{\mathcal{N}}$ is called *adjointable* whenever there exists $\mathrm{S}: \mathcal{H}_{\mathcal{N}} \to \mathcal{H}_{\mathcal{N}}$ such that

$$\langle x, \mathrm{T} y \rangle = \langle \mathrm{S} x, y \rangle \qquad \text{for all} \qquad x, y \in \mathcal{H}_{\mathcal{N}}.$$

Let us recall how elements in A_k act on $\mathcal{H}_{\mathcal{N}}$. We decompose any $a_k \in \mathsf{A}_k$ as

$$a_k = \overset{\circ}{a}_k + \mathsf{E}_k(a_k).$$

Any element in $\mathcal{N}$ acts on $\mathcal{H}_{\mathcal{N}}$ by left multiplication. Therefore, it suffices to define the action of mean-zero elements. The *-homomorphism $\pi_k : \mathsf{A}_k \to \mathcal{L}(\mathcal{H}_{\mathcal{N}})$ is then defined as follows

$$\pi_k(\overset{\circ}{a}_k)(b_{j_1} \otimes \cdots \otimes b_{j_m}) = \begin{cases} \overset{\circ}{a}_k \otimes b_{j_1} \otimes \cdots \otimes b_{j_m}, & \text{if } k \neq j_1 \\ \\ \mathsf{E}_k(\overset{\circ}{a}_k b_{j_1})\, b_{j_2} \otimes \cdots \otimes b_{j_m} \oplus \\ \big(\overset{\circ}{a}_k b_{j_1} - \mathsf{E}_k(\overset{\circ}{a}_k b_{j_1})\big) \otimes b_{j_2} \otimes \cdots \otimes b_{j_m}, & \text{if } k = j_1. \end{cases}$$

This definition also applies for the empty word. Now, since the algebra $\mathcal{L}(\mathcal{H}_{\mathcal{N}})$ is a C*-algebra (*c.f.* [**34**]), we can define the *reduced $\mathcal{N}$-amalgamated free product* C*$(*_{\mathcal{N}} \mathsf{A}_k)$ as the C*-closure of linear combinations of operators of the form

$$\pi_{j_1}(a_1) \pi_{j_2}(a_2) \cdots \pi_{j_m}(a_m).$$

Now we consider the case in which $\mathcal{N}$ and $A_1, A_2, \ldots, A_n$ are von Neumann algebras. Let $\varphi : \mathcal{N} \to \mathbb{C}$ be a n.f. state on $\mathcal{N}$. This provides us with the induced states $\varphi_k : \mathsf{A}_k \to \mathbb{C}$ given by $\varphi_k = \varphi \circ \mathsf{E}_k$. The Hilbert space

$$L_2\big(\overset{\circ}{\mathsf{A}}_{j_1} \otimes \overset{\circ}{\mathsf{A}}_{j_2} \otimes \cdots \otimes \overset{\circ}{\mathsf{A}}_{j_m}, \varphi\big)$$

is obtained from $\overset{\circ}{\mathsf{A}}_{j_1} \otimes \overset{\circ}{\mathsf{A}}_{j_2} \otimes \cdots \otimes \overset{\circ}{\mathsf{A}}_{j_m}$ by considering the inner product

$$\langle a_1 \otimes \cdots \otimes a_m, b_1 \otimes \cdots \otimes b_m \rangle_\varphi = \varphi\Big(\langle a_1 \otimes \cdots \otimes a_m, b_1 \otimes \cdots \otimes b_m \rangle\Big).$$

Then we define the orthogonal direct sum

$$\mathcal{H}_\varphi = L_2(\mathcal{N}) \oplus \bigoplus_{m \geq 1} \bigoplus_{j_1 \neq j_2 \neq \cdots \neq j_m} L_2\big(\overset{\circ}{\mathsf{A}}_{j_1} \otimes \overset{\circ}{\mathsf{A}}_{j_2} \otimes \cdots \otimes \overset{\circ}{\mathsf{A}}_{j_m}, \varphi\big).$$

Let us consider the $*$-representation $\lambda : \mathcal{L}(\mathcal{H}_\mathcal{N}) \to \mathcal{B}(\mathcal{H}_\varphi)$ defined by $\lambda(\mathrm{T})x = \mathrm{T}x$. The faithfulness of λ is implied by the fact that φ is also faithful. Let $\mathcal{M}(\mathcal{H}_\mathcal{N})$ be the von Neumann algebra generated in $\mathcal{B}(\mathcal{H}_\varphi)$ by $\mathcal{L}(\mathcal{H}_\mathcal{N})$. Then, we define the *reduced $\mathcal{N}$-amalgamated free product* $*_\mathcal{N} \mathsf{A}_k$ as the weak* closure of $\mathrm{C}^*(*_\mathcal{N}\mathsf{A}_k)$ in $\mathcal{M}(\mathcal{H}_\mathcal{N})$. After decomposing

$$a_k = \overset{\circ}{a}_k + \mathsf{E}_k(a_k)$$

and identifying $\overset{\circ}{\mathsf{A}}_k$ with $\lambda\big(\pi_k(\overset{\circ}{\mathsf{A}}_k)\big)$, we can think of $*_\mathcal{N}\mathsf{A}_k$ as

$$*_\mathcal{N}\mathsf{A}_k = \Big(\mathcal{N} \oplus \bigoplus_{m \geq 1} \bigoplus_{j_1 \neq j_2 \neq \cdots \neq j_m} \overset{\circ}{\mathsf{A}}_{j_1}\overset{\circ}{\mathsf{A}}_{j_2}\cdots \overset{\circ}{\mathsf{A}}_{j_m}\Big)''.$$

Let us consider the orthogonal projections

$$\begin{aligned}\mathcal{Q}_\emptyset : \mathcal{H}_\varphi &\to L_2(\mathcal{N}), \\ \mathcal{Q}_{j_1\cdots j_m} : \mathcal{H}_\varphi &\to L_2\big(\overset{\circ}{\mathsf{A}}_{j_1} \otimes \overset{\circ}{\mathsf{A}}_{j_2} \otimes \cdots \otimes \overset{\circ}{\mathsf{A}}_{j_m}, \varphi\big).\end{aligned}$$

If we also consider the projection $\mathcal{Q}_{\mathsf{A}_k} = \mathcal{Q}_\emptyset + \mathcal{Q}_k$, the following mappings

$$\begin{aligned}\mathsf{E}_\mathcal{N} : x \in *_\mathcal{N}\mathsf{A}_k &\mapsto \mathcal{Q}_\emptyset x \mathcal{Q}_\emptyset \in \mathcal{N}, \\ \mathcal{E}_{\mathsf{A}_k} : x \in *_\mathcal{N}\mathsf{A}_k &\mapsto \mathcal{Q}_{\mathsf{A}_k} x \mathcal{Q}_{\mathsf{A}_k} \in \mathsf{A}_k,\end{aligned}$$

are faithful conditional expectations. Then, it turns out that $\mathsf{A}_1, \mathsf{A}_2, \ldots, \mathsf{A}_n$ are von Neumann subalgebras of $*_\mathcal{N}\mathsf{A}_k$ freely independent over $\mathsf{E}_\mathcal{N}$. Reciprocally, if $\mathsf{A}_1, \mathsf{A}_2, \ldots, \mathsf{A}_n$ is a collection of von Neumann subalgebras of $\mathcal{A}$ freely independent over $\mathsf{E}_\mathcal{N} : \mathcal{A} \to \mathcal{N}$ and generating $\mathcal{A}$, then $\mathcal{A}$ is isomorphic to $*_\mathcal{N}\mathsf{A}_k$.

REMARK 5.1. Let $\mathcal{N}_1$ and $\mathcal{N}_2$ be von Neumann algebras and assume that $\mathcal{N}_2$ is equipped with a *n.f.* state φ_2. A relevant example of the construction outlined above is the following. Let $\mathcal{A} = \mathcal{N}_1 \otimes \mathcal{N}_2$ and let us consider the conditional expectation $\mathsf{E}_{\mathcal{N}_1} : \mathcal{A} \to \mathcal{N}_1$ defined by

$$\mathsf{E}_{\mathcal{N}_1}(x_1 \otimes x_2) = x_1 \otimes \varphi_2(x_2)1.$$

Assume that $\mathsf{A}_1, \mathsf{A}_2, \ldots, \mathsf{A}_n$ are freely independent subalgebras of $\mathcal{N}_2$ over φ_2. Then, it is well-known that $\mathcal{N}_1 \otimes \mathsf{A}_1, \mathcal{N}_1 \otimes \mathsf{A}_2, \ldots, \mathcal{N}_1 \otimes \mathsf{A}_n$ is a family of freely independent subalgebras of $\mathcal{A}$ over $\mathsf{E}_{\mathcal{N}_1}$, see e.g. Section 7 of [**17**]. In particular, if $\mathsf{A}_1, \mathsf{A}_2, \ldots, \mathsf{A}_n$ generate $\mathcal{N}_2$, we obtain

$$(5.3) \qquad \mathcal{A} = \mathcal{N}_1 \otimes \big(\underset{k=1}{\overset{n}{*}} \mathsf{A}_k\big) = *_{\mathcal{N}_1}(\mathcal{N}_1 \otimes \mathsf{A}_k).$$

5.1.2. A Rosenthal/Voiculescu type inequality. In this paragraph we recall the free analogue [**26**] of Rosenthal inequalities [**56**] for mean-zero random variables and prove a simple consequence of it. Let $\mathsf{A}_1, \mathsf{A}_2, \ldots, \mathsf{A}_n$ be a family of von Neumann algebras having $\mathcal{N}$ as a common von Neumann subalgebra and let $*_\mathcal{N}\mathsf{A}_k$ be the corresponding amalgamated free product. Given a family $a_1, a_2, \ldots, a_n$ in $*_\mathcal{N}\mathsf{A}_k$ we consider the row and column conditional square functions

$$\begin{aligned}\mathcal{S}^r_{\mathrm{cond}}(a) &= \Big(\sum_{k=1}^n \mathsf{E}_\mathcal{N}(a_k a_k^*)\Big)^{1/2}, \\ \mathcal{S}^c_{\mathrm{cond}}(a) &= \Big(\sum_{k=1}^n \mathsf{E}_\mathcal{N}(a_k^* a_k)\Big)^{1/2}.\end{aligned}$$

THEOREM 5.2. *If $2 \leq p \leq \infty$ and $a_1, a_2, \ldots, a_n \in L_p(*_\mathcal{N} \mathsf{A}_k)$ with $a_k \in L_p(\overset{\circ}{\mathsf{A}}_k)$ for $1 \leq k \leq n$, the following equivalence of norms holds with relevant constants independent of p or n*

$$\Big\| \sum_{k=1}^n a_k \Big\|_p \sim \Big(\sum_{k=1}^n \|a_k\|_p^p \Big)^{1/p} + \big\| \mathcal{S}_{\mathrm{cond}}^r(a) \big\|_p + \big\| \mathcal{S}_{\mathrm{cond}}^c(a) \big\|_p.$$

This is the operator-valued/free analogue of Rosenthal's original result. On the other hand, a noncommutative analogue was obtained in [**30**] for general algebras (non necessarily free products), see also [**68**] for the notion of noncommutative independence (called order independence) employed in it. Recalling that freeness implies order independence, Theorem 5.2 follows from [**30**] for $2 \leq p < \infty$. However, the constants there are not uniformly bounded as $p \to \infty$, in sharp contrast with the situation in Theorem 5.2. This is another example of an L_p inequality involving independent random variables which only holds in the limit case as $p \to \infty$ when considering their free analogue. This constitutes a significant difference in this paper. Theorem 5.2 for $p = \infty$ was proved in [**17**] and constitutes the operator valued extension of Voiculescu's inequality [**66**]. Finally, we refer the reader to [**26**] for a generalization of Theorem 5.2, where $a_1, a_2, \ldots, a_n$ are replaced by certain words of a fixed degree $d \geq 1$.

The following result is a standard application for positive random variables.

COROLLARY 5.3. *If $2 \leq p \leq \infty$ and $a_1, \ldots, a_n \in L_p(*_\mathcal{N} \mathsf{A}_k)$ with $a_k \in L_p(\mathsf{A}_k)$ for $1 \leq k \leq n$, the following equivalence of norms holds with relevant constants independent of p or n*

$$\Big\| \Big(\sum_{k=1}^n a_k a_k^* \Big)^{1/2} \Big\|_p \sim \Big(\sum_{k=1}^n \|a_k\|_p^p \Big)^{1/p} + \big\| \mathcal{S}_{\mathrm{cond}}^r(a) \big\|_p,$$

$$\Big\| \Big(\sum_{k=1}^n a_k^* a_k \Big)^{1/2} \Big\|_p \sim \Big(\sum_{k=1}^n \|a_k\|_p^p \Big)^{1/p} + \big\| \mathcal{S}_{\mathrm{cond}}^c(a) \big\|_p.$$

PROOF. We we clearly have

$$\Big(\sum_{k=1}^n \|a_k\|_p^p \Big)^{1/p} \leq \Big\| \Big(\sum_{k=1}^n a_k a_k^* \Big)^{1/2} \Big\|_p.$$

Indeed, our claim follows by complex interpolation from the trivial case $p = \infty$ and the case $p = 2$, where the equality clearly holds. This, together with the fact that $\mathsf{E}_\mathcal{N}$ is a contraction on $L_{p/2}(*_\mathcal{N} \mathsf{A}_k)$, proves the lower estimate with constant 2. For the upper estimate, we begin with the triangle inequality in $L_p(*_\mathcal{N} \mathsf{A}_k; R_p^n)$

$$\Big\| \Big(\sum_{k=1}^n a_k a_k^* \Big)^{1/2} \Big\|_p \leq \Big\| \sum_{k=1}^n \mathsf{E}_\mathcal{N}(a_k) \otimes e_{1k} \Big\|_p + \Big\| \sum_{k=1}^n \overset{\circ}{a}_k \otimes e_{1k} \Big\|_p = \mathrm{A} + \mathrm{B}.$$

To estimate A, we apply Kadison's inequality (see Lemma 1.4 i))

$$\mathrm{A} = \Big\| \Big(\sum_{k=1}^n \mathsf{E}_\mathcal{N}(a_k) \mathsf{E}_\mathcal{N}(a_k)^* \Big)^{1/2} \Big\|_p \leq \Big\| \Big(\sum_{k=1}^n \mathsf{E}_\mathcal{N}(a_k a_k^*) \Big)^{1/2} \Big\|_p = \big\| \mathcal{S}_{\mathrm{cond}}^r(a) \big\|_p.$$

On the other hand, according to (5.3) we can regard $\overset{\circ}{a}_k \otimes e_{1k}$ as an element of

$$S_p^n \big(L_p(*_\mathcal{N} \mathsf{A}_k) \big) = L_p \Big(*_{S_\infty^n(\mathcal{N})} S_\infty^n(\mathsf{A}_k) \Big),$$

where $\mathsf{E}_\mathcal{N}$ is replaced by $1_{M_n} \otimes \mathsf{E}_\mathcal{N}$. Then, writing x_k for $\mathring{a}_k \otimes e_{1k}$ and $\mathbf{E}_\mathcal{N}$ for $1_{M_n} \otimes \mathsf{E}_\mathcal{N}$, we estimate B using the free Rosenthal inequalities in this bigger space. Indeed, using $\|x_k\|_p = \|a_k\|_p$ we obtain

$$\mathrm{B} \sim \Big(\sum_{k=1}^n \|a_k\|_p^p\Big)^{1/p} + \Big\|\Big(\sum_{k=1}^n \mathbf{E}_\mathcal{N}(x_k x_k^*)\Big)^{1/2}\Big\|_p + \Big\|\Big(\sum_{k=1}^n \mathbf{E}_\mathcal{N}(x_k^* x_k)\Big)^{1/2}\Big\|_p.$$

Let us note that

$$\sum_{k=1}^n \mathbf{E}_\mathcal{N}(x_k x_k^*) = \sum_{k=1}^n e_{11} \otimes \mathsf{E}_\mathcal{N}(\mathring{a}_k \mathring{a}_k^*) \leq \sum_{k=1}^n e_{11} \otimes \mathsf{E}_\mathcal{N}(a_k a_k^*),$$

$$\sum_{k=1}^n \mathbf{E}_\mathcal{N}(x_k^* x_k) = \sum_{k=1}^n e_{kk} \otimes \mathsf{E}_\mathcal{N}(\mathring{a}_k^* \mathring{a}_k) \leq \sum_{k=1}^n e_{kk} \otimes \mathsf{E}_\mathcal{N}(a_k^* a_k).$$

This implies

$$\Big\|\Big(\sum_{k=1}^n \mathbf{E}_\mathcal{N}(x_k x_k^*)\Big)^{1/2}\Big\|_p \leq \Big\|\Big(\sum_{k=1}^n \mathsf{E}_\mathcal{N}(a_k a_k^*)\Big)^{1/2}\Big\|_p.$$

On the other hand, the third term is controlled by

$$\Big\|\Big(\sum_{k=1}^n \mathbf{E}_\mathcal{N}(x_k^* x_k)\Big)^{1/2}\Big\|_p = \Big\|\sum_{k=1}^n \mathbf{E}_\mathcal{N}(x_k^* x_k)\Big\|_{p/2}^{1/2}$$

$$\leq \Big\|\sum_{k=1}^n e_{kk} \otimes \mathsf{E}_\mathcal{N}(a_k^* a_k)\Big\|_{p/2}^{1/2}$$

$$= \Big(\sum_{k=1}^n \|\mathsf{E}_\mathcal{N}(a_k^* a_k)\|_{p/2}^{p/2}\Big)^{1/p} \leq \Big(\sum_{k=1}^n \|a_k\|_p^p\Big)^{1/p}.$$

Combining the inequalities above we have the upper estimate with constant 2. □

5.2. Estimates for BMO type norms

Apart from the free Rosenthal inequalities, our second key tool in the proof of (5.1) and (5.2) will be certain estimates that we develop in this section. Let us recall the definition of several noncommutative Hardy spaces. Xu's survey [68] contains a systematic exposition of these notions. Let $\mathcal{M}$ be a von Neumann algebra equipped with a n.f. state φ. Let $\mathcal{M}_1, \mathcal{M}_2, \ldots$ be an increasing filtration of von Neumann subalgebras of $\mathcal{M}$ which are invariant under the modular automorphism group on $\mathcal{M}$ associated to φ. This allows us to consider the corresponding conditional expectations $\mathcal{E}_n : \mathcal{M} \to \mathcal{M}_n$ and noncommutative L_p martingales $x = (x_n)_{n \geq 1}$ with martingale differences $dx_1, dx_2, \ldots$ adapted to this filtration. The row and column Hardy spaces $\mathcal{H}_p^r(\mathcal{M})$ and $\mathcal{H}_p^c(\mathcal{M})$ are defined respectively as the closure of the space of finite L_p martingales with respect to the following norms

$$\|x\|_{\mathcal{H}_p^r(\mathcal{M})} = \Big\|\Big(\sum_k dx_k dx_k^*\Big)^{\frac{1}{2}}\Big\|_p,$$

$$\|x\|_{\mathcal{H}_p^c(\mathcal{M})} = \Big\|\Big(\sum_k dx_k^* dx_k\Big)^{\frac{1}{2}}\Big\|_p.$$

On the other hand, the space $\mathcal{H}_p^p(\mathcal{M})$ measures the p-variation

$$\|x\|_{\mathcal{H}_p^p(\mathcal{M})} = \Big(\sum_k \|dx_k\|_p^p\Big)^{1/p}.$$

Finally, the conditional row and column Hardy spaces for martingales $h_p^r(\mathcal{M})$ and $h_p^c(\mathcal{M})$ are defined as the closure of the space of finite L_p martingales with respect to the following norms

$$\|x\|_{h_p^r(\mathcal{M})} = \left\|\left(\sum_k \mathcal{E}_{k-1}(dx_k dx_k^*)\right)^{\frac{1}{2}}\right\|_p,$$

$$\|x\|_{h_p^c(\mathcal{M})} = \left\|\left(\sum_k \mathcal{E}_{k-1}(dx_k^* dx_k)\right)^{\frac{1}{2}}\right\|_p.$$

A further tool are maximal functions. The notion of maximal function was introduced in [16] via the spaces $L_p(\mathcal{M}; \ell_\infty)$. Here we are using variations of these from Musat's paper [38]. Given $2 \le p \le \infty$ we define the spaces $L_p^r(\mathcal{M}; \ell_\infty)$ and $L_p^c(\mathcal{M}; \ell_\infty)$ as the space of bounded sequences in $L_p(\mathcal{M})$ with respect to the following norms

$$\|(x_n)\|_{L_p^r(\mathcal{M};\ell_\infty)} = \left\|\sup_{n\ge 1} x_n x_n^*\right\|_{p/2}^{1/2},$$

$$\|(x_n)\|_{L_p^c(\mathcal{M};\ell_\infty)} = \left\|\sup_{n\ge 1} x_n^* x_n\right\|_{p/2}^{1/2}.$$

This definition requires some explanation. Indeed, in the noncommutative setting there is no obvious analogue for the pointwise supremum of a family of positive operators. Therefore, the above is to be understood in the sense of the suggestive notation introduced in [16]. Among several characterizations of the norm in $L_p(\mathcal{M}; \ell_\infty)$ of a sequence $(z_n)_{n\ge 1}$ of positive operators, we outline the following obtained by duality

(5.4) $$\left\|\sup_{n\ge 1} z_n\right\|_p = \sup\left\{\sum_n \text{tr}(z_n w_n) \,\Big|\, w_n \ge 0, \left\|\sum_n w_n\right\|_{p'} \le 1\right\}.$$

The reader is also referred to [29] for a rather complete exposition. One of the fundamental properties obtained in [38] of the spaces $L_p^r(\mathcal{M}; \ell_\infty)$ and $L_p^c(\mathcal{M}; \ell_\infty)$ is that they form interpolation families. Indeed, given $2 \le p_0, p_1 \le \infty$, we have the following isometric isomorphisms for $1/p_\theta = (1-\theta)/p_0 + \theta/p_1$

(5.5)
$$\begin{aligned}
\left[L_{p_0}^r(\mathcal{M};\ell_\infty), L_{p_1}^r(\mathcal{M};\ell_\infty)\right]_\theta &= L_{p_\theta}^r(\mathcal{M};\ell_\infty), \\
\left[L_{p_0}^c(\mathcal{M};\ell_\infty), L_{p_1}^c(\mathcal{M};\ell_\infty)\right]_\theta &= L_{p_\theta}^c(\mathcal{M};\ell_\infty).
\end{aligned}$$

Here it is also interesting to point out that these spaces can easily be identified as one-sided amalgamated L_p spaces. In particular, the interpolation formula (5.5) follows from Theorem 3.2.

5.2.1. One-sided estimates. We will simplify our arguments considerably by assuming that $\mathcal{M}$ is finite and the density D associated to the state φ satisfies $c_1 1_\mathcal{M} \le D \le c_2 1_\mathcal{M}$. The general case will follow one more time from Haagerup's approximation theorem.

LEMMA 5.4. *Let $1 \le p \le 2 \le q \le \infty$ be such that $1/p = 1/2 + 1/q$. Given $0 < \theta < 1$, let x be a norm one element in $[\mathcal{H}_p^p(\mathcal{M}), \mathcal{H}_p^r(\mathcal{M})]_\theta$ and let us consider the indices $1 \le u \le 2 \le v \le \infty$ defined as follows*

$$1/u = 1/p - \theta/q \quad \text{and} \quad 1/v = \theta/q.$$

Then we may find a positive element $a \in L_{v/2}(\mathcal{M})$ and $b_k \in L_u(\mathcal{M}_k)$ such that
$$dx_k = \mathcal{E}_k(a)^{\frac{1}{2}} b_k \quad \text{and} \quad \max\left\{\|a\|_{v/2}, \left(\sum_{k\geq 1}\|b_k\|_u^u\right)^{1/u}\right\} \leq \sqrt{2}.$$

If x belongs to $[\mathcal{H}_p^p(\mathcal{M}), \mathcal{H}_p^c(\mathcal{M})]_\theta$, the same conclusion holds with $dx_k = b_k^* \mathcal{E}_k(a)^{\frac{1}{2}}$.

PROOF. The last assertion follows from the first part of the statement by taking adjoints. To prove the first assertion, we may assume by approximation that x is a finite martingale in $L_p(\mathcal{M}_m)$ for some $m \geq 1$. Therefore, since x is of norm 1 there exists an analytic function $f : \mathcal{S} \to L_p(\mathcal{M}_m)$ of the form
$$f(z) = \sum_{k=1}^{m} d_k(z),$$
which satisfies $f(\theta) = x$ and the estimate
$$\max\left\{\sup_{z \in \partial_0}\left(\sum_{k=1}^{m}\|d_k(z)\|_p^p\right)^{\frac{1}{p}}, \sup_{z \in \partial_1}\left\|\left(\sum_{k=1}^{m} d_k(z)d_k(z)^*\right)^{\frac{1}{2}}\right\|_p\right\} \leq 1.$$

Note that we are assuming that the d_k's are also analytic where $d_1(z), d_2(z), \ldots$ denote the martingale differences of $f(z)$. Now we consider the following functions on the strip for $1 \leq k \leq m$
$$g_k(z) = \begin{cases} 1, & \text{if } z \in \partial_0, \\ \left(\sum_{j=1}^k d_j(z)d_j(z)^* + \delta \mathrm{D}^{\frac{2}{p}}\right)^{\frac{p}{q}}, & \text{if } z \in \partial_1. \end{cases}$$

According to our original assumption on the finiteness of $\mathcal{M}$ and the invertibility of D, we are in position to apply Devinatz's theorem. Indeed, here we need Xu's modification, which can be found in Section 8 of Pisier and Xu's survey [52]. This provides us with analytic function h_k with analytic inverse and such that

(5.6) $\qquad h_k(z)h_k(z)^* = g_k(z) \quad \text{for all} \quad z \in \partial\mathcal{S}.$

STEP 1. We claim that

(5.7) $\qquad \left\|\sum_{k=1}^{m} \delta_k \otimes h_k(\theta)\right\|_{L_v^r(\mathcal{M};\ell_\infty)} \leq \left(1+\delta^{\frac{p}{2}}\right)^{\frac{\theta}{q}},$

where δ appears in the definition of g_k. Indeed, according to the interpolation isometries (5.5) and the three lines lemma, it suffices to see that the following estimates hold

(5.8) $\qquad \sup_{z \in \partial_0}\left\|\sum_{k=1}^{m} \delta_k \otimes h_k(z)\right\|_{L_\infty^r(\mathcal{M};\ell_\infty)} \leq 1,$

(5.9) $\qquad \sup_{z \in \partial_1}\left\|\sum_{k=1}^{m} \delta_k \otimes h_k(z)\right\|_{L_q^r(\mathcal{M};\ell_\infty)} \leq \left(1+\delta^{\frac{p}{2}}\right)^{\frac{1}{q}}.$

To prove (5.8) we first recall from (5.6) and the definition of g_k that $h_k(z)$ is a unitary for any $z \in \partial_0$ and any $1 \leq k \leq m$. Therefore, since Fubini's theorem gives $L_\infty^r(\mathcal{M};\ell_\infty) = L_\infty(\ell_\infty(\mathcal{M}))$, we conclude
$$\sup_{z \in \partial_0}\left\|\sum_{k=1}^{m} \delta_k \otimes h_k(z)\right\|_{L_\infty^r(\mathcal{M};\ell_\infty)} = \sup_{z \in \partial_0}\sup_{1 \leq k \leq m}\|h_k(z)\|_\infty = 1.$$

On the other hand, if $z \in \partial_1$ we have the following estimate from (5.6)

$$\begin{aligned}
\Big\| \sum_{k=1}^{m} \delta_k \otimes h_k(z) \Big\|_{L_q^r(\mathcal{M};\ell_\infty)}^q &= \Big\| \sup_{1 \le k \le m} \Big(\sum_{j=1}^{k} d_j(z)d_j(z)^* + \delta \mathrm{D}^{\frac{2}{p}} \Big)^{p/q} \Big\|_{q/2}^{q/2} \\
&\le \Big\| \Big(\sum_{j=1}^{m} d_j(z)d_j(z)^* + \delta \mathrm{D}^{\frac{2}{p}} \Big)^{p/q} \Big\|_{q/2}^{q/2} \\
&= \Big\| \sum_{j=1}^{m} d_j(z)d_j(z)^* + \delta \mathrm{D}^{\frac{2}{p}} \Big\|_{p/2}^{p/2} \le 1 + \delta^{\frac{p}{2}},
\end{aligned}$$

where the last inequality follows from the fact that $L_{p/2}(\mathcal{M})$ is a $p/2$-normed space and also from the right boundary estimate for the function f given above. Thus, inequality (5.9) follows and the proof of (5.7) is completed.

STEP 2. Let us consider the functions $w_k(z) = h_k(z)^{-1} d_k(z)$. Now we claim that

(5.10) $$\Big(\sum_{k=1}^{m} \|w_k(\theta)\|_u^u \Big)^{1/u} \le \sqrt{2}^\theta \big(1 + \delta^{\frac{p}{2}}\big)^{\theta/2}.$$

According to (5.6), we can write $h_k(z) = g_k(z)^{\frac{1}{2}} u_k(z)$ for some unitary $u_k(z)$. Thus, we deduce that for $z \in \partial_0$ we have $w_k(z) = u_k(z)^* d_k(z)$. This and the left boundary estimate for f yield

$$\sup_{z \in \partial_0} \Big(\sum_{k=1}^{m} \|w_k(z)\|_p^p \Big)^{1/p} \le 1.$$

The interesting argument, based on the classical Fefferman-Stein duality theorem, appears for $z \in \partial_1$. In that case we have $h_k(z)^{-1} = u_k(z)^* g_k(z)^{-\frac{1}{2}}$ and we find the following estimate for any $z \in \partial_1$

$$\begin{aligned}
\sum_{k=1}^{m} \|w_k(z)\|_2^2 &= \sum_{k=1}^{m} \mathrm{tr}\Big(h_k(z)^{-1} d_k(z) d_k(z)^* h_k(z)^{-1*} \Big) \\
&= \sum_{k=1}^{m} \mathrm{tr}\Big(u_k(z)^* g_k(z)^{-\frac{1}{2}} d_k(z) d_k(z)^* g_k(z)^{-\frac{1}{2}} u_k(z) \Big) \\
&= \sum_{k=1}^{m} \mathrm{tr}\Big(g_k(z)^{-\frac{1}{2}} d_k(z) d_k(z)^* g_k(z)^{-\frac{1}{2}} \Big).
\end{aligned}$$

Now we define the positive operators

$$\begin{aligned}
\alpha_k &= \Big(\sum_{j=1}^{k-1} d_j(z) d_j(z)^* + \delta \mathrm{D}^{\frac{2}{p}} \Big)^{\frac{p}{2}}, \\
\beta_k &= \Big(\sum_{j=1}^{k} d_j(z) d_j(z)^* + \delta \mathrm{D}^{\frac{2}{p}} \Big)^{\frac{p}{2}},
\end{aligned}$$

and the indices $(s,t) = (2/q, 2/p)$. Lemma 7.2 in [**28**] gives

$$\mathrm{tr}\Big(\beta_k^{-s/2} (\beta_k^t - \alpha_k^t) \beta_k^{-s/2} \Big) \le 2 \, \mathrm{tr}\big(\beta_k - \alpha_k \big).$$

According to our choice of (s,t), we may rewrite this inequality as follows
$$\mathrm{tr}\Big(g_k(z)^{-\frac{1}{2}}d_k(z)d_k(z)^*g_k(z)^{-\frac{1}{2}}\Big) \leq 2\,\mathrm{tr}\big(\beta_k - \beta_{k-1}\big).$$
Summing up, we deduce that
$$\sum_{k=1}^m \|w_k(z)\|_2^2 \leq 2 \Big\|\sum_{j=1}^m d_j(z)d_j(z)^* + \delta \mathrm{D}^{\frac{2}{p}}\Big\|_{p/2}^{p/2} \leq 2\big(1+\delta^{\frac{p}{2}}\big).$$

Finally, recalling that the functions $w_k : \mathcal{S} \to L_p(\mathcal{M})$ are analytic since are products of analytic functions, inequality (5.10) follows from the estimates given above and the three lines lemma.

STEP 3. For the moment, we have seen that
$$dx_k = h_k(\theta)w_k(\theta).$$
Now, recalling that
$$g_k : \partial \mathcal{S} \to L_{q/2}(\mathcal{M}_k) \quad \text{for each} \quad 1 \leq k \leq m,$$
we conclude that $h(\theta) = (h_1(\theta), h_2(\theta), \ldots, h_m(\theta))$ is an *adapted* sequence. That is, we have $h_k \in L_v(\mathcal{M}_k)$ for all $1 \leq k \leq m$. On the other hand, according to the definition of the space $L_v^r(\mathcal{M}; \ell_\infty)$, we may find for any $\delta > 0$ a positive operator a such that $h_k(\theta)h_k(\theta)^* \leq a$ for $1 \leq k \leq m$ and
$$(5.11) \qquad \|a\|_{v/2}^{1/2} \leq (1+\delta) \Big\|\sum_{k=1}^m \delta_k \otimes h_k(\theta)\Big\|_{L_v^r(\mathcal{M};\ell_\infty)} \leq \big(1+\delta^{\frac{p}{2}}\big)^{1+\frac{\theta}{q}}.$$
Moreover, since $h(\theta)$ is adapted, we have $h_k(\theta)h_k(\theta)^* = \mathcal{E}_k(h_k(\theta)h_k(\theta)^*) \leq \mathcal{E}_k(a)$. This gives a contraction $\gamma_k \in \mathcal{M}_k$ with $h_k(\theta) = \mathcal{E}_k(a)^{\frac{1}{2}}\gamma_k$. In particular, we deduce
$$(5.12) \qquad dx_k = \mathcal{E}_k(a)^{\frac{1}{2}}\gamma_k w_k(\theta) = \mathcal{E}_k(a)^{\frac{1}{2}}b_k.$$
Finally, since γ_k is a contraction
$$(5.13) \qquad \Big(\sum_{k=1}^m \|b_k\|_u^u\Big)^{1/u} \leq \Big(\sum_{k=1}^m \|w_k(\theta)\|_u^u\Big)^{1/u} \leq \sqrt{2}^\theta \big(1+\delta^{\frac{p}{2}}\big)^{\theta/2}.$$
Therefore, the assertion follows from (5.11), (5.12) and (5.13) by letting $\delta \to 0^+$. □

Applying the anti-linear duality bracket $\langle x, y\rangle = \mathrm{tr}(x^*y)$, we consider in the following result an immediate application of Lemma 5.4 for the following dual spaces
$$\begin{aligned} \mathcal{Z}_p^r(\mathcal{M},\theta) &= \big[\mathcal{H}_p^p(\mathcal{M}), \mathcal{H}_p^r(\mathcal{M})\big]_\theta^*, \\ \mathcal{Z}_p^c(\mathcal{M},\theta) &= \big[\mathcal{H}_p^p(\mathcal{M}), \mathcal{H}_p^c(\mathcal{M})\big]_\theta^*. \end{aligned}$$

LEMMA 5.5. *If p, q, u, v are as above and $1/u + 1/s = 1$, we have*
$$\|x\|_{\mathcal{Z}_p^r(\mathcal{M},\theta)} \leq 2 \sup_{\substack{\|a\|_{v/2} \leq 1 \\ a \geq 0}} \Big(\sum_k \big\|\mathcal{E}_k(a)^{\frac{1}{2}}dx_k\big\|_s^s\Big)^{1/s},$$
$$\|x\|_{\mathcal{Z}_p^c(\mathcal{M},\theta)} \leq 2 \sup_{\substack{\|a\|_{v/2} \leq 1 \\ a \geq 0}} \Big(\sum_k \big\|dx_k \mathcal{E}_k(a)^{\frac{1}{2}}\big\|_s^s\Big)^{1/s}.$$

PROOF. There exists y in the unit ball of $[\mathcal{H}_p^p(\mathcal{M}), \mathcal{H}_p^r(\mathcal{M})]_\theta$ such that
$$\|x\|_{\mathcal{Z}_p^r(\mathcal{M},\theta)} = |\text{tr}(xy^*)|.$$
On the other hand, according to Lemma 5.4 and using homogeneity, we may write $dy_k = \mathcal{E}_k(a)^{1/2} b_k$ with a being a positive operator in the unit ball of $L_{v/2}(\mathcal{M})$ and $b_1, b_2, \ldots$ satisfying

(5.14) $$\Big(\sum_{k \geq 1} \|b_k\|_u^u\Big)^{1/u} \leq 2.$$

With this decomposition we obtain the following estimate
$$\begin{aligned}
\|x\|_{\mathcal{Z}_p^r(\mathcal{M},\theta)} &= \Big|\sum_k \text{tr}(dx_k dy_k^*)\Big| \\
&= \Big|\sum_k \text{tr}(dx_k b_k^* \mathcal{E}_k(a)^{\frac{1}{2}})\Big| \\
&\leq \Big(\sum_k \|b_k\|_u^u\Big)^{1/u} \Big(\sum_k \big\|\mathcal{E}_k(a)^{\frac{1}{2}} dx_k\big\|_s^s\Big)^{1/s}.
\end{aligned}$$
The assertion follows from (5.14). The proof of the second estimate is similar. □

In the following result and in the rest of this paper we shall write $A \lesssim B$ to denote the existence of an absolute constant c such that $A \leq c\, B$ holds.

LEMMA 5.6. *If p, q, u, v are as above and $1/u + 1/s = 1$, we have*
$$\sup_{\substack{\|a\|_{v/2} \leq 1 \\ a \geq 0}} \Big(\sum_k \big\|\mathcal{E}_k(a)^{\frac{1}{2}} dx_k\big\|_s^s\Big)^{1/s} \lesssim \|x\|_{[\mathcal{H}_{p'}^{p'}(\mathcal{M}), \mathcal{H}_{p'}^r(\mathcal{M})]_\theta},$$
$$\sup_{\substack{\|a\|_{v/2} \leq 1 \\ a \geq 0}} \Big(\sum_k \big\|dx_k \mathcal{E}_k(a)^{\frac{1}{2}}\big\|_s^s\Big)^{1/s} \lesssim \|x\|_{[\mathcal{H}_{p'}^{p'}(\mathcal{M}), \mathcal{H}_{p'}^c(\mathcal{M})]_\theta}.$$

PROOF. We may find an analytic function
$$f: \mathcal{S} \to \mathcal{H}_{p'}^{p'}(\mathcal{M}) + \mathcal{H}_{p'}^r(\mathcal{M})$$
such that $f(\theta) = x$ and
$$\max\Big\{\sup_{z \in \partial_0} \|f(z)\|_{\mathcal{H}_{p'}^{p'}(\mathcal{M})}, \sup_{z \in \partial_1} \|f(z)\|_{\mathcal{H}_{p'}^r(\mathcal{M})}\Big\} \lesssim \|x\|_{[\mathcal{H}_{p'}^{p'}(\mathcal{M}), \mathcal{H}_{p'}^r(\mathcal{M})]_\theta}.$$
By homogeneity, let us assume that the right hand side above is 1. Now we take positive element a in the unit ball of $L_{v/2}(\mathcal{M})$ so that we may consider an analytic function $g: \mathcal{S} \to L_\infty(\mathcal{M}) + L_{q/2}(\mathcal{M})$ satisfying $g(\theta) = a$ and
$$\max\Big\{\sup_{z \in \partial_0} \|g(z)\|_\infty, \sup_{z \in \partial_1} \|g(z)\|_{q/2}\Big\} \leq 1.$$
Then we construct the analytic function
$$h(z) = \sum_k \delta_k \otimes d_k(f(\bar{z}))^* \mathcal{E}_k(g(z)) d_k(f(z)) \in \ell_\infty(L_1(\mathcal{M})).$$
For $z \in \partial_0$ we have
$$\|h(z)\|_{p'/2} \leq \Big(\sum_{k \geq 1} \|d_k(f(\bar{z}))\|_{p'}^{p'}\Big)^{1/p'} \sup_{k \geq 1} \|\mathcal{E}_k(g(z))\|_\infty \Big(\sum_{k \geq 1} \|d_k(f(z))\|_{p'}^{p'}\Big)^{1/p'} \lesssim 1.$$

In the case $z \in \partial_1$, we choose a factorization $g(z) = g_1(z)g_2(z)$ such that
$$\|g_1(z)\|_q = \|g_2(z)\|_q = \sqrt{\|g(z)\|_{q/2}} \leq 1.$$
Then, Hölder inequality provides the following estimate
$$\begin{aligned}
\|h(z)\|_1 &= \sum_{k \geq 1} \left\| \mathcal{E}_k\big(d_k(f(\bar{z}))^* g_1(z) g_2(z) d_k(f(z))\big) \right\|_1 \\
&\leq \left(\sum_{k \geq 1} \operatorname{tr}\big(d_k(f(\bar{z}))^* g_1(z) g_1(z)^* d_k(f(\bar{z}))\big) \right)^{\frac{1}{2}} \\
&\quad \times \left(\sum_{k \geq 1} \operatorname{tr}\big(d_k(f(z))^* g_2(z)^* g_2(z) d_k(f(z))\big) \right)^{\frac{1}{2}} \\
&\leq \left\| \sum_{k \geq 1} d_k(f(\bar{z})) d_k(f(\bar{z}))^* \right\|_{p'/2}^{1/2} \left\| \sum_{k \geq 1} d_k(f(z)) d_k(f(z))^* \right\|_{p'/2}^{1/2} \lesssim 1.
\end{aligned}$$
Indeed, in order to apply Hölder inequality in the first inequality above, we factorize the conditional expectation $\mathcal{E}_{k-1}(a^*b)$ as the product $u_{k-1}(a)^* u_{k-1}(b)$ by a right $\mathcal{M}_{k-1}$-module map $u_{k-1} : \mathcal{M} \to C_\infty(\mathcal{M}_{k-1})$, see e.g. [**16, 26**]. By complex interpolation $(2/s = 2(1-\theta)/p' + \theta)$, we conclude that
$$\left(\sum_k \left\| \mathcal{E}_k(a)^{\frac{1}{2}} dx_k \right\|_s^s \right)^{1/s} = \|h(\theta)\|_{s/2}^{1/2} \lesssim 1.$$
The column version of this inequality follows by taking adjoints. $\square$

REMARK 5.7. When $1 < p \leq 2$ we have
$$\|x\|_{\mathcal{Z}_p^r(\mathcal{M},\theta)} \sim \sup_{\substack{\|a\|_{v/2} \leq 1 \\ a \geq 0}} \left(\sum_k \left\| \mathcal{E}_k(a)^{\frac{1}{2}} dx_k \right\|_s^s \right)^{1/s},$$
$$\|x\|_{\mathcal{Z}_p^c(\mathcal{M},\theta)} \sim \sup_{\substack{\|a\|_{v/2} \leq 1 \\ a \geq 0}} \left(\sum_k \left\| dx_k \mathcal{E}_k(a)^{\frac{1}{2}} \right\|_s^s \right)^{1/s}.$$
Indeed, since anti-linear duality is compatible with complex interpolation via the analytic function $\operatorname{tr}(f(\bar{z})^* g(z))$, we find by reflexivity in the case $1 < p \leq 2$ the following isomorphisms
$$(5.15) \quad \begin{aligned} \mathcal{Z}_p^r(\mathcal{M},\theta) &\simeq [\mathcal{H}_{p'}^{p'}(\mathcal{M}), \mathcal{H}_{p'}^r(\mathcal{M})]_\theta, \\ \mathcal{Z}_p^c(\mathcal{M},\theta) &\simeq [\mathcal{H}_{p'}^{p'}(\mathcal{M}), \mathcal{H}_{p'}^c(\mathcal{M})]_\theta. \end{aligned}$$
Therefore, the result follows from Lemma 5.5 and Lemma 5.6.

REMARK 5.8. Lemmas 5.4, 5.5, 5.6 and Remark 5.7 immediately generalize for the row and column conditional Hardy spaces. Indeed, if we replace the row and column Hardy spaces $\mathcal{H}_p^r(\mathcal{M})$ and $\mathcal{H}_p^c(\mathcal{M})$ by their conditional analogues $h_p^r(\mathcal{M})$ and $h_p^c(\mathcal{M})$ and the conditional expectation $\mathcal{E}_k$ by $\mathcal{E}_{k-1}$, it can be easily checked that the same arguments can be adapted to obtain the conditioned results.

We shall also need to generalize the norms of $h_p^r(\mathcal{M})$ and $h_p^c(\mathcal{M})$ for arbitrary (non-necessarily adapted) sequences $z_1, z_2, \ldots$ in $L_p(\mathcal{M})$ as follows
$$\left\| \sum_k \delta_k \otimes z_k \right\|_{L_{\text{cond}}^p(\mathcal{M}; \ell_2^r)} = \left\| \left(\sum_k \mathcal{E}_{k-1}(z_k z_k^*) \right)^{\frac{1}{2}} \right\|_p,$$

$$\Big\|\sum_k \delta_k \otimes z_k\Big\|_{L^p_{\mathrm{cond}}(\mathcal{M};\ell_2^c)} = \Big\|\Big(\sum_k \mathcal{E}_{k-1}(z_k^* z_k)\Big)^{\frac{1}{2}}\Big\|_p.$$

LEMMA 5.9. *Let p,q be as above and let us consider the indices (s,t) given by $1/s = (1-\eta)/p' + \eta/2$ and $1/t = \eta/q$ for some parameter $0 < \eta < 1$. Then, the following estimates hold for every martingale difference sequence $dx_1, dx_2, \ldots$ in $L_{p'}(\mathcal{M})$*

$$\Big\|\sum_k \delta_k \otimes dx_k a_k\Big\|_{L^s_{\mathrm{cond}}(\mathcal{M};\ell_2^r)} \leq \Big(\sum_{k\geq 1} \|a_k\|_t^t\Big)^{1/t} \|x\|_{[h^r_{p'}(\mathcal{M}), \mathcal{H}^{p'}_{p'}(\mathcal{M})]_\eta},$$

$$\Big\|\sum_k \delta_k \otimes a_k dx_k\Big\|_{L^s_{\mathrm{cond}}(\mathcal{M};\ell_2^c)} \leq \Big(\sum_{k\geq 1} \|a_k\|_t^t\Big)^{1/t} \|x\|_{[h^c_{p'}(\mathcal{M}), \mathcal{H}^{p'}_{p'}(\mathcal{M})]_\eta}.$$

PROOF. We recall that

$$\big[L^{p_0}_{\mathrm{cond}}(\mathcal{M};\ell_2^r), L^{p_1}_{\mathrm{cond}}(\mathcal{M};\ell_2^r)\big]_\theta \subset L^{p_\theta}_{\mathrm{cond}}(\mathcal{M};\ell_2^r),$$
$$\big[L^{p_0}_{\mathrm{cond}}(\mathcal{M};\ell_2^c), L^{p_1}_{\mathrm{cond}}(\mathcal{M};\ell_2^c)\big]_\theta \subset L^{p_\theta}_{\mathrm{cond}}(\mathcal{M};\ell_2^c),$$

hold isometrically. Indeed, we recall the factorization

$$\mathcal{E}_{k-1}(a^* b) = u_{k-1}(a)^* u_{k-1}(b)$$

used in the proof of Lemma 5.6 and the resulting isometric embeddings

$$L^p_{\mathrm{cond}}(\mathcal{M};\ell_2^r) \subset L_p(\mathcal{M}; R_p(\mathbb{N}^2)),$$
$$L^p_{\mathrm{cond}}(\mathcal{M};\ell_2^c) \subset L_p(\mathcal{M}; C_p(\mathbb{N}^2)).$$

We refer the reader to [**16**] for a more detailed explanation. According to our claim and by bilinear interpolation, it suffices to prove the assertion for the extremal cases. When $\eta = 0$ we have $(s,t) = (p', \infty)$ and the following estimate holds

$$\Big\|\Big(\sum_k \mathcal{E}_{k-1}(dx_k a_k a_k^* dx_k^*)\Big)^{\frac{1}{2}}\Big\|_{p'} \leq \sup_{k\geq 1} \|a_k\|_\infty \Big\|\Big(\sum_k \mathcal{E}_{k-1}(dx_k dx_k^*)\Big)^{\frac{1}{2}}\Big\|_{p'}.$$

On the other hand, for $\eta = 1$ we have $(s,t) = (2, q)$ so that

$$\Big\|\Big(\sum_k \mathcal{E}_{k-1}(dx_k a_k a_k^* dx_k^*)\Big)^{\frac{1}{2}}\Big\|_2 = \Big(\sum_k \mathrm{tr}(a_k a_k^* dx_k^* dx_k)\Big)^{1/2}$$
$$\leq \Big(\sum_k \|a_k\|_q^2 \|dx_k\|_{p'}^2\Big)^{1/2}$$
$$\leq \Big(\sum_k \|a_k\|_q^q\Big)^{1/q} \Big(\sum_k \|dx_k\|_{p'}^{p'}\Big)^{1/p'}.$$

This proves the first inequality. The second one follows by taking adjoints. □

The following is the main result of this paragraph.

PROPOSITION 5.10. *If p, q, u, v are as above and $1/u + 1/s = 1$, we have*

$$\|x\|_{\mathcal{Z}^r_p(\mathcal{M},\theta)} \leq c(p,\theta) \max\Big\{\|x\|_{\mathcal{H}^{p'}_{p'}(\mathcal{M})}, \|x\|_{[\mathcal{H}^{p'}_{p'}(\mathcal{M}), h^r_{p'}(\mathcal{M})]_\theta}\Big\},$$

$$\|x\|_{\mathcal{Z}^c_p(\mathcal{M},\theta)} \leq c(p,\theta) \max\Big\{\|x\|_{\mathcal{H}^{p'}_{p'}(\mathcal{M})}, \|x\|_{[\mathcal{H}^{p'}_{p'}(\mathcal{M}), h^c_{p'}(\mathcal{M})]_\theta}\Big\}.$$

Here the constant $c(p,\theta)$ satisfies $c(p,\theta) \sim 1$ as $v \to \infty$ and

$$c(p,\theta) \sim \Big(\frac{p}{(2-p)\theta} - 1\Big)^{-1/2} = \sqrt{\frac{2}{v-2}} \quad as \quad v \to 2.$$

PROOF. According to Lemma 5.5 we have
$$\|x\|_{\mathcal{Z}_p^r(\mathcal{M},\theta)} \leq 2 \sup_{\substack{\|a\|_{v/2}\leq 1 \\ a\geq 0}} \Big(\sum_k \|\mathcal{E}_k(a)^{\frac{1}{2}} dx_k\|_s^s\Big)^{1/s}.$$

However, since $2 \leq s \leq \infty$ we have

(5.16) $\Big(\sum_k \|\mathcal{E}_k(a)^{\frac{1}{2}} dx_k\|_s^s\Big)^{1/s}$
$= \Big(\sum_k \|dx_k^*(da_k + \mathcal{E}_{k-1}(a))dx_k\|_{s/2}^{s/2}\Big)^{1/s}$
$\leq \Big(\sum_k \|dx_k^* da_k dx_k\|_{s/2}^{s/2}\Big)^{1/s} + \Big(\sum_k \|dx_k^* \mathcal{E}_{k-1}(a) dx_k\|_{s/2}^{s/2}\Big)^{1/s}.$

As we pointed above, Lemma 5.6 and Remark 5.7 generalize to conditional Hardy spaces after replacing $\mathcal{E}_k$ by $\mathcal{E}_{k-1}$. In other words, the inequality below holds with absolute constants for $1 \leq p \leq 2$

(5.17) $\sup_{\substack{\|a\|_{v/2}\leq 1 \\ a\geq 0}} \Big(\sum_k \|dx_k^* \mathcal{E}_{k-1}(a) dx_k\|_{s/2}^{s/2}\Big)^{1/s} \lesssim \|x\|_{[\mathcal{H}_{p'}^{p'}(\mathcal{M}), h_{p'}^r(\mathcal{M})]_\theta}.$

Therefore, it suffices to estimate the first term on the right of (5.16).

STEP 1. We first assume $4 \leq v \leq \infty$. Since $1/s = 1/v + 1/p'$
$$\Big(\sum_k \|dx_k da_k dx_k\|_{s/2}^{s/2}\Big)^{1/s} \leq \Big(\sum_k \|da_k\|_{v/2}^{v/2}\Big)^{1/v} \Big(\sum_k \|dx_k\|_{p'}^{p'}\Big)^{1/p'}.$$

Then, complex interpolation gives
$$\Big(\sum_k \|da_k\|_{v/2}^{v/2}\Big)^{1/v} \leq \sqrt{2} \|a\|_{v/2}^{1/2} \leq \sqrt{2} \quad \text{for} \quad 4 \leq v \leq \infty.$$

Indeed, our claim is trivial for the extremal cases.

STEP 2. The case $2 < v < 4$ is a little more complicated. By the noncommutative Burkholder inequality [28], we may find a decomposition $da_k = d\alpha_k + d\beta_k + d\gamma_k$ into three martingales satisfying the following estimates

(5.18)
$$\Big(\sum_k \|d\alpha_k\|_{v/2}^{v/2}\Big)^{2/v} \leq c_v \|a\|_{v/2},$$
$$\Big\|\Big(\sum_k \mathcal{E}_{k-1}(d\beta_k d\beta_k^*)\Big)^{\frac{1}{2}}\Big\|_{v/2} \leq c_v \|a\|_{v/2},$$
$$\Big\|\Big(\sum_k \mathcal{E}_{k-1}(d\gamma_k^* d\gamma_k)\Big)^{\frac{1}{2}}\Big\|_{v/2} \leq c_v \|a\|_{v/2},$$

where we know from [54] that
$$c_v \lesssim \frac{1}{v-2} \quad \text{for} \quad 2 \leq v \leq 4.$$

Since we need to estimate the first term on the right of (5.16), we decompose it into three terms according to the martingale decomposition above. The resulting term associated to α can be estimated as in Step 1. For the second term (associated to β) we may find a norm one element (b_k) in $L_{(s/2)'}(\mathcal{M}; \ell_{(s/2)'})$ such that

(5.19) $\Big(\sum_k \|dx_k^* d\beta_k dx_k\|_{s/2}^{s/2}\Big)^{1/s} = \sqrt{\Big|\sum_k \operatorname{tr}(b_k dx_k^* d\beta_k dx_k)\Big|}.$

However, we have

$$\begin{aligned}
\left|\sum_k \mathrm{tr}\big(b_k dx_k^* d\beta_k dx_k\big)\right| &= \left|\sum_k \mathrm{tr}\big(d\beta_k dx_k b_k dx_k^*\big)\right| \\
&= \left|\sum_k \mathrm{tr}\big(\mathcal{E}_{k-1}(d\beta_k dx_k b_k dx_k^*)\big)\right| \\
&\le \left\|\big(\sum_k \mathcal{E}_{k-1}(d\beta_k d\beta_k^*)\big)^{\frac{1}{2}}\right\|_{v/2} \\
&\quad \times \left\|\big(\sum_k \mathcal{E}_{k-1}(dx_k b_k^* dx_k^* dx_k b_k dx_k^*)\big)^{\frac{1}{2}}\right\|_{(v/2)'} \\
&\le c_v \|a\|_{v/2} \left\|\big(\sum_k \mathcal{E}_{k-1}(dx_k b_k^* dx_k^* dx_k b_k dx_k^*)\big)^{\frac{1}{2}}\right\|_{(v/2)'}.
\end{aligned}$$

Indeed, the first inequality above follows from Hölder inequality after representing $\mathcal{E}_{k-1}(a^*b)$ as $u_{k-1}(a)^* u_{k-1}(b)$ via the right module map u_{k-1} considered in the proof of Lemma 5.6. The estimate for the term associated to γ is similar and yields the same term with b_k^* and b_k exchanged. Now we have to estimate the term

$$\left\|\big(\sum_k \mathcal{E}_{k-1}(dx_k b_k^* dx_k^* dx_k b_k dx_k^*)\big)^{\frac{1}{2}}\right\|_{(v/2)'}.$$

Writing each b_k as a linear combination

$$b_k = (b_{k1} - b_{k2}) + i(b_{k3} - b_{k4})$$

of positive elements and allowing an additional constant 2, we may assume that the operators $b_1, b_2, \ldots$ are positive. This consideration allows us to construct the positive elements $z_k = (dx_k b_k dx_k^*)^{\frac{1}{2}}$. Then, recalling our assumption $2 < v < 4$, we have $2 < (v/2)' < \infty$ and Lemma 5.2 of [**28**] gives for $t = \frac{1}{2}(v/2)'$

$$\left\|\sum_k \mathcal{E}_{k-1}(z_k^4)\right\|_t \le \left\|\sum_k \mathcal{E}_{k-1}(z_k^2)\right\|_{2t}^{\frac{2(t-1)}{2t-1}} \left(\sum_k \|z_k\|_{4t}^{4t}\right)^{\frac{1}{2t-1}}.$$

In our situation, this implies

(5.20) $\left\|\big(\sum_k \mathcal{E}_{k-1}(dx_k b_k^* dx_k^* dx_k b_k dx_k^*)\big)^{\frac{1}{2}}\right\|_{(v/2)'}$

$$\le \left\|\sum_k \mathcal{E}_{k-1}(dx_k b_k dx_k^*)\right\|_{(v/2)'}^{\frac{t-1}{2t-1}} \left(\sum_k \|dx_k b_k dx_k^*\|_{(v/2)'}^{(v/2)'}\right)^{\frac{1}{4t-2}}.$$

Using $\frac{1}{(v/2)'} - \frac{1}{(s/2)'} = \frac{2}{s} - \frac{2}{v} = \frac{2}{p'}$, we find

$$\left(\sum_k \|dx_k b_k dx_k^*\|_{(v/2)'}^{(v/2)'}\right)^{\frac{1}{(v/2)'}} \le \|dx\|_{p'} \|b\|_{(s/2)'} \|dx^*\|_{p'} \le \left(\sum_k \|dx_k\|_{p'}^{p'}\right)^{\frac{2}{p'}}.$$

In other words, we have

$$\left(\sum_k \|dx_k b_k dx_k^*\|_{(v/2)'}^{(v/2)'}\right)^{\frac{1}{4t-2}} \le \|x\|_{\mathcal{H}_{p'}^{p'}(\mathcal{M})}^{\frac{4t}{4t-2}}.$$

For the first term on the right of (5.20) we use Lemma 5.9 with

$$(\eta, s, t) = \big(1-\theta, 2(v/2)', 2(s/2)'\big).$$

This yields

$$\begin{aligned}
\left\|\sum_k \mathcal{E}_{k-1}(dx_k b_k dx_k^*)\right\|_{(v/2)'} &= \left\|\sum_k \delta_k \otimes dx_k b_k^{\frac{1}{2}}\right\|_{L^{2(v/2)'}_{\mathrm{cond}}(\mathcal{M};\ell_2^r)}^2 \\
&\le \left(\sum_k \|b_k\|_{(s/2)'}^{(s/2)'}\right)^{\frac{1}{(s/2)'}} \|x\|_{[\mathcal{H}_{p'}^{p'}(\mathcal{M}), h_{p'}^r(\mathcal{M})]_\theta}^2.
\end{aligned}$$

Hence, since (b_k) is in the unit ball of $L_{(s/2)'}(\mathcal{M}; \ell_{(s/2)'})$, we conclude

$$\Big\| \sum_k \mathcal{E}_{k-1}(dx_k b_k dx_k^*) \Big\|_{(v/2)'}^{\frac{t-1}{2t-1}} \le \|x\|_{[\mathcal{H}_{p'}^{p'}(\mathcal{M}), h_{p'}^r(\mathcal{M})]_\theta}^{\frac{4t-4}{4t-2}}$$

The estimates above and (5.20) give rise to

$$\Big\| \Big(\sum_k \mathcal{E}_{k-1}(dx_k b_k^* dx_k^* dx_k b_k dx_k^*) \Big)^{\frac{1}{2}} \Big\|_{(v/2)'} \le \max\Big\{ \|x\|_{\mathcal{H}_{p'}^{p'}(\mathcal{M})}^2, \|x\|_{[\mathcal{H}_{p'}^{p'}(\mathcal{M}), h_{p'}^r(\mathcal{M})]_\theta}^2 \Big\}.$$

Taking square roots as imposed by (5.19) and keeping track of the constants, we obtain the assertion for $\mathcal{Z}_p^r(\mathcal{M}, \theta)$. The column case follows by taking adjoints. □

At the beginning of this paragraph, we assumed that the von Neumann algebra $\mathcal{M}$ was finite and equipped with a n.f. state φ with respect to which the associated density satisfied $c_1 1_\mathcal{M} \le D \le c_2 1_\mathcal{M}$. This assumption was only needed in Lemma 5.4 (and its conditional version) in order to apply a variation of Devinatz's theorem. On the other hand, this result has been only applied to prove Lemma 5.5. Therefore, if we are able to show that Lemma 5.5 (and its conditional version) holds for arbitrary σ-finite von Neumann algebras, the same will hold for Proposition 5.10. As we shall see, this is relevant for our aims since we shall use these results in the context of free products. Let us indicate how to derive Lemma 5.5 for σ-finite von Neumann algebras. As expected, we apply Haagerup's construction and consider $\mathcal{R} = \mathcal{M} \rtimes_\sigma G$ for the discrete group

$$G = \bigcup_{n \in \mathbb{N}} 2^{-n} \mathbb{Z}.$$

The crossed product $\mathcal{R}$ is a direct limit of a family of finite von Neumann algebras $\mathcal{R}_1, \mathcal{R}_2, \ldots$ (we change our usual notation here since in this chapter $\mathcal{M}_1, \mathcal{M}_2, \ldots$ stand for a filtration of $\mathcal{M}$) which are obtained as centralizers constructed from the modular action for $\varphi \circ \mathsf{E}_\mathcal{M}$, where

$$\mathsf{E}_\mathcal{M} : \sum_{g \in G} x_g \lambda(g) \in \mathcal{R} \mapsto x_0 \in \mathcal{M}$$

denotes the natural conditional expectation onto $\mathcal{M}$. The trace in $\mathcal{R}_n$ is given by $\tau_n(x) = \varphi \circ \mathsf{E}_\mathcal{M}(d_n x)$ where $c_{1n} 1_{\mathcal{R}_n} \le d_n \le c_{2n} 1_{\mathcal{R}_n}$. It is then easily checked that

$$\hat{\mathcal{M}}_n = \mathcal{M}_n \rtimes_\sigma G,$$
$$\hat{\mathcal{M}}_n(m) = \hat{\mathcal{M}}_n \cap \mathcal{R}_m,$$

are increasing filtrations in $\mathcal{R}$ and $\mathcal{R}_m$ respectively. Thus, Lemma 5.5 (and its conditional version) holds for $\mathcal{R}_m$ and the filtration $\hat{\mathcal{M}}_1(m), \hat{\mathcal{M}}_2(m), \ldots$ for fixed $m \ge 1$. Moreover, according to (2.3) and a simple density argument, Lemma 5.5 remains valid for $\mathcal{R}$ and the filtration $\hat{\mathcal{M}}_1, \hat{\mathcal{M}}_2, \ldots$ Finally, it remains to see that $\mathsf{E}_\mathcal{M}$ extends to a contraction on $\mathcal{H}_p^s(\mathcal{R})$ and $h_p^s(\mathcal{R})$ where $s \in \{r, c, p\}$. This is obvious for $s = p$, see e.g. [28] for the convention $h_p^p(\mathcal{R}) = \mathcal{H}_p^p(\mathcal{R})$. For $s = r, c$ we recall that $\mathsf{E}_\mathcal{M}$ and $\mathcal{E}_{k-1}$ commute. In the case $2 \le p \le \infty$, this implies

$$\Big\| \sum_k \mathcal{E}_{k-1}\big(d_k(\mathsf{E}_\mathcal{M}(x)) d_k(\mathsf{E}_\mathcal{M}(x))^*\big) \Big\|_{p/2} = \Big\| \sum_k \mathcal{E}_{k-1}\big(\mathsf{E}_\mathcal{M}(dx_k) \mathsf{E}_\mathcal{M}(dx_k^*)\big) \Big\|_{p/2}$$
$$\le \Big\| \sum_k \mathcal{E}_{k-1}\big(\mathsf{E}_\mathcal{M}(dx_k dx_k^*)\big) \Big\|_{p/2}$$
$$= \Big\| \mathsf{E}_\mathcal{M}\Big(\sum_k \mathcal{E}_{k-1}(dx_k dx_k^*)\Big) \Big\|_{p/2}$$

$$\leq \Big\|\sum_k \mathcal{E}_{k-1}(dx_k dx_k^*)\Big\|_{p/2}.$$

The arguments for $\mathcal{H}_p^r(\mathcal{R})$, $p \geq 2$ (as well as the analogues for the column spaces) are the same. For $1 \leq p < 2$ we have to argue differently. By duality, it suffices to prove the assertion for $L_q^r \mathcal{MO}$ with $2 < q \leq \infty$, see Theorem 4.1 of [**28**] for this duality result. Indeed, the norm in that space is given by

$$\|x\|_{L_q^r \mathcal{MO}} = \Big\|\sup_{n\geq 1} \sum_{k\geq n} \mathcal{E}_n(dx_k dx_k^*)\Big\|_{q/2}^{\frac{1}{2}}.$$

Therefore, using the inequality

$$\Big\|\sup_{n\geq 1} \mathsf{E}_\mathcal{M}(z_n z_n^*)\Big\|_{q/2} \leq \Big\|\sup_{n\geq 1} z_n z_n^*\Big\|_{q/2}$$

which follows easily from (5.4), we see that the same argument above applies. On the other hand, it is easily checked (as in [**28**], Theorem 4.1) that the corresponding dual in the conditional case is given by

$$\|x\|_{L_q^r mo} = \Big\|\sup_{n\geq 1} \sum_{k>n} \mathcal{E}_n(dx_k dx_k^*)\Big\|_{q/2}^{\frac{1}{2}}.$$

5.2.2. Two-sided estimates. Now we will perform a similar task considering two-sided terms. Since most of the arguments are the same, we shall only sketch the main ideas in the proofs. Again, we begin by assuming that $\mathcal{M}$ is finite and the density D associated to the state φ satisfies $c_1 1_\mathcal{M} \leq D \leq c_2 1_\mathcal{M}$. The above argument via Haagerup's construction leads to the σ-finite case.

LEMMA 5.11. *Let $1 \leq p \leq 2 \leq q \leq \infty$ be such that $1/p = 1/2 + 1/q$. Given $0 < \theta < 1$, let x be a norm one element in $[\mathcal{H}_p^r(\mathcal{M}), \mathcal{H}_p^c(\mathcal{M})]_\theta$ and let us consider the indices $2 \leq w_r, w_c \leq \infty$ defined as follows*

$$1/w_r = (1-\theta)/q \quad \text{and} \quad 1/w_c = \theta/q.$$

Then there exists $(a_r, a_c) \in L_{w_r/2}(\mathcal{M})_+ \times L_{w_c/2}(\mathcal{M})_+$ and $b_k \in L_2(\mathcal{M}_k)$ such that

$$dx_k = \mathcal{E}_k(a_r)^{\frac{1}{2}} b_k \mathcal{E}_k(a_c)^{\frac{1}{2}} \quad \text{and} \quad \max\Big\{\|a_r\|_{w_r/2}, \Big(\sum_{k\geq 1} \|b_k\|_2^2\Big)^{\frac{1}{2}}, \|a_c\|_{w_c/2}\Big\} \leq 2.$$

PROOF. Assume by approximation that x is a finite martingale in $L_p(\mathcal{M}_m)$ for some integer $m \geq 1$. Therefore, since x is of norm 1 there exists an analytic function $f : \mathcal{S} \to L_p(\mathcal{M}_m)$ of the form

$$f(z) = \sum_{k=1}^m d_k(z),$$

which satisfies $f(\theta) = x$ and the estimate

$$\max\Big\{\sup_{z\in\partial_0} \Big\|\Big(\sum_{k=1}^m d_k(z) d_k(z)^*\Big)^{\frac{1}{2}}\Big\|_p, \sup_{z\in\partial_1} \Big\|\Big(\sum_{k=1}^m d_k(z)^* d_k(z)\Big)^{\frac{1}{2}}\Big\|_p\Big\} \leq 1.$$

Now we define

$$g_k^r(z) = \begin{cases} \big(\sum_{j=1}^k d_j(z) d_j(z)^* + \delta \mathrm{D}^{\frac{2}{p}}\big)^{\frac{p}{q}}, & \text{if } z \in \partial_0, \\ 1, & \text{if } z \in \partial_1. \end{cases}$$

$$g_k^c(z) = \begin{cases} 1, & \text{if } z \in \partial_0, \\ \bigl(\sum_{j=1}^k d_j(z)^* d_j(z) + \delta \mathrm{D}^{\frac{2}{p}}\bigr)^{\frac{p}{q}}, & \text{if } z \in \partial_1. \end{cases}$$

As in Lemma 5.4, we are now in position to apply Devinatz's theorem. This provides us with two analytic functions h_k^r and h_k^c with analytic inverses and satisfying the following relations

(5.21)
$$h_k^r(z) h_k^r(z)^* = g_k^r(z) \quad \text{for all} \quad z \in \partial \mathcal{S},$$
$$h_k^c(z)^* h_k^c(z) = g_k^c(z) \quad \text{for all} \quad z \in \partial \mathcal{S}.$$

According to the argument in Step 1 of the proof of Lemma 5.4, we have

(5.22)
$$\Bigl\| \sum_{k=1}^m \delta_k \otimes h_k^r(\theta) \Bigr\|_{L_{w_r}^r(\mathcal{M}; \ell_\infty)} \leq \bigl(1 + \delta^{\frac{p}{2}}\bigr)^{\frac{1-\theta}{q}},$$
$$\Bigl\| \sum_{k=1}^m \delta_k \otimes h_k^c(\theta) \Bigr\|_{L_{w_c}^c(\mathcal{M}; \ell_\infty)} \leq \bigl(1 + \delta^{\frac{p}{2}}\bigr)^{\frac{\theta}{q}}.$$

Now we consider the functions

$$w_k(z) = h_k^r(z)^{-1} d_k(z) h_k^c(z)^{-1}.$$

Note that we have unitaries $u_k^r(z)$ and $u_k^c(z)$ for which

$$h_k^r(z) = g_k^r(z)^{\frac{1}{2}} u_k^r(z) \quad \text{and} \quad h_k^c(z) = u_k^c(z) g_k^c(z)^{\frac{1}{2}}.$$

Thus, w_k can be rewritten as follows on $\partial \mathcal{S}$

$$w_k(z) = u_k^r(z)^* g_k^r(z)^{-\frac{1}{2}} d_k(z) u_k^c(z)^* \quad \text{on} \quad \partial_0,$$
$$w_k(z) = u_k^r(z)^* d_k(z) g_k^c(z)^{-\frac{1}{2}} u_k^c(z)^* \quad \text{on} \quad \partial_1.$$

In particular, the same argument as in Lemma 5.4 (second part of Step 2) yields

$$\sup_{z \in \partial \mathcal{S}} \Bigl(\sum_{k=1}^m \|w_k(z)\|_2^2 \Bigr)^{\frac{1}{2}} \leq \sqrt{2\bigl(1 + \delta^{\frac{p}{2}}\bigr)}.$$

Therefore, the same bound holds for $z = \theta$ by the three lines lemma. For the moment, we have seen that $dx_k = h_k^r(\theta) w_k(\theta) h_k^c(\theta)$. Now, recalling that g_k^r and g_k^c take values in $L_{q/2}(\mathcal{M}_k)$, we deduce that the sequence $h_1^r(\theta), h_2^r(\theta), \ldots$ as well as $h_1^c(\theta), h_2^c(\theta), \ldots$ are adapted. In particular, the argument in Step 3 of Lemma 5.4 gives rise to

$$h_k^r(\theta) = \mathcal{E}_k(a_r)^{\frac{1}{2}} \gamma_k^r \quad \text{and} \quad h_k^c(\theta) = \gamma_k^c \mathcal{E}_k(a_c)^{\frac{1}{2}}$$

for some contractions $\gamma_k^r, \gamma_k^c \in \mathcal{M}_k$ and some positive elements a_r, a_c satisfying

$$\max\bigl\{ \|a_r\|_{w_r/2}, \|a_c\|_{w_c/2} \bigr\} \leq \bigl(1 + \delta^{\frac{p}{2}}\bigr)^{1+\frac{1}{q}}.$$

The proof is completed by taking the elements b_k to be $\gamma_k^r w_k \gamma_k^c$ for $1 \leq k \leq m$. □

Exactly as we did in Lemma 5.5, the following result is a direct application of Lemma 5.11 for the dual space

$$\mathcal{Z}_p(\mathcal{M}, \theta) = \bigl[\mathcal{H}_p^r(\mathcal{M}), \mathcal{H}_p^c(\mathcal{M})\bigr]_\theta^*.$$

LEMMA 5.12. *If p, q, w_r, w_c are as above, we have*

$$\|x\|_{\mathcal{Z}_p(\mathcal{M},\theta)} \leq 8 \sup_{\substack{\|a_r\|_{w_r/2}, \|a_c\|_{w_c/2} \leq 1 \\ a_r, a_c \geq 0}} \Bigl(\sum_k \bigl\| \mathcal{E}_k(a_r)^{\frac{1}{2}} dx_k \mathcal{E}_k(a_c)^{\frac{1}{2}} \bigr\|_2^2 \Bigr)^{1/2}.$$

In Proposition 5.10 we found the constant $c(p,\theta)$. Now we define
$$c'(p,\theta) = \max\Big\{c(p,\theta), c(p,1-\theta)\Big\}.$$

PROPOSITION 5.13. *If p, q, w_r, w_c are as above, we have*
$$\|x\|_{\mathcal{Z}_p(\mathcal{M},\theta)} \leq c'(p,\theta) \max\Big\{\|x\|_{\mathcal{H}_{p'}^{p'}}, \|x\|_{[h_{p'}^r, \mathcal{H}_{p'}^{p'}]_\theta}, \|x\|_{[\mathcal{H}_{p'}^{p'}, h_{p'}^c]_\theta}, \|x\|_{[h_{p'}^r, h_{p'}^c]_\theta}\Big\}.$$

PROOF. According to Lemma 5.12 we have to estimate the term
$$A = \Big(\sum_k \big\|\mathcal{E}_k(a_r)^{\frac{1}{2}} dx_k \mathcal{E}_k(a_c)^{\frac{1}{2}}\big\|_2^2\Big)^{1/2}$$
for any pair (a_r, a_c) of positive elements satisfying $\|a_r\|_{w_r/2} \leq 1$ and $\|a_c\|_{w_c/2} \leq 1$. To that aim, we decompose it into the following three terms
$$(5.23) \quad A^2 = \sum_k \operatorname{tr}\big(dx_k \mathcal{E}_k(a_c) dx_k^* \mathcal{E}_k(a_r)\big)$$
$$\leq \Big|\sum_k \operatorname{tr}\big(dx_k d_k(a_c) dx_k^* d_k(a_r)\big)\Big|$$
$$+ \Big|\sum_k \operatorname{tr}\big(dx_k \mathcal{E}_{k-1}(a_c) dx_k^* d_k(a_r)\big)\Big|$$
$$+ \Big|\sum_k \operatorname{tr}\big(dx_k d_k(a_c) dx_k^* \mathcal{E}_{k-1}(a_r)\big)\Big|$$
$$+ \Big|\sum_k \operatorname{tr}\big(dx_k \mathcal{E}_{k-1}(a_c) dx_k^* \mathcal{E}_{k-1}(a_r)\big)\Big| = A_1^2 + A_2^2 + A_3^2 + A_4^2.$$

In particular, we have $A \leq A_1 + A_2 + A_3 + A_4$. The estimate for A_4 is rather simple. Indeed, arguing as in Remark 5.8, the conditional version of Lemma 5.12 follows after replacing $\mathcal{E}_k$ by $\mathcal{E}_{k-1}$. Moreover, as in (5.17) the argument in Lemma 5.6 gives
$$\sup_{\substack{\|a_r\|_{w_r/2}, \|a_c\|_{w_c/2} \leq 1 \\ a_r, a_c \geq 0}} \Big(\sum_k \big\|\mathcal{E}_{k-1}(a_r)^{\frac{1}{2}} dx_k \mathcal{E}_{k-1}(a_c)^{\frac{1}{2}}\big\|_2^2\Big)^{1/2} \lesssim \|x\|_{[h_{p'}^r(\mathcal{M}), h_{p'}^c(\mathcal{M})]_\theta},$$
with absolute constants. Therefore, we find
$$A_4 = \Big(\sum_k \big\|\mathcal{E}_{k-1}(a_r)^{\frac{1}{2}} dx_k \mathcal{E}_{k-1}(a_c)^{\frac{1}{2}}\big\|_2^2\Big)^{1/2} \lesssim \|x\|_{[h_{p'}^r(\mathcal{M}), h_{p'}^c(\mathcal{M})]_\theta}.$$
It remains to estimate the terms A_1, A_2 and A_3.

STEP 1. We first estimate the term A_1. Recalling that $1/w_r + 1/w_c = 1/q \leq 1/2$, we must have $4 \leq \max(w_r, w_c) \leq \infty$. Moreover, since both cases can be argued in the same way, we assume without lost of generality that $4 \leq w_r \leq \infty$. In this case we have
$$A_1^2 = \Big|\sum_k \operatorname{tr}\big(dx_k d_k(a_c) dx_k^* d_k(a_r)\big)\Big|$$
$$\leq \Big(\sum_k \|d_k(a_r)\|_{w_r/2}^{w_r/2}\Big)^{2/w_r} \Big(\sum_k \|dx_k d_k(a_c) dx_k^*\|_{(w_r/2)'}^{(w_r/2)'}\Big)^{1/(w_r/2)'}.$$

The first term on the right is controlled by $\|a_r\|_{w_r/2}$ since we are assuming that $w_r \geq 4$. For the second term on the right, we observe that $2(w_r/2)'$ is determined by the following relation
$$\frac{1}{2(w_r/2)'} = \frac{1}{2} - \frac{1}{w_r} = \frac{1}{2} - \frac{1-\theta}{q} = 1 - \frac{1}{p} + \frac{\theta}{q} = 1 - \frac{1}{u} = \frac{1}{s}.$$

Thus, this term is estimated as follows

$$\Big(\sum_k \|dx_k d_k(a_c) dx_k^*\|_{s/2}^{s/2}\Big)^{2/s} \leq \Big(\sum_k \|dx_k \mathcal{E}_k(a_c)^{\frac{1}{2}}\|_s^s\Big)^{2/s}$$
$$+ \Big(\sum_k \|dx_k \mathcal{E}_{k-1}(a_c)^{\frac{1}{2}}\|_s^s\Big)^{2/s} = A_{11}^2 + A_{12}^2.$$

By (the proof of) Proposition 5.10 we conclude that

$$A_{11} \leq c(p,\theta) \max\Big\{\|x\|_{\mathcal{H}_{p'}^{p'}(\mathcal{M})}, \|x\|_{[\mathcal{H}_{p'}^{p'}(\mathcal{M}), h_{p'}^c(\mathcal{M})]_\theta}\Big\}.$$

On the other hand, (5.17) yields the estimate

$$A_{12} \lesssim \|x\|_{[\mathcal{H}_{p'}^{p'}(\mathcal{M}), h_{p'}^c(\mathcal{M})]_\theta}.$$

The case $4 \leq w_c \leq \infty$ is similar and yields

$$A_1 \leq c(p,\theta) \max\Big\{\|x\|_{\mathcal{H}_{p'}^{p'}(\mathcal{M})}, \|x\|_{[h_{p'}^r(\mathcal{M}), \mathcal{H}_{p'}^{p'}(\mathcal{M})]_\theta}\Big\}.$$

Therefore, in the general case we conclude

$$A_1 \leq c(p,\theta) \max\Big\{\|x\|_{\mathcal{H}_{p'}^{p'}(\mathcal{M})}, \|x\|_{[h_{p'}^r(\mathcal{M}), \mathcal{H}_{p'}^{p'}(\mathcal{M})]_\theta}, \|x\|_{[\mathcal{H}_{p'}^{p'}(\mathcal{M}), h_{p'}^c(\mathcal{M})]_\theta}\Big\}.$$

STEP 2. The same arguments as in Step 1 yield the right estimate for A_2 in the case $4 \leq w_r \leq \infty$ and for A_3 in the case $4 \leq w_c \leq \infty$. Of course, there is an obvious symmetry between both cases so that we only prove the estimate for A_3 in the case $4 \leq w_c \leq \infty$. We have

$$A_3^2 \leq \Big(\sum_k \|d_k(a_c)\|_{w_c/2}^{w_c/2}\Big)^{2/w_c} \Big(\sum_k \|\mathcal{E}_{k-1}(a_r)^{\frac{1}{2}} dx_k\|_{2(w_c/2)'}^{2(w_c/2)'}\Big)^{1/(w_c/2)'}.$$

The first term on the right is controlled by $\|a_c\|_{w_c/2}$ while

$$\frac{1}{2(w_c/2)'} = \frac{1}{2} - \frac{1}{w_c} = \frac{1}{2} - \frac{\theta}{q} = 1 - \frac{1}{p} + \frac{1-\theta}{q}.$$

That is, the roles of θ and $1-\theta$ have exchanged with respect to the situation in Step 1 above. Therefore, according to the equivalence (5.17) we easily conclude that

$$A_3 \lesssim \|x\|_{[h_{p'}^r(\mathcal{M}), \mathcal{H}_{p'}^{p'}(\mathcal{M})]_\theta}.$$

When $4 \leq w_r \leq \infty$ we obtain the estimate

$$A_2 \lesssim \|x\|_{[\mathcal{H}_{p'}^{p'}(\mathcal{M}), h_{p'}^c(\mathcal{M})]_\theta}.$$

STEP 3. Now we estimate A_2 for $2 < w_r < 4$ and A_3 for $2 < w_c < 4$. Again by symmetry, we only prove the estimate for A_2. The proof of this estimate follows the argument given in Step 2 of Proposition 5.10. By the noncommutative Burkholder inequality, we may find a decomposition $d_k(a_r) = d_k(\alpha_r) + d_k(\beta_r) + d_k(\gamma_r)$ into three martingales satisfying (5.18) with $(\alpha, \beta, \gamma) \rightsquigarrow (\alpha_r, \beta_r, \gamma_r)$ and $v \rightsquigarrow w_r$. Then we have $A_2^2 \leq A_2(\alpha)^2 + A_2(\beta)^2 + A_2(\gamma)^2$ with

$$A_2(\alpha)^2 = \Big|\sum_k \mathrm{tr}(dx_k \mathcal{E}_{k-1}(a_c) dx_k^* d_k(\alpha_r))\Big|,$$
$$A_2(\beta)^2 = \Big|\sum_k \mathrm{tr}(dx_k \mathcal{E}_{k-1}(a_c) dx_k^* d_k(\beta_r))\Big|,$$
$$A_2(\gamma)^2 = \Big|\sum_k \mathrm{tr}(dx_k \mathcal{E}_{k-1}(a_c) dx_k^* d_k(\gamma_r))\Big|.$$

The term $A_2(\alpha)$ is estimated as in Step 2, due to the first inequality in (5.18). The terms $A_2(\beta)$ and $A_2(\gamma)$ are estimated in the same way so that we only show how to estimate $A_2(\beta)$. We proceed as in Proposition 5.10 again and obtain

$$\begin{aligned} A_2(\beta)^2 &= \Big|\sum_k \mathrm{tr}\big(\mathcal{E}_{k-1}(d_k(\beta_r)dx_k\mathcal{E}_{k-1}(a_c)dx_k^*)\big)\Big| \\ &\le \Big\|\big(\sum_k \mathcal{E}_{k-1}(d_k(\beta_r)d_k(\beta_r)^*)\big)^{\frac{1}{2}}\Big\|_{w_r/2} \\ &\quad\times \Big\|\big(\sum_k \mathcal{E}_{k-1}(dx_k\mathcal{E}_{k-1}(a_c)dx_k^*)^2\big)^{\frac{1}{2}}\Big\|_{(w_r/2)'} \\ &\le c_{w_r}\|a_r\|_{w_r/2}\Big\|\big(\sum_k \mathcal{E}_{k-1}(dx_k\mathcal{E}_{k-1}(a_c)dx_k^*)^2\big)^{\frac{1}{2}}\Big\|_{(w_r/2)'}. \end{aligned}$$

To estimate the last term on the right we define $z_k = (dx_k\mathcal{E}_{k-1}(a_c)dx_k^*)^{\frac{1}{2}}$. Recalling, our assumption $2 < w_r < 4$, we have $2 < (w_r/2)' < \infty$ and Lemma 5.2 of [**28**] gives for $t = \frac{1}{2}(w_r/2)'$

$$\Big\|\sum_k \mathcal{E}_{k-1}(z_k^4)\Big\|_t \le \Big\|\sum_k \mathcal{E}_{k-1}(z_k^2)\Big\|_{2t}^{\frac{2(t-1)}{2t-1}} \Big(\sum_k \|z_k\|_{4t}^{4t}\Big)^{\frac{1}{2t-1}}.$$

In our situation, this implies

$$\Big\|\big(\sum_k \mathcal{E}_{k-1}(dx_k\mathcal{E}_{k-1}(a_c)dx_k^*)^2\big)^{\frac{1}{2}}\Big\|_{(\frac{w_r}{2})'} \le \Big\|\sum_k \mathcal{E}_{k-1}(dx_k\mathcal{E}_{k-1}(a_c)dx_k^*)\Big\|_{(w_r/2)'}^{\frac{t-1}{2t-1}}$$

$$\times \Big(\sum_k \|dx_k\mathcal{E}_{k-1}(a_c)dx_k^*\|_{(w_r/2)'}^{(w_r/2)'}\Big)^{\frac{1}{4t-2}}.$$

If we take

$$M_\theta = \max\Big\{\|x\|_{\mathcal{H}_{p'}^{p'}},\, \|x\|_{[h_{p'}^r,\mathcal{H}_{p'}^{p'}]_\theta},\, \|x\|_{[\mathcal{H}_{p'}^{p'},h_{p'}^c]_\theta},\, \|x\|_{[h_{p'}^r,h_{p'}^c]_\theta}\Big\},$$

we have already seen in Step 1 above that

$$\Big(\sum_k \|dx_k\mathcal{E}_{k-1}(a_c)dx_k^*\|_{(w_r/2)'}^{(w_r/2)'}\Big)^{\frac{1}{4t}} \lesssim M_\theta.$$

We claim that

$$(5.24) \qquad \Big\|\sum_k \mathcal{E}_{k-1}(dx_k\mathcal{E}_{k-1}(a_c)dx_k^*)\Big\|_{(w_r/2)'} \le \|x\|_{[h_{p'}^r,h_{p'}^c]_\theta}^2 \le M_\theta^2.$$

If we prove (5.24) then it is easy to see that $A_2(\beta) \lesssim M$ and the estimate for A_2 will be completed. Arguing as in Lemma 5.6, it is not difficult to check that the left hand side of (5.24) does interpolate. Hence, it suffices to estimate the extremal cases. When $\theta = 0$, we have $(w_r, w_c) = (q, \infty)$ and $(w_r/2)' = p'/2$. Consequently, we find

$$\Big\|\sum_k \mathcal{E}_{k-1}(dx_k\mathcal{E}_{k-1}(a_c)dx_k^*)\Big\|_{p'/2} \le \|a_c\|_\infty \Big\|\sum_k \mathcal{E}_{k-1}(dx_kdx_k^*)\Big\|_{p'/2} \le \|x\|_{h_{p'}^r(\mathcal{M})}^2.$$

When $\theta = 1$ we have $(w_r, w_c) = (\infty, q)$ and $(w_r/2)' = 1$ so that

$$\begin{aligned}\Big\|\sum_k \mathcal{E}_{k-1}(dx_k\mathcal{E}_{k-1}(a_c)dx_k^*)\Big\|_1 &= \sum_k \mathrm{tr}\big(\mathcal{E}_{k-1}(a_c)dx_k^*dx_k\big) \\ &= \sum_k \mathrm{tr}\big(a_c\mathcal{E}_{k-1}(dx_k^*dx_k)\big) \\ &\le \|a_c\|_{\frac{q}{2}} \Big\|\sum_k \mathcal{E}_{k-1}(dx_k^*dx_k)\Big\|_{\frac{p'}{2}} \le \|x\|_{h_{p'}^c(\mathcal{M})}^2.\end{aligned}$$

Note that the first identity assumes that a_c is positive and this is not necessarily true on the boundary. However, decomposing into a linear combination of four positive elements and allowing an additional constant 2, we may and do assume positivity. Therefore, (5.24) follows from the tree lines lemma. A detailed reading of the proof gives now the constant $c'(p,\theta)$ stated above. This completes the proof. $\square$

5.3. Interpolation of 2-term intersections

Let us fix some notation which will be used in the sequel. As usual, we begin by fixing a von Neumann algebra $\mathcal{M}$ and a von Neumann subalgebra $\mathcal{N}$ with conditional expectation $\mathsf{E}: \mathcal{M} \to \mathcal{N}$. Given $1 \le q \le p \le \infty$ and a positive integer $n \ge 1$, the main spaces in this paragraph will be the following

$$\mathcal{R}^n_{2p,q}(\mathcal{M}, \mathsf{E}) = n^{\frac{1}{2p}} L_{2p}(\mathcal{M}) \cap n^{\frac{1}{2q}} L^{2p}_{(\frac{2pq}{p-q},\infty)}(\mathcal{M}, \mathsf{E}),$$

$$\mathcal{C}^n_{2p,q}(\mathcal{M}, \mathsf{E}) = n^{\frac{1}{2p}} L_{2p}(\mathcal{M}) \cap n^{\frac{1}{2q}} L^{2p}_{(\infty,\frac{2pq}{p-q})}(\mathcal{M}, \mathsf{E}).$$

In order to study these spaces we need to introduce some terminology. We set A_k to be $\mathcal{M} \oplus \mathcal{M}$ for $1 \le k \le n$. Then we consider the reduced amalgamated free product $\mathcal{A} = *_\mathcal{N} \mathsf{A}_k$ where the conditional expectation $\mathsf{E}_\mathcal{N} : \mathcal{A} \to \mathcal{N}$, defined in Paragraph 5.1.1 as $\mathsf{E}_\mathcal{N}(a) = \mathcal{Q}_\emptyset a \mathcal{Q}_\emptyset$, has the following form when restricted to A_k

$$\mathsf{E}_\mathcal{N}(x_1, x_2) = \frac{1}{2}\big(\mathsf{E}(x_1) + \mathsf{E}(x_2)\big).$$

Given a n.f. state $\varphi : \mathcal{N} \to \mathbb{C}$, let $\varphi_2 : \mathcal{M} \oplus \mathcal{M} \to \mathbb{C}$ be the n.f. state

$$\varphi_2(x_1, x_2) = \frac{1}{2}\big(\varphi(\mathsf{E}(x_1)) + \varphi(\mathsf{E}(x_2))\big) = \varphi\big(\mathsf{E}_\mathcal{N}(x_1, x_2)\big).$$

We shall write $\mathcal{A}_{\oplus n}$ for the direct sum $\mathcal{A} \oplus \mathcal{A} \oplus \ldots \oplus \mathcal{A}$ with n terms. If ϕ stands for the free product state on $\mathcal{A}$, we consider the n.f. state $\phi_n : \mathcal{A}_{\oplus n} \to \mathbb{C}$ and the conditional expectation $\mathcal{E}_n : \mathcal{A}_{\oplus n} \to \mathcal{A}$ given by

$$\phi_n\big(\sum_{k=1}^n x_k \otimes \delta_k\big) = \frac{1}{n}\sum_{k=1}^n \phi(x_k) \quad \text{and} \quad \mathcal{E}_{\oplus n}\big(\sum_{k=1}^n x_k \otimes \delta_k\big) = \frac{1}{n}\sum_{k=1}^n x_k.$$

Let $\pi_k : \mathsf{A}_k \to \mathcal{A}$ denote the embedding of A_k into $\mathcal{A}$ as defined at the beginning of this chapter. Moreover, given $x \in \mathcal{M}$ we shall write x_k as an abbreviation of $\pi_k(x, -x)$. Note that x_k is a mean-zero element for $1 \le k \le n$. In the following we shall use with no further comment the identities

$$\mathsf{E}_\mathcal{N}(x_k x_k^*) = \mathsf{E}(xx^*) \quad \text{and} \quad \mathsf{E}_\mathcal{N}(x_k^* x_k) = \mathsf{E}(x^*x).$$

Let us consider the following map

(5.25) $$u : x \in \mathcal{M} \mapsto \sum_{k=1}^n x_k \otimes \delta_k \in \mathcal{A}_{\oplus n}.$$

Moreover, if $d_{\widehat{\varphi}}$ denotes the density associated to the n.f. state $\widehat{\varphi} = \varphi \circ \mathsf{E}$ on $\mathcal{M}$ and d_{ϕ_n} stands for the density associated to the n.f. state ϕ_n on $\mathcal{A}_{\oplus n}$, we may extend the definition of u to other indices by taking

$$u\big(d_{\widehat{\varphi}}^{\frac{1}{p}} x\big) = d_{\phi_n}^{\frac{1}{p}} u(x) \quad \text{and} \quad u\big(x d_{\widehat{\varphi}}^{\frac{1}{p}}\big) = u(x) d_{\phi_n}^{\frac{1}{p}}.$$

In the following we shall consider the filtration on the von Neumann algebra $\mathcal{A}$ given by $\mathcal{A}_k = \mathsf{A}_1 *_\mathcal{N} \mathsf{A}_2 *_\mathcal{N} \cdots *_\mathcal{N} \mathsf{A}_k$. In particular, for any $x \in L_p(\mathcal{M})$ we obtain that $u(x) = (x_1, x_2, \ldots, x_n)$ is the sequence of martingale differences of $\sum_k x_k$.

LEMMA 5.14. *If $1 \leq p < \infty$, the following mappings are isomorphisms onto complemented subspaces*
$$u: \mathcal{R}^n_{2p,1}(\mathcal{M},\mathsf{E}) \to \mathcal{H}^r_{2p}(\mathcal{A}),$$
$$u: \mathcal{C}^n_{2p,1}(\mathcal{M},\mathsf{E}) \to \mathcal{H}^c_{2p}(\mathcal{A}).$$
Moreover, the constants are independent of n and remain bounded as $p \to 1$.

PROOF. Let us observe that
$$\mathcal{R}^n_{2p,1}(\mathcal{M},\mathsf{E}) = n^{\frac{1}{2p}} L_{2p}(\mathcal{M}) \cap \sqrt{n} L^r_{2p}(\mathcal{M},\mathsf{E}),$$
$$\mathcal{C}^n_{2p,1}(\mathcal{M},\mathsf{E}) = n^{\frac{1}{2p}} L_{2p}(\mathcal{M}) \cap \sqrt{n} L^c_{2p}(\mathcal{M},\mathsf{E}).$$
Given $x \in \mathcal{R}^n_{2p,1}(\mathcal{M},\mathsf{E})$, Corollary 5.3 gives
$$\|u(x)\|_{\mathcal{H}^r_{2p}(\mathcal{A})} = \Big\|\Big(\sum_{k=1}^n x_k x_k^*\Big)^{1/2}\Big\|_{2p} \sim \Big(\sum_{k=1}^n \|x_k\|_{2p}^{2p}\Big)^{2p} + \Big\|\Big(\sum_{k=1}^n \mathsf{E}_\mathcal{N}(x_k x_k^*)\Big)^{\frac{1}{2}}\Big\|_{2p}.$$
In other words, we have
$$\|u(x)\|_{\mathcal{H}^r_{2p}(\mathcal{A})} \sim n^{\frac{1}{2p}} \|x\|_{L_{2p}(\mathcal{M})} + \sqrt{n} \|x\|_{L^r_{2p}(\mathcal{M},\mathsf{E})}.$$
This proves that $u: \mathcal{R}^n_{2p,1}(\mathcal{M},\mathsf{E}) \to \mathcal{H}^r_{2p}(\mathcal{A})$ is an isomorphism onto its image with relevant constants independent of p,n. A similar argument yields to the same conclusion for the column case. To prove the complementation, we recall that
$$\mathcal{H}^r_{2p}(\mathcal{A})^* \simeq \mathcal{H}^r_{(2p)'}(\mathcal{A}) \quad \text{for} \quad 1 \leq p < \infty$$
(with relevant constants which remain bounded as $p \to 1$) and consider the map
$$\omega: x \in \frac{1}{n^{\frac{1}{2p}}} L_{(2p)'}(\mathcal{M}) + \frac{1}{\sqrt{n}} L^r_{(2p)'}(\mathcal{M},\mathsf{E}) \longmapsto \frac{1}{n} \sum_{k=1}^n x_k \otimes \delta_k \in \mathcal{H}^r_{(2p)'}(\mathcal{A}).$$
Let $d_{\widehat{\varphi}}$ be the density associated to $\varphi \circ \mathsf{E}$. Assume by approximation that
$$x = \alpha d_{\widehat{\varphi}}^{1/(2p)'} a$$
for some $(\alpha, a) \in \mathcal{N} \times \mathcal{M}$. Then, taking d_ϕ to be the density associated to the state ϕ on $\mathcal{A}$ and defining $a_k = \pi_k(a, -a)$, the following estimate holds by Theorem 7.1 in [**28**]
$$\|\omega(x)\|_{\mathcal{H}^r_{(2p)'}(\mathcal{A})} = \frac{1}{n}\Big\|\alpha d_\phi^{1/(2p)'}\Big(\sum_{k=1}^n a_k a_k^*\Big) d_\phi^{1/(2p)'} \alpha^*\Big\|_{L_{(2p)'/2}(\mathcal{A})}^{1/2}$$
$$\leq \frac{1}{n}\Big\|\alpha d_\varphi^{1/(2p)'}\Big(\sum_{k=1}^n \mathsf{E}_\mathcal{N}(a_k a_k^*)\Big) d_\varphi^{1/(2p)'} \alpha^*\Big\|_{L_{(2p)'/2}(\mathcal{N})}^{1/2}.$$
This gives
$$(5.26) \qquad \|\omega(x)\|_{\mathcal{H}^r_{(2p)'}(\mathcal{A})} \leq \frac{1}{\sqrt{n}} \|x\|_{L^r_{(2p)'}(\mathcal{M},\mathsf{E})}.$$
On the other hand, the inequality
$$(5.27) \qquad \|\omega(x)\|_{\mathcal{H}^r_{(2p)'}(\mathcal{A})} \leq \frac{1}{n^{\frac{1}{2p}}} \|x\|_{L_{(2p)'}(\mathcal{M})}$$

follows by the complex interpolation method between the (trivial) extremal cases for $p = 1$ and $p = \infty$. The estimates (5.26) and (5.27) show that the map ω is a contraction. Note also that
$$\langle u(x_1), \omega(x_2)\rangle = \frac{1}{n}\sum_{k=1}^{n} \mathrm{tr}_{\mathcal{A}}(x_{1k}^* x_{2k}) = \frac{1}{n}\sum_{k=1}^{n} \mathrm{tr}_{\mathcal{M}}(x_1^* x_2) = \langle x_1, x_2\rangle.$$
In particular, since we have
$$\mathcal{R}_{2p,1}^n(\mathcal{M},\mathsf{E}) = \Big(\frac{1}{n^{\frac{1}{2p}}}L_{(2p)'}(\mathcal{M}) + \frac{1}{\sqrt{n}}L_{(2p)'}^r(\mathcal{M},\mathsf{E})\Big)^*,$$
it turns out that the map $\omega^* u$ is the identity on $\mathcal{R}_{2p,1}^n(\mathcal{M},\mathsf{E})$ and $u\omega^*$ is a bounded projection onto the image of u with constants independent of n and bounded for $p \sim 1$. This completes the proof in the row case. The column case is the same. $\square$

Before proving our interpolation result, we need to consider a variation of Lemma 5.14. Namely, we know from [28] that $\mathcal{H}_{2p}^r(\mathcal{A}) \simeq L_{2p}^r\mathcal{MO}(\mathcal{A})$ for $1 < p < \infty$ and with constants depending on p which diverge as $p \to \infty$. We claim however that Lemma 5.14 still holds in this setting with bounded constants as $p \to \infty$.

LEMMA 5.15. *If $1 < p \leq \infty$, the following mappings are isomorphisms onto complemented subspaces*
$$u: \mathcal{R}_{2p,1}^n(\mathcal{M},\mathsf{E}) \to L_{2p}^r\mathcal{MO}(\mathcal{A}),$$
$$u: \mathcal{C}_{2p,1}^n(\mathcal{M},\mathsf{E}) \to L_{2p}^c\mathcal{MO}(\mathcal{A}).$$
Moreover, the constants are independent of n and remain bounded as $p \to \infty$.

PROOF. The noncommutative Doob's inequality [16] gives
$$\|u(x)\|_{L_{2p}^r\mathcal{MO}} = \Big\|\sup_{1 \leq m \leq n} \mathcal{E}_m\Big(\sum_{k=m}^n x_k x_k^*\Big)\Big\|_p^{\frac{1}{2}} \leq \gamma_p \Big\|\sum_{k=1}^n x_k x_k^*\Big\|_p^{\frac{1}{2}}.$$
Note that $\gamma_p \to \infty$ as $p \to 1$ but $\gamma_p \leq 2$ for $p \geq 2$. On the other hand, we may estimate the term on the right by using the free Rosenthal inequality (see Corollary 5.3) one more time
$$\Big\|\sum_{k=1}^n x_k x_k^*\Big\|_p^{\frac{1}{2}} \sim \Big(\sum_{k=1}^n \|x_k\|_{2p}^{2p}\Big)^{2p} + \Big\|\Big(\sum_{k=1}^n \mathcal{E}_{\mathcal{N}}(x_k x_k^*)\Big)^{\frac{1}{2}}\Big\|_{2p} = \|x\|_{\mathcal{R}_{2p,1}^n(\mathcal{M},\mathsf{E})}.$$
This shows that $u: \mathcal{R}_{2p,1}^n(\mathcal{M},\mathsf{E}) \to L_{2p}^r\mathcal{MO}(\mathcal{A})$ is bounded with constant γ_p. To prove complementation and the boundedness of the inverse we proceed by duality as in Lemma 5.14. Indeed, using the map w one more time and recalling that
$$L_{2p}^r\mathcal{MO}(\mathcal{A}) \simeq \mathcal{H}_{(2p)'}^r(\mathcal{A})^*$$
with constants which remain bounded as $p \to \infty$ (*c.f.* Theorem 4.1 in [28]), we may follows verbatim the proof of Lemma 5.14 to conclude that the map $\omega^* u$ is the identity on $\mathcal{R}_{2p,1}^n(\mathcal{M},\mathsf{E})$ and $u\omega^*$ is a bounded projection onto the image of u with constants independent of n and bounded for $p \sim \infty$. This completes the proof in the row case. The column case follows in the same way. $\square$

THEOREM 5.16. *If $1 \leq p \leq \infty$ and $1/q = 1 - \theta + \theta/p$, we have*
$$\big[\mathcal{R}_{2p,1}^n(\mathcal{M},\mathsf{E}), \mathcal{R}_{2p,p}^n(\mathcal{M},\mathsf{E})\big]_\theta \simeq \mathcal{R}_{2p,q}^n(\mathcal{M},\mathsf{E}),$$
$$\big[\mathcal{C}_{2p,1}^n(\mathcal{M},\mathsf{E}), \mathcal{C}_{2p,p}^n(\mathcal{M},\mathsf{E})\big]_\theta \simeq \mathcal{C}_{2p,q}^n(\mathcal{M},\mathsf{E}),$$

isomorphically with relevant constant c(p, q) independent of n and such that

$$c(p,q) \lesssim \sqrt{\frac{p-q}{pq+q-p}} \quad as \quad (p,q) \to (\infty, 1).$$

PROOF. By Corollary 4.7 i), we have contractive inclusions

$$\big[\mathcal{R}^n_{2p,1}(\mathcal{M}, \mathsf{E}), \mathcal{R}^n_{2p,p}(\mathcal{M}, \mathsf{E})\big]_\theta \subset \mathcal{R}^n_{2p,q}(\mathcal{M}, \mathsf{E}),$$
$$\big[\mathcal{C}^n_{2p,1}(\mathcal{M}, \mathsf{E}), \mathcal{C}^n_{2p,p}(\mathcal{M}, \mathsf{E})\big]_\theta \subset \mathcal{C}^n_{2p,q}(\mathcal{M}, \mathsf{E}).$$

To prove the converse, we consider again the map given in (5.25). It is clear that

$$\|u(x)\|_{\mathcal{H}^{2p}_{2p}(\mathcal{A})} = \Big(\sum_{k=1}^n \|x_k\|_{2p}^{2p}\Big)^{\frac{1}{2p}} = n^{\frac{1}{2p}} \|x\|_{2p}.$$

This shows that $u : n^{\frac{1}{2p}} L_{2p}(\mathcal{M}) \to \mathcal{H}^{2p}_{2p}(\mathcal{A})$ is an isometric isomorphism. Moreover, arguing as in the proof of Lemma 5.14 we easily obtain that the image of u is contractively complemented. This observation together with Lemmas 5.14 and 5.15 give rise to the following equivalences

$$\|x\|_{[\mathcal{R}^n_{2p,1}(\mathcal{M},\mathsf{E}),\mathcal{R}^n_{2p,p}(\mathcal{M},\mathsf{E})]_\theta} \sim \|u(x)\|_{[\mathcal{H}^r_{2p}(\mathcal{A}),\mathcal{H}^{2p}_{2p}(\mathcal{A})]_\theta} \quad \text{for small } p,$$

$$\|x\|_{[\mathcal{R}^n_{2p,1}(\mathcal{M},\mathsf{E}),\mathcal{R}^n_{2p,p}(\mathcal{M},\mathsf{E})]_\theta} \sim \|u(x)\|_{[L^r_{2p}\mathcal{MO}(\mathcal{A}),\mathcal{H}^{2p}_{2p}(\mathcal{A})]_\theta} \quad \text{for large } p,$$

with constants independent of p, q, n. On the other hand, Berg's theorem gives isometric inclusions

$$\big[\mathcal{H}^r_{2p}(\mathcal{A}), \mathcal{H}^{2p}_{2p}(\mathcal{A})\big]_\theta \subset \big[\mathcal{H}^r_{2p}(\mathcal{A}), \mathcal{H}^{2p}_{2p}(\mathcal{A})\big]^\theta,$$
$$\big[L^r_{2p}\mathcal{MO}(\mathcal{A}), \mathcal{H}^{2p}_{2p}(\mathcal{A})\big]_\theta \subset \big[L^r_{2p}\mathcal{MO}(\mathcal{A}), \mathcal{H}^{2p}_{2p}(\mathcal{A})\big]^\theta.$$

Now we can use duality and obtain

$$\big[\mathcal{H}^r_{2p}(\mathcal{A}), \mathcal{H}^{2p}_{2p}(\mathcal{A})\big]^\theta \simeq \big[\mathcal{H}^r_{(2p)'}(\mathcal{A}), \mathcal{H}^{(2p)'}_{(2p)'}(\mathcal{A})\big]^*_\theta \quad \text{for } 1 \leq p < \infty,$$

$$\big[L^r_{2p}\mathcal{MO}(\mathcal{A}), \mathcal{H}^{2p}_{2p}(\mathcal{A})\big]^\theta \simeq \big[\mathcal{H}^r_{(2p)'}(\mathcal{A}), \mathcal{H}^{(2p)'}_{(2p)'}(\mathcal{A})\big]^*_\theta \quad \text{for } 1 < p \leq \infty,$$

where the constants in the first isomorphism remain bounded as $p \to 1$ and the constants in the second one remain bounded as $p \to \infty$. Therefore, recalling the terminology used in the previous section

$$\big[\mathcal{H}^r_{(2p)'}(\mathcal{A}), \mathcal{H}^{(2p)'}_{(2p)'}(\mathcal{A})\big]^*_\theta = \mathcal{Z}^r_{(2p)'}(\mathcal{A}, 1-\theta)$$

and taking adjoints, we obtain

$$\|x\|_{[\mathcal{R}^n_{2p,1}(\mathcal{M},\mathsf{E}),\mathcal{R}^n_{2p,p}(\mathcal{M},\mathsf{E})]_\theta} \sim \|u(x)\|_{\mathcal{Z}^r_{(2p)'}(\mathcal{A},1-\theta)} = \mathrm{A}^r_p(\mathcal{A},\theta),$$

$$\|x\|_{[\mathcal{C}^n_{2p,1}(\mathcal{M},\mathsf{E}),\mathcal{C}^n_{2p,p}(\mathcal{M},\mathsf{E})]_\theta} \sim \|u(x)\|_{\mathcal{Z}^c_{(2p)'}(\mathcal{A},1-\theta)} = \mathrm{A}^c_p(\mathcal{A},\theta),$$

with constants independent of p, q, n. According to Proposition 5.10 we have

$$\mathrm{A}^r_p(\mathcal{A},\theta) \leq c(p,q) \max \Big\{ \|u(x)\|_{\mathcal{H}^{2p}_{2p}(\mathcal{A})}, \|u(x)\|_{[\mathcal{H}^{2p}_{2p}(\mathcal{A}), h^r_{2p}(\mathcal{A})]_{1-\theta}} \Big\}.$$

Let us estimate the two terms on the right

$$\|u(x)\|_{\mathcal{H}^{2p}_{2p}(\mathcal{A})} = \Big(\sum_{k=1}^n \|x_k\|_{2p}^{2p}\Big)^{\frac{1}{2p}} = n^{\frac{1}{2p}} \|x\|_{2p} \leq \|x\|_{\mathcal{R}^n_{2p,q}(\mathcal{M},\mathsf{E})}.$$

For the second term we observe that

$$\|u(x)\|_{h^r_{2p}(\mathcal{A})} = \Big\|\Big(\sum_{k=1}^n \mathcal{E}_{k-1}(x_k x_k^*)\Big)^{\frac12}\Big\|_{2p}$$

$$= \Big\|\Big(\sum_{k=1}^n \mathsf{E}_\mathcal{N}(x_k x_k^*)\Big)^{\frac12}\Big\|_{2p}$$

$$= \Big\|\Big(\sum_{k=1}^n \mathsf{E}(xx^*)\Big)^{\frac12}\Big\|_{2p} = \sqrt{n}\,\|\mathsf{E}(xx^*)\|_{2p},$$

where the second inequality follows by freeness. By complex interpolation and Corollary 4.7 we conclude

$$\|u(x)\|_{[h^r_{2p}(\mathcal{A}),\mathcal{H}^{2p}_{2p}(\mathcal{A})]_\theta} \leq n^{\frac{1-\theta}{2}+\frac{\theta}{2p}} \|x\|_{L^{2p}_{(\frac{2p'}{1-\theta},\infty)}(\mathcal{M},\mathsf{E})}$$

$$= n^{\frac{1}{2q}} \|x\|_{L^{2p}_{(\frac{2pq}{p-q},\infty)}(\mathcal{M},\mathsf{E})} \leq \|x\|_{\mathcal{R}^n_{2p,q}(\mathcal{M},\mathsf{E})}.$$

In summary, we have proved that

$$\|x\|_{\mathcal{R}^n_{2p,q}(\mathcal{M},\mathsf{E})} \leq \|x\|_{[\mathcal{R}^n_{2p,1}(\mathcal{M},\mathsf{E}),\mathcal{R}^n_{2p,p}(\mathcal{M},\mathsf{E})]_\theta} \leq c(p,q)\,\|x\|_{\mathcal{R}^n_{2p,q}(\mathcal{M},\mathsf{E})}$$

where (recalling that $p \rightsquigarrow (2p)'$ and $\theta \rightsquigarrow 1-\theta$), it follows from Proposition 5.10 that

$$c(p,q) \sim \sqrt{\frac{p-q}{pq+q-p}} \quad \text{as} \quad (p,q) \to (\infty,1).$$

This proves the assertion for rows. The column case follows by taking adjoints. □

REMARK 5.17. At the time of this writing, we do not know whether or not the relevant constants in Theorem 5.16 are uniformly bounded in p and q. Our constants are not uniformly bounded due to fact that we use the noncommutative Burkholder inequality from [**28**] in our approach. We take this opportunity to pose this question as a problem for the interested reader. The same question applies to Theorems 5.18 and 7.2 below.

5.4. Interpolation of 4-term intersections

In this section we study the interpolation spaces between $\mathcal{R}^n_{2p,1}(\mathcal{M},\mathsf{E})$ and $\mathcal{C}^n_{2p,1}(\mathcal{M},\mathsf{E})$. Of course, as it is to be expected, our main tools will be the free Rosenthal inequalities and the two-sided estimates for BMO type norms. We shall use below the constant $c'(p,\theta)$ in Proposition 5.13.

THEOREM 5.18. *If $1 \leq p \leq \infty$, we have*

$$\big[\mathcal{R}^n_{2p,1}(\mathcal{M},\mathsf{E}), \mathcal{C}^n_{2p,1}(\mathcal{M},\mathsf{E})\big]_\theta \simeq \bigcap_{u,v\in\{2p',\infty\}} n^{\frac{1-\theta}{u}+\frac{1}{2p}+\frac{\theta}{v}} L^{2p}_{(\frac{u}{1-\theta},\frac{v}{\theta})}(\mathcal{M},\mathsf{E})$$

isomorphically with relevant constant controlled by $c'(p,\theta)$ and independent of n.

PROOF. According to Corollary 4.7, we have

$$\big[\sqrt{n}L^r_{2p}(\mathcal{M},\mathsf{E}), n^{\frac{1}{2p}}L_{2p}(\mathcal{M})\big]_\theta = n^{\frac{1-\theta}{2}+\frac{\theta}{2p}} L^{2p}_{(\frac{2p'}{1-\theta},\infty)}(\mathcal{M},\mathsf{E}),$$

$$\big[n^{\frac{1}{2p}}L_{2p}(\mathcal{M}), \sqrt{n}L^c_{2p}(\mathcal{M},\mathsf{E})\big]_\theta = n^{\frac{1-\theta}{2p}+\frac{\theta}{2}} L^{2p}_{(\infty,\frac{2p'}{\theta})}(\mathcal{M},\mathsf{E}),$$

for $1 \leq p \leq \infty$. Moreover, if $1 \leq p < \infty$ the same result gives
$$\left[\sqrt{n}\, L_p^r(\mathcal{M}, \mathsf{E}), \sqrt{n}\, L_p^c(\mathcal{M}, \mathsf{E})\right]_\theta = \sqrt{n}\, L_{(\frac{2p'}{1-\theta}, \frac{2p'}{\theta})}^{2p}(\mathcal{M}, \mathsf{E}).$$

In the extremal case we claim that we have a contractive inclusion
$$\left[L_\infty^r(\mathcal{M}, \mathsf{E}), L_\infty^c(\mathcal{M}, \mathsf{E})\right]_\theta \subset L_{(\frac{2}{1-\theta}, \frac{2}{\theta})}^\infty(\mathcal{M}, \mathsf{E}).$$

Indeed, let us consider the multi-linear mappings
$$\mathsf{T}_1: (\alpha, x, \beta) \in L_2(\mathcal{M}) \times L_\infty^r(\mathcal{M}, \mathsf{E}) \times L_\infty(\mathcal{M}) \mapsto \alpha x \beta \in L_2(\mathcal{M}),$$
$$\mathsf{T}_2: (\alpha, x, \beta) \in L_\infty(\mathcal{M}) \times L_\infty^c(\mathcal{M}, \mathsf{E}) \times L_2(\mathcal{M}) \mapsto \alpha x \beta \in L_2(\mathcal{M}).$$

By the definition of $L_\infty^r(\mathcal{M}, \mathsf{E})$ and $L_\infty^c(\mathcal{M}, \mathsf{E})$ it is clear that both T_1 and T_2 are contractions. In particular, it is easily checked that our claim follows by multi-linear interpolation, details are left to the reader. Therefore, according to the observation above and Corollary 4.7, we obtain the lower estimate with constant 1. In other words, there exists a contractive inclusion
$$\left[\mathcal{R}_{2p,1}^n(\mathcal{M}, \mathsf{E}), \mathcal{C}_{2p,1}^n(\mathcal{M}, \mathsf{E})\right]_\theta \subset \bigcap_{u,v \in \{2p', \infty\}} n^{\frac{1-\theta}{u} + \frac{1}{2p} + \frac{\theta}{v}} L_{(\frac{u}{1-\theta}, \frac{v}{\theta})}^{2p}(\mathcal{M}, \mathsf{E}).$$

To prove the converse, we consider again the map given in (5.25). Arguing as in the proof of Theorem 5.16 and according to Lemmas 5.14 and 5.15, we obtain the following equivalences
$$\|x\|_{[\mathcal{R}_{2p,1}^n(\mathcal{M},\mathsf{E}), \mathcal{C}_{2p,1}^n(\mathcal{M},\mathsf{E})]_\theta} \sim \|u(x)\|_{[\mathcal{H}_{2p}^r(\mathcal{A}), \mathcal{H}_{2p}^c(\mathcal{A})]_\theta} \quad \text{for } p \leq 2,$$
$$\|x\|_{[\mathcal{R}_{2p,1}^n(\mathcal{M},\mathsf{E}), \mathcal{C}_{2p,1}^n(\mathcal{M},\mathsf{E})]_\theta} \sim \|u(x)\|_{[L_{2p}^r \mathcal{MO}(\mathcal{A}), L_{2p}^c \mathcal{MO}(\mathcal{A})]_\theta} \quad \text{for } p \geq 2,$$

with constants independent of p, q, n. On the other hand, Berg's theorem gives isometric inclusions
$$\left[\mathcal{H}_{2p}^r(\mathcal{A}), \mathcal{H}_{2p}^c(\mathcal{A})\right]_\theta \subset \left[\mathcal{H}_{2p}^r(\mathcal{A}), \mathcal{H}_{2p}^c(\mathcal{A})\right]^\theta,$$
$$\left[L_{2p}^r \mathcal{MO}(\mathcal{A}), L_{2p}^c \mathcal{MO}(\mathcal{A})\right]_\theta \subset \left[L_{2p}^r \mathcal{MO}(\mathcal{A}), L_{2p}^c \mathcal{MO}(\mathcal{A})\right]^\theta.$$

Now we can use duality an obtain
$$\left[\mathcal{H}_{2p}^r(\mathcal{A}), \mathcal{H}_{2p}^c(\mathcal{A})\right]^\theta \simeq \left[\mathcal{H}_{(2p)'}^r(\mathcal{A}), \mathcal{H}_{(2p)'}^c(\mathcal{A})\right]_\theta^* \quad \text{for } 1 \leq p < \infty,$$
$$\left[L_{2p}^r \mathcal{MO}(\mathcal{A}), L_{2p}^c \mathcal{MO}(\mathcal{A})\right]^\theta \simeq \left[\mathcal{H}_{(2p)'}^r(\mathcal{A}), \mathcal{H}_{(2p)'}^c(\mathcal{A})\right]_\theta^* \quad \text{for } 1 < p \leq \infty,$$

where the constants in the first isomorphism remain bounded as $p \to 1$ and the constants in the second one remain bounded as $p \to \infty$. Therefore, recalling the terminology used above
$$\left[\mathcal{H}_{(2p)'}^r(\mathcal{A}), \mathcal{H}_{(2p)'}^c(\mathcal{A})\right]_\theta^* = \mathcal{Z}_{(2p)'}(\mathcal{A}, \theta),$$

we deduce the following equivalence
$$\|x\|_{[\mathcal{R}_{2p,1}^n(\mathcal{M},\mathsf{E}), \mathcal{C}_{2p,1}^n(\mathcal{M},\mathsf{E})]_\theta} \sim \|u(x)\|_{\mathcal{Z}_{(2p)'}(\mathcal{A},\theta)}.$$

According to Proposition 5.13, the right hand side is controlled by
$$c'(p, \theta) \max \left\{ \|u(x)\|_{\mathcal{H}_{2p}^{2p}}, \|u(x)\|_{[h_{2p}^r, \mathcal{H}_{2p}^{2p}]_\theta}, \|u(x)\|_{[\mathcal{H}_{2p}^{2p}, h_{2p}^c]_\theta}, \|u(x)\|_{[h_{2p}^r, h_{2p}^c]_\theta} \right\}.$$

The first two terms are estimated as in Theorem 5.16
$$\|u(x)\|_{\mathcal{H}_{2p}^{2p}(\mathcal{A})} = n^{\frac{1}{2p}} \|x\|_{2p},$$

5.4. INTERPOLATION OF 4-TERM INTERSECTIONS

$$\|u(x)\|_{[h^r_{2p}(\mathcal{A}), \mathcal{H}^{2p}_{2p}(\mathcal{A})]_\theta} \leq n^{\frac{1}{2q}} \|x\|_{L^{2p}_{(\frac{2pq}{p-q}, \infty)}(\mathcal{M}, \mathsf{E})}.$$

Note that the latter term is the norm of x in

$$n^{\frac{1-\theta}{u} + \frac{1}{2p} + \frac{\theta}{v}} L^{2p}_{(\frac{u}{1-\theta}, \frac{v}{\theta})}(\mathcal{M}, \mathsf{E}) \quad \text{with} \quad (u,v) = (2p', \infty).$$

Taking adjoints and replacing θ by $1 - \theta$, we obtain

$$\|u(x)\|_{[\mathcal{H}^{2p}_{2p}(\mathcal{A}), h^c_{2p}(\mathcal{A})]_\theta} \leq n^{\frac{1}{2p} + \frac{\theta}{2p'}} \|x\|_{L^{2p}_{(\infty, \frac{2p'}{\theta})}(\mathcal{M}, \mathsf{E})}.$$

It remains to estimate the last term in the maximum. As in Theorem 5.16

$$\|u(x)\|_{h^r_{2p}(\mathcal{A})} = \sqrt{n} \, \|\mathsf{E}(xx^*)\|_{2p},$$
$$\|u(x)\|_{h^c_{2p}(\mathcal{A})} = \sqrt{n} \, \|\mathsf{E}(x^*x)\|_{2p}.$$

Thus, by complex interpolation we conclude

$$\|u(x)\|_{[h^r_{2p}, h^c_{2p}]_\theta} \leq \sqrt{n} \, \|x\|_{L^{2p}_{(\frac{2p'}{1-\theta}, \frac{2p'}{\theta})}(\mathcal{M}, \mathsf{E})}.$$

Combining the estimates obtained above, the assertion follows. □

REMARK 5.19. As we shall see in Chapter 7 below, another useful way to write the intersection space appearing on the right side of Theorem 5.18 is in the following form

$$\bigcap_{\alpha, \beta \in \{2p, 2q\}} n^{\frac{1-\eta}{\alpha} + \frac{\eta}{\beta}} L^{2p}_{(\frac{4pq}{(1-\eta)(2p-\alpha)}, \frac{4pq}{\eta(2p-\beta)})}(\mathcal{M}, \mathsf{E}).$$

REMARK 5.20. It might be of independent interest to mention that our methods immediately imply that the spaces $\mathcal{R}^n_{2p,1}(\mathcal{M}, \mathsf{E})$ and $\mathcal{C}^n_{2p,1}(\mathcal{M}, \mathsf{E})$ form interpolation families with respect to the index p. In other words, for any $1 \leq p \leq \infty$

$$\left[\mathcal{R}^n_{\infty,1}(\mathcal{M}, \mathsf{E}), \mathcal{R}^n_{2,1}(\mathcal{M}, \mathsf{E})\right]_{1/p} \simeq \mathcal{R}^n_{2p,1}(\mathcal{M}, \mathsf{E}),$$
$$\left[\mathcal{C}^n_{\infty,1}(\mathcal{M}, \mathsf{E}), \mathcal{C}^n_{2,1}(\mathcal{M}, \mathsf{E})\right]_{1/p} \simeq \mathcal{C}^n_{2p,1}(\mathcal{M}, \mathsf{E}).$$

Moreover, using anti-linear duality we may replace intersections by sums and extend our results to the whole range $1 \leq 2p \leq \infty$. These generalizations of Theorems 5.16 and 5.18 are out of the scope of this paper.

CHAPTER 6

Factorization of $\mathcal{J}_{p,q}^n(\mathcal{M}, \mathsf{E})$

Let (X_1, X_2) be a pair of operator spaces containing a von Neumann algebra $\mathcal{M}$ as a common two-sided ideal. We define the *amalgamated* Haagerup tensor product $X_1 \otimes_{\mathcal{M},h} X_2$ as the quotient of the Haagerup tensor product $X_1 \otimes_h X_2$ by the closed subspace $\mathcal{I}$ generated by the differences

$$x_1\gamma \otimes x_2 - x_1 \otimes \gamma x_2 \quad \text{with} \quad \gamma \in \mathcal{M}.$$

We shall be interested only in the Banach space structure of the operator spaces $X_1 \otimes_{\mathcal{M},h} X_2$. In particular, we shall write $X_1 \otimes_\mathcal{M} X_2$ to denote the underlying Banach space of $X_1 \otimes_{\mathcal{M},h} X_2$. According to the definition of the Haagerup tensor product and recalling the isometric embeddings $X_j \subset \mathcal{B}(\mathcal{H}_j)$, we have

$$(6.1) \quad \|x\|_{X_1 \otimes_\mathcal{M} X_2} = \inf \left\{ \left\|\left(\sum_k x_{1k} x_{1k}^*\right)^{1/2}\right\|_{\mathcal{B}(\mathcal{H}_1)} \left\|\left(\sum_k x_{2k}^* x_{2k}\right)^{1/2}\right\|_{\mathcal{B}(\mathcal{H}_2)} \right\},$$

where the infimum runs over all possible decompositions of $x + \mathcal{I}$ into a finite sum

$$x = \sum_k x_{1k} \otimes x_{2k} + \mathcal{I}.$$

REMARK 6.1. Our definition (6.1) of the norm in $X_1 \otimes_\mathcal{M} X_2$ uses the operator space structure of X_1 and X_2 since the row and column square functions live in $\mathcal{B}(\mathcal{H}_j)$ but not necessarily in X_j. However, in the sequel it will be important to note that much less structure on (X_1, X_2) is needed to define the norm in $X_1 \otimes_\mathcal{M} X_2$. Indeed, we just need to impose conditions under which the row and column square functions become closed operations in X_1 and X_2 respectively. In particular, this is guaranteed if X_1 is a right $\mathcal{M}$-module and X_2 is a left $\mathcal{M}$-module. On the other hand, note that this structure does not provide us with a natural operator space structure on $X_1 \otimes_\mathcal{M} X_2$, as we did above with $X_1 \otimes_{\mathcal{M},h} X_2$.

Let us consider any pair of indices $1 \leq p, q \leq \infty$ satisfying $q \leq p$ and let us define $1/r = 1/q - 1/p$. Then, given a positive integer n, the rest of this paper will be devoted to study the following intersection spaces

$$\mathcal{J}_{p,q}^n(\mathcal{M}, \mathsf{E}) = \bigcap_{u,v \in \{2r, \infty\}} n^{\frac{1}{u} + \frac{1}{p} + \frac{1}{v}} L_{(u,v)}^p(\mathcal{M}, \mathsf{E}).$$

The aim of this chapter is the following *factorization* result for the spaces $\mathcal{J}_{pq}^n(\mathcal{M}, \mathsf{E})$.

THEOREM 6.2. *If* $1 \leq q \leq p \leq \infty$, *we have*

$$\mathcal{J}_{p,q}^n(\mathcal{M}, \mathsf{E}) \simeq \mathcal{R}_{2p,q}^n(\mathcal{M}, \mathsf{E}) \otimes_\mathcal{M} \mathcal{C}_{2p,q}^n(\mathcal{M}, \mathsf{E}),$$

isomorphically. Moreover, the constants are independent of p, q, n.

6.1. Amalgamated tensors

Before any other consideration, let us simplify the expression (6.1) for the amalgamated Haagerup tensor product in Theorem 6.2 above. Since $\mathcal{R}_{2p,q}^n(\mathcal{M}, \mathsf{E})$ and $\mathcal{C}_{2p,q}^n(\mathcal{M}, \mathsf{E})$ coincide with $L_{2p}(\mathcal{M})$ as a set, the product ab of any two elements $(a, b) \in \mathcal{R}_{2p,q}^n(\mathcal{M}, \mathsf{E}) \times \mathcal{C}_{2p,q}^n(\mathcal{M}, \mathsf{E})$ is well-defined and the amalgamation over $\mathcal{M}$ allows us to identify finite sums

$$\sum_k a_k b_k \simeq \sum_k a_k \otimes b_k + \mathcal{I}.$$

Moreover, it is easily seen that

$$a_1, a_2, \ldots, a_n \in L_{(u,\infty)}^{2p}(\mathcal{M}, \mathsf{E}) \quad \Rightarrow \quad \Big(\sum_k a_k a_k^*\Big)^{\frac{1}{2}} \in L_{(u,\infty)}^{2p}(\mathcal{M}, \mathsf{E}),$$

$$b_1, b_2, \ldots, b_n \in L_{(\infty,v)}^{2p}(\mathcal{M}, \mathsf{E}) \quad \Rightarrow \quad \Big(\sum_k b_k^* b_k\Big)^{\frac{1}{2}} \in L_{(\infty,v)}^{2p}(\mathcal{M}, \mathsf{E}).$$

In particular, it turns out that (6.1) simplifies in this case as follows

$$\|x\|_{\mathcal{R}_{2p,q}^n(\mathcal{M},\mathsf{E}) \otimes_\mathcal{M} \mathcal{C}_{2p,q}^n(\mathcal{M},\mathsf{E})} = \inf \Big\{ \Big\|\Big(\sum_k a_k a_k^*\Big)^{1/2}\Big\|_{\mathcal{R}_{2p,q}^n} \Big\|\Big(\sum_k b_k^* b_k\Big)^{1/2}\Big\|_{\mathcal{C}_{2p,q}^n} \Big\},$$

where the infimum runs over all possible decompositions

$$x = \sum_k a_k b_k.$$

Of course, this argument holds in a more general context. Indeed, arguing as in Proposition 4.5, we deduce that the conditional L_p space $L_{(u,v)}^p(\mathcal{M}, \mathsf{E})$ embeds contractively into $L_s(\mathcal{M})$ with

$$1/s = 1/u + 1/p + 1/v.$$

Thus, the same arguments lead to the same simplification for the spaces

$$\begin{aligned} \mathsf{A} &= L_{2p}(\mathcal{M}) \otimes_\mathcal{M} L_{2p}(\mathcal{M}), \\ \mathsf{B} &= L_{2p}(\mathcal{M}) \otimes_\mathcal{M} L_{(\infty, \frac{2pq}{p-q})}^{2p}(\mathcal{M}, \mathsf{E}), \\ \mathsf{C} &= L_{(\frac{2pq}{p-q}, \infty)}^{2p}(\mathcal{M}, \mathsf{E}) \otimes_\mathcal{M} L_{2p}(\mathcal{M}), \\ \mathsf{D} &= L_{(\frac{2pq}{p-q}, \infty)}^{2p}(\mathcal{M}, \mathsf{E}) \otimes_\mathcal{M} L_{(\infty, \frac{2pq}{p-q})}^{2p}(\mathcal{M}, \mathsf{E}). \end{aligned}$$

Our first step in the proof of Theorem 6.2 is the following.

LEMMA 6.3. *We have*

$$\mathcal{R}_{2p,q}^n(\mathcal{M}, \mathsf{E}) \otimes_\mathcal{M} \mathcal{C}_{2p,q}^n(\mathcal{M}, \mathsf{E}) \simeq n^{\frac{1}{p}} \mathsf{A} \cap n^{\frac{1}{2p}+\frac{1}{2q}} \mathsf{B} \cap n^{\frac{1}{2p}+\frac{1}{2q}} \mathsf{C} \cap n^{\frac{1}{q}} \mathsf{D},$$

where the relevant constants in the isomorphism above are independent of p, q, n.

PROOF. It is clear that

$$\mathcal{R}_{2p,q}^n(\mathcal{M}, \mathsf{E}) \otimes_\mathcal{M} \mathcal{C}_{2p,q}^n(\mathcal{M}, \mathsf{E}) \subset n^{\frac{1}{p}} \mathsf{A} \cap n^{\frac{1}{2p}+\frac{1}{2q}} \mathsf{B} \cap n^{\frac{1}{2p}+\frac{1}{2q}} \mathsf{C} \cap n^{\frac{1}{q}} \mathsf{D}$$

contractively. To prove the reverse inclusion it suffices to see that

$$(6.2) \quad n^{\frac{1}{p}} \mathsf{A} \cap n^{\frac{1}{2p}+\frac{1}{2q}} \mathsf{C} \subset \mathcal{R}_{2p,q}^n(\mathcal{M}, \mathsf{E}) \otimes_\mathcal{M} n^{\frac{1}{2p}} L_{2p}(\mathcal{M}) = \mathcal{X}_1,$$

$$(6.3) \quad n^{\frac{1}{2p}+\frac{1}{2q}} \mathsf{B} \cap n^{\frac{1}{q}} \mathsf{D} \subset \mathcal{R}_{2p,q}^n(\mathcal{M}, \mathsf{E}) \otimes_\mathcal{M} n^{\frac{1}{2q}} L_{(\infty, \frac{2pq}{p-q})}^{2p}(\mathcal{M}, \mathsf{E}) = \mathcal{X}_2,$$

with constants independent of p, q, n and also

(6.4) $$\mathcal{X}_1 \cap \mathcal{X}_2 \subset \mathcal{R}_{2p,q}^n(\mathcal{M}, \mathsf{E}) \otimes_\mathcal{M} \mathcal{C}_{2p,q}^n(\mathcal{M}, \mathsf{E})$$

with absolute constants. In fact, the three inclusions can be proved using the same principle, which obviously works in a much more general setting. Indeed, let us prove (6.2). If x is a norm one element in $n^{\frac{1}{p}}\mathsf{A} \cap n^{\frac{1}{2p}+\frac{1}{2q}}\mathsf{C}$ and $\delta > 0$, we may find decompositions $x = \sum_k a_{1k}a_{2k}$ and $x = \sum_k c_{1k}c_{2k}$ satisfying the following estimates

$$\max\left\{n^{\frac{1}{2p}}\left\|\left(\sum_k a_{1k}a_{1k}^*\right)^{\frac{1}{2}}\right\|_{2p}, n^{\frac{1}{2p}}\left\|\left(\sum_k a_{2k}^*a_{2k}\right)^{\frac{1}{2}}\right\|_{2p}\right\} \leq 1+\delta,$$

$$\max\left\{n^{\frac{1}{2q}}\left\|\left(\sum_k c_{1k}c_{1k}^*\right)^{\frac{1}{2}}\right\|_{L_{(\frac{2pq}{p-q},\infty)}^{2p}(\mathcal{M},\mathsf{E})}, n^{\frac{1}{2p}}\left\|\left(\sum_k c_{2k}^*c_{2k}\right)^{\frac{1}{2}}\right\|_{2p}\right\} \leq 1+\delta.$$

Let us consider the following element in $L_{2p}(\mathcal{M})$

$$\gamma = \left(\sum_k a_{2k}^*a_{2k} + \sum_k c_{2k}^*c_{2k} + \delta \mathrm{D}^{1/p}\right)^{1/2}.$$

Since γ is invertible as a measurable operator, we may define ξ by $x = \xi\gamma$. Moreover, we also define (a_{2k}', c_{2k}') by $a_{2k} = a_{2k}'\gamma$ and $c_{2k} = c_{2k}'\gamma$. This gives rise to the following expressions for x that will be used below

$$x = \xi\gamma = \left(\sum_k a_{1k}a_{2k}'\right)\gamma,$$
$$x = \xi\gamma = \left(\sum_k c_{1k}c_{2k}'\right)\gamma.$$

The norm of x in $\mathcal{X}_1$ is estimated as follows

$$\|x\|_{\mathcal{X}_1} \leq \|\xi\|_{\mathcal{R}_{2p,q}^n(\mathcal{M},\mathsf{E})}\|\gamma\|_{n^{\frac{1}{2p}}L_{2p}(\mathcal{M})},$$

where the norm of ξ in the space $\mathcal{R}_{2p,q}^n(\mathcal{M}, \mathsf{E})$ is given by

$$\max\left\{n^{\frac{1}{2p}}\left\|\sum_k a_{1k}a_{2k}'\right\|_{2p}, n^{\frac{1}{2q}}\left\|\sum_k c_{1k}c_{2k}'\right\|_{L_{(\frac{2pq}{p-q},\infty)}^{2p}(\mathcal{M},\mathsf{E})}\right\} = \max\{\mathrm{S}_1, \mathrm{S}_2\}.$$

However, since

$$\sum_k a_{2k}'^*a_{2k}' = \gamma^{-1}\Big(\sum_k a_{2k}^*a_{2k}\Big)\gamma^{-1} \leq 1,$$

we obtain

$$\mathrm{S}_1 = n^{\frac{1}{2p}}\left\|\sum_k a_{1k}a_{2k}'\right\|_{2p} \leq n^{\frac{1}{2p}}\left\|\left(\sum_k a_{1k}a_{1k}^*\right)^{\frac{1}{2}}\right\|_{2p}\left\|\left(\sum_k a_{2k}'^*a_{2k}'\right)^{\frac{1}{2}}\right\|_\infty \leq 1+\delta.$$

Similarly, we have

$$\sum_k c_{2k}'^*c_{2k}' = \gamma^{-1}\Big(\sum_k c_{2k}^*c_{2k}\Big)\gamma^{-1} \leq 1$$

and therefore we deduce

$$\mathrm{S}_2 = n^{\frac{1}{2q}} \sup\left\{\left\|\alpha\sum_k c_{1k}c_{2k}'\right\|_{L_{2q}(\mathcal{M})} \mid \|\alpha\|_{L_{\frac{2pq}{p-q}}(\mathcal{N})} \leq 1\right\}$$
$$\leq n^{\frac{1}{2q}} \sup\left\{\left\|\sum_k \alpha c_{1k}c_{1k}^*\alpha^*\right\|_{L_q(\mathcal{M})}^{1/2} \mid \|\alpha\|_{L_{\frac{2pq}{p-q}}(\mathcal{N})} \leq 1\right\}$$
$$= n^{\frac{1}{2q}} \sup\left\{\left\|\alpha\Big(\sum_k c_{1k}c_{1k}^*\Big)^{\frac{1}{2}}\right\|_{L_{2q}(\mathcal{M})} \mid \|\alpha\|_{L_{\frac{2pq}{p-q}}(\mathcal{N})} \leq 1\right\} \leq 1+\delta.$$

Thus, we have proved

$$\|\xi\|_{\mathcal{R}_{2p,q}^n(\mathcal{M},\mathsf{E})} \leq 1+\delta.$$

It remains to estimate the norm of γ in $n^{\frac{1}{2p}} L_{2p}(\mathcal{M})$

$$n^{\frac{1}{2p}} \|\gamma\|_{2p} \le n^{\frac{1}{2p}} \left(\left\| \sum_k a_{2k}^* a_{2k} \right\|_p + \left\| \sum_k c_{2k}^* c_{2k} \right\|_p + \delta \right)^{\frac{1}{2}} \le \sqrt{2}(1+\delta) + \delta^{\frac{1}{2}} n^{\frac{1}{2p}}.$$

In conclusion, letting $\delta \to 0^+$ we obtain

$$\|x\|_{\mathcal{X}_1} \le \sqrt{2}.$$

This proves (6.2) and inclusion (6.3) is proved in a similar way. To prove (6.4) we just need to observe that the *common factor* $\mathcal{R}_{2p,q}^n(\mathcal{M}, \mathsf{E})$ in $\mathcal{X}_1 \cap \mathcal{X}_2$ is on the left, in contrast with the previous situations where the common factor was on the right. The only consequence is that the roles of ξ and γ above must be exchanged. $\square$

LEMMA 6.4. *We have*

$$\begin{aligned} \mathsf{A} &= L_p(\mathcal{M}), \\ \mathsf{B} &= L^p_{(\infty, \frac{2pq}{p-q})}(\mathcal{M}, \mathsf{E}), \\ \mathsf{C} &= L^p_{(\frac{2pq}{p-q}, \infty)}(\mathcal{M}, \mathsf{E}), \\ \mathsf{D} &= L^p_{(\frac{2pq}{p-q}, \frac{2pq}{p-q})}(\mathcal{M}, \mathsf{E}), \end{aligned}$$

isometrically whenever the indices p and q satisfy $1 \le q \le p < \infty$ or $1 < q \le p \le \infty$.

PROOF. Let us define

$$1/s = 1/u + 1/p + 1/v = 1/s_1 + 1/s_2$$

with the indices s_1 and s_2 given by

$$\begin{aligned} 1/s_1 &= 1/u + 1/2p, \\ 1/s_2 &= 1/2p + 1/v. \end{aligned}$$

Then we obtain the following estimate

$$\begin{aligned} \left\| \sum_k a_k b_k \right\|_{L^p_{(u,v)}(\mathcal{M},\mathsf{E})} &= \sup_{\alpha,\beta} \left\| \sum_k \alpha a_k b_k \beta \right\|_s \\ &\le \sup_{\alpha,\beta} \left\| \alpha \Big(\sum_k a_k a_k^* \Big)^{\frac{1}{2}} \right\|_{s_1} \left\| \Big(\sum_k b_k^* b_k \Big)^{\frac{1}{2}} \beta \right\|_{s_2} \\ &= \left\| \Big(\sum_k a_k a_k^* \Big)^{\frac{1}{2}} \right\|_{L^{2p}_{(u,\infty)}(\mathcal{M},\mathsf{E})} \left\| \Big(\sum_k b_k^* b_k \Big)^{\frac{1}{2}} \right\|_{L^{2p}_{(\infty,v)}(\mathcal{M},\mathsf{E})}, \end{aligned}$$

where the supremum runs over all α in the unit ball in $L_u(\mathcal{N})$ and all β in the unit ball of $L_v(\mathcal{N})$. Therefore, taking infima on the right we obtain a contractive inclusion

$$L^{2p}_{(u,\infty)}(\mathcal{M}, \mathsf{E}) \otimes_\mathcal{M} L^{2p}_{(\infty,v)}(\mathcal{M}, \mathsf{E}) \subset L^p_{(u,v)}(\mathcal{M}, \mathsf{E}).$$

This shows at once the lower estimate for all isometries and for $1 \le q \le p \le \infty$. That is, with no restriction on the indices. In order to show the reverse inequalities, we restrict the indices u and v to be either $\frac{2pq}{p-q}$ or ∞. We shall obtain

$$L^p_{(u,v)}(\mathcal{M}, \mathsf{E}) \subset L^{2p}_{(u,\infty)}(\mathcal{M}, \mathsf{E}) \otimes_\mathcal{M} L^{2p}_{(\infty,v)}(\mathcal{M}, \mathsf{E})$$

contractively. To do so we begin with some remarks. First, the isometry $\mathsf{A} = L_p(\mathcal{M})$ is very well-known and there is nothing to prove. Thus, the case $q = p$ is trivial since the spaces on the left collapse into $L_{2p}(\mathcal{M}) \otimes_\mathcal{M} L_{2p}(\mathcal{M})$ while the spaces

on the right coincide with $L_p(\mathcal{M})$. Therefore, we just need to consider the cases $1 \leq q < p < \infty$ and $1 < q < p \leq \infty$. In both cases we have
$$2 < \frac{2pq}{p-q} < \infty.$$
In other words, we may assume that $2 < \min(u,v) < \infty$. In particular, we are in position to apply the standard Grothendieck-Pietsch separation argument as in Theorem 3.16 and Observation 3.17. This will be our main tool in the proof. Let us consider the following norm on $L_p(\mathcal{M})$
$$|||x||| = \inf \left\{ \left\|\left(\sum_k a_k a_k^*\right)^{\frac12}\right\|_{L^{2p}_{(u,\infty)}(\mathcal{M},\mathsf{E})} \left\|\left(\sum_k b_k^* b_k\right)^{\frac12}\right\|_{L^{2p}_{(\infty,v)}(\mathcal{M},\mathsf{E})} \right\}$$
where the infimum runs over all decompositions of x into a finite sum $\sum_k a_k b_k$ with $a_k, b_k \in L_{2p}(\mathcal{M})$. Since this norm majorizes that of $L^{2p}_{(u,\infty)}(\mathcal{M},\mathsf{E}) \otimes_{\mathcal{M}} L^{2p}_{(\infty,v)}(\mathcal{M},\mathsf{E})$ in $L_p(\mathcal{M})$, it suffices by density to see that the norm in $L^p_{(u,v)}(\mathcal{M},\mathsf{E})$ controls $|||\ |||$ from above. To that aim, given $x \in L_p(\mathcal{M})$, we consider a norm one functional
$$\phi: (L_p(\mathcal{M}), |||\ |||) \to \mathbb{C} \quad \text{satisfying} \quad |\phi(x)| = |||x|||.$$
Then, the reverse inequalities will follow from

(6.5) $\qquad |\phi(x)| \leq \sup \left\{ \|\alpha x \beta\|_{L_s(\mathcal{M})} \,\big|\, \alpha \in \mathsf{B}_{L_u(\mathcal{N})}, \beta \in \mathsf{B}_{L_v(\mathcal{N})} \right\}.$

Applying the standard Grothendieck-Pietsch separation argument, we may find positive elements $\alpha \in \mathsf{B}_{L_u(\mathcal{N})}$ and $\beta \in \mathsf{B}_{L_v(\mathcal{N})}$ satisfying
$$|\phi(ab)| \leq \|\alpha a\|_{L_{s_1}(\mathcal{M})} \|b\beta\|_{L_{s_2}(\mathcal{M})}.$$
Then, taking q_α and q_β to be the support projections of α and β respectively, we obtain after the usual arguments (see e.g. Theorem 3.16) a right $\mathcal{M}$-module map $\Psi: q_\alpha L_{s_1}(\mathcal{M}) \to q_\beta L_{s_2'}(\mathcal{M})$ determined by
$$\phi(ab) = \mathrm{tr}_{\mathcal{M}}\Big(\Psi(\alpha a) b \beta\Big).$$
In particular, there exists $m_\Psi \in \mathsf{B}_{L_{s'}(\mathcal{M})}$ satisfying $\Psi(\alpha a) = m_\Psi \alpha a$ so that
$$|\phi(x)| = \big|\mathrm{tr}_{\mathcal{M}}\big(m_\Psi \alpha x \beta\big)\big| \leq \sup \left\{ \|\alpha x \beta\|_{L_s(\mathcal{M})} \,\big|\, \alpha \in \mathsf{B}_{L_u(\mathcal{N})}, \beta \in \mathsf{B}_{L_v(\mathcal{N})} \right\}.$$
This completes the proof of inequality (6.5) and thus the proof is concluded. $\square$

Let us observe that Lemmas 6.3 and 6.4 give Theorem 6.2 for every pair of indices (p,q) except for the case of $\mathcal{J}^n_{\infty,1}(\mathcal{M},\mathsf{E})$. However, this is for several reasons one of the most important factorization results that we need. In order to factorize the space $\mathcal{J}^n_{\infty,1}(\mathcal{M},\mathsf{E})$, we note that Lemma 6.3 is still valid. Moreover, due to the obvious isometries
$$\mathsf{X}_1 \otimes_{\mathcal{M}} \mathcal{M} = \mathsf{X}_1 \quad \text{and} \quad \mathcal{M} \otimes_{\mathcal{M}} \mathsf{X}_2 = \mathsf{X}_2$$
for any right (resp. left) $\mathcal{M}$-module, we deduce that the first three isometries in Lemma 6.4 are trivial in the limit case $p = \infty$. Hence, we just need to show that the isometry for D still holds in the case $(p,q) = (\infty, 1)$. Again, we recall that the lower estimate can be proved as in Lemma 6.4 so that it suffices to prove the upper estimate. Unfortunately, the application of Grothendieck-Pietsch separation argument in this case is much more delicate and we need some preparation. After some auxiliary results in the next paragraph, we will go back to this question.

6.2. Conditional expectations and ultraproducts

We study certain ultraproduct von Neumann algebras and the corresponding conditional expectations. These auxiliary results will be used to factorize the norm of $\mathcal{J}^n_{\infty,1}(\mathcal{M}, \mathsf{E})$ as explained above.

LEMMA 6.5. *Let F be a finite dimensional subspace of $\mathcal{M}^*$ and let G be a finite dimensional subspace of $\mathcal{M}$. Then, given any $\delta > 0$ there exists a linear mapping*

$$\omega : \mathsf{F} \to \mathcal{M}_*$$

satisfying the following properties:

i) $\|\omega\|_{cb} \le 1 + \delta$.
ii) *The space $\omega(\mathsf{F} \cap \mathcal{N}^*)$ is contained in $\mathcal{N}_*$.*
iii) *The following estimate holds for any $f \in \mathsf{F}$ and $g \in \mathsf{G}$*

$$|g(\omega(f)) - f(g)| \le \delta \|f\|_\mathsf{F} \|g\|_\mathsf{G}.$$

PROOF. Let $(f_1, f_2, \ldots f_k; \mathsf{f}_1^*, \mathsf{f}_2^*, \ldots, \mathsf{f}_k^*)$ be an Auerbach basis of $\mathsf{F} \cap \mathcal{N}^*$. That is, $(f_1, f_2, \ldots, f_k)$ is a basis of $\mathsf{F} \cap \mathcal{N}^*$ with $\|f_j\| = 1$ for $1 \le j \le k$ and the f_j^*'s are functionals on $\mathsf{F} \cap \mathcal{N}^*$ satisfying $\mathsf{f}_i^*(f_j) = \delta_{ij}$. Let us take Hahn-Banach extensions $f_1^*, \ldots, f_k^* : \mathsf{F} \to \mathbb{C}$ of $\mathsf{f}_1^*, \ldots, \mathsf{f}_k^*$ respectively. Then we may define the projection

$$\mathsf{P} : f \in \mathsf{F} \mapsto \sum_{j=1}^k f_j^*(f) f_j \in \mathsf{F} \cap \mathcal{N}^*.$$

Now, using P we may also consider an Auerbach basis $(f_{k+1}, \ldots, f_n; \mathsf{f}_{k+1}^*, \ldots, \mathsf{f}_n^*)$ of $(1_\mathsf{F} - \mathsf{P})(\mathsf{F})$. Then, given any $k+1 \le j \le n$ we consider the linear functional $f_j^* : \mathsf{F} \to \mathbb{C}$ defined by the relation

$$f_j^*(f) = \mathsf{f}_j^*\Big(f - \mathsf{P}(f)\Big).$$

Finally we consider an Auerbach basis $(g_1, g_2, \ldots, g_m; g_1^*, g_2^*, \ldots, g_m^*)$ of G. This allows us to define the following set for any $\varepsilon > 0$

$$\mathsf{C}(\varepsilon) = \mathrm{conv}\Big\{\omega : \mathsf{F} \to \mathcal{M}_* \,\Big|\, \omega(\mathsf{F} \cap \mathcal{N}^*) \subset \mathcal{N}_*,\ |g_k(\omega(f_j)) - f_j(g_k)| \le \varepsilon\Big\}.$$

Let us assume that

$$(1+\delta)\,\mathsf{B}_{\mathcal{CB}(\mathsf{F}, \mathcal{M}_*)} \cap \overline{\mathsf{C}(\varepsilon)} = \emptyset,$$

where $\overline{\mathsf{C}(\varepsilon)}$ denotes the closure of $\mathsf{C}(\varepsilon)$ in the $\sigma(\mathcal{CB}(\mathsf{F}, \mathcal{M}_*), \mathsf{F} \otimes \mathcal{M})$ topology. We will show below that $\mathsf{C}(\varepsilon)$ is not empty. Thus, by the Hahn-Banach theorem we may find a linear functional $\xi : \mathcal{CB}(\mathsf{F}, \mathcal{M}_*) \to \mathbb{C}$ such that

$$\mathrm{Re}\big(\xi(\omega_1)\big) \le 1 \le \mathrm{Re}\big(\xi(\omega_2)\big)$$

for all $\omega_1 \in (1+\delta)\,\mathsf{B}_{\mathcal{CB}(\mathsf{F},\mathcal{M}_*)}$ and $\omega_2 \in \mathsf{C}(\varepsilon)$. This implies $\|\xi\|_{\mathcal{CB}(\mathsf{F},\mathcal{M}_*)^*} \le (1+\delta)^{-1}$. After identifying the space $\mathcal{CB}(\mathsf{F}, \mathcal{M}_*)$ with the minimal tensor product $\mathsf{F}^* \otimes_{\min} \mathcal{M}_*$, we consider the associated linear map $T_\xi : \mathcal{M}_* \to \mathsf{F}$ defined by

$$f^*(T_\xi(m_*)) = \xi(f^* \otimes m_*).$$

Then, taking $m_j = f_j^* \circ T_\xi \in \mathcal{M}$ it turns out that

$$\xi = \sum_{j=1}^n m_j \otimes f_j.$$

We claim that
$$\|\xi\|_{N(\mathcal{M}_*,F)} = \|\xi\|_{I(\mathcal{M}_*,F)} \leq (1+\delta)^{-1},$$
where $N(\mathcal{M}_*,F)$ (resp. $I(\mathcal{M}_*,F)$) denotes the space of completely nuclear maps (resp. completely integral maps) from $\mathcal{M}_*$ to F. Indeed, according to the main result of [10], $\mathcal{M}_*$ is locally reflexive and so the first identity follows from Proposition 4.4 in [10]. The inequality following it holds by Corollary 12.3.4 of [11] and the fact that $\|\xi\|_{\mathcal{CB}(F,\mathcal{M}_*)^*} \leq (1+\delta)^{-1}$. Moreover, since F is finite dimensional
$$N(\mathcal{M}_*,F) = F \hat{\otimes} \mathcal{M} \quad \text{and} \quad N(\mathcal{M}_*,F)^* = \mathcal{CB}(F,\mathcal{M}^*).$$
This means that given any linear map $\omega : F \to \mathcal{M}^*$, we have
$$|\xi(\omega)| = \Big| \sum_{j=1}^n \omega(f_j)(m_j) \Big| \leq \|\xi\|_{N(\mathcal{M}_*,F)} \|\omega\|_{cb} \leq (1+\delta)^{-1} \|\omega\|_{cb}.$$
In particular, the inclusion map $j : F \to \mathcal{M}^*$ satisfies
$$|\xi(j)| \leq (1+\delta)^{-1}.$$
On the other hand we can write
$$j = \sum_{j=1}^n f_j^* \otimes f_j.$$
Given $1 \leq j \leq k$, let $(f_{j,\alpha})_{\alpha \in \Lambda} \subset \mathcal{N}_*$ be a net converging to f_j in the $\sigma(\mathcal{N}^*, \mathcal{N})$ topology. Thus $f_{j,\alpha} \circ E$ converges to $f_j \circ E$ in the $\sigma(\mathcal{M}^*, \mathcal{M})$ topology. When $j = k+1, k+2, \ldots, n$ we may also fix a net $(f_{j,\alpha})_{\alpha \in \Lambda} \subset \mathcal{M}_*$ converging to f_j in the $\sigma(\mathcal{M}^*, \mathcal{M})$ topology. This implies that the maps $\omega_\alpha : F \to \mathcal{M}_*$ defined by
$$\omega_\alpha(f) = \sum_{j=1}^n f_j^*(f) f_{j,\alpha} \quad \text{satisfy} \quad \omega_\alpha(F \cap \mathcal{N}^*) \subset \mathcal{N}_*$$
since $f_{j,\alpha} \in \mathcal{N}_*$ for $j = 1, 2, \ldots, k$ and $f_j^*(P(F)) = 0$ for $j = k+1, k+2, \ldots, n$. Moreover, for large enough α we clearly find
$$\big| g_k(\omega_\alpha(f_j)) - f_j(g_k) \big| \leq \varepsilon.$$
This shows that there exists α for which $\omega_\alpha \in C(\varepsilon)$. Finally, we have
$$\lim_\alpha \xi(\omega_\alpha) = \lim_\alpha \sum_{j=1}^n m_j(\omega_\alpha(f_j)) = \lim_\alpha \sum_{j=1}^n m_j(f_{j,\alpha}) = \xi(j).$$
Therefore we may find α such that
$$\omega_\alpha \in C(\varepsilon) \quad \text{and} \quad |\xi(\omega_\alpha)| < 1.$$
However, any $\omega \in C(\varepsilon)$ satisfies $\operatorname{Re}(\xi(\omega)) \geq 1$. Therefore, we have a contradiction so that we can find
$$\omega \in (1+\delta) B_{\mathcal{CB}(F,\mathcal{M}_*)} \cap C(\varepsilon).$$
Such a mapping clearly satisfies
$$\big| g(\omega(f)) - g(f) \big| = \Big| \sum_{j,k} f_j^*(f) g_k^*(g) \big(g_k(\omega(f_j)) - f_j(g_k) \big) \Big| \leq \varepsilon \sum_{j=1}^n |f_j^*(f)| \sum_{k=1}^m |g_k^*(g)|.$$
Then, recalling the definition of f_j^* and g_k^* we easily obtain
$$\big| g(\omega(f)) - g(f) \big| \leq \varepsilon m n \|1_F - P\| \|f\|_F \|g\|_G \leq 2\varepsilon m n^2 \|f\|_F \|g\|_G$$

for all $f \in \mathsf{F}$ and $g \in \mathsf{G}$. Therefore, taking $\varepsilon < \delta/2mn^2$ the assertion follows. □

In the following we use the notation
$$(x_i)^\bullet = \text{Equivalence class of } (x_i) \text{ in } \prod\nolimits_{i,\mathcal{U}} X_i.$$

LEMMA 6.6. *There exist an ultrafilter $\mathcal{U}$ on an index set I and a linear map*
$$\alpha : \mathcal{M}^* \to \prod\nolimits_{i,\mathcal{U}} \mathcal{M}_*$$
satisfying the following properties:
   i) *The map α is a complete contraction.*
   ii) *The space $\alpha(\mathcal{N}^*)$ is contained in $\prod_{i,\mathcal{U}} \mathcal{N}_*$.*
   iii) *The following identity holds for all $\psi \in \mathcal{M}^*$ and $m \in \mathcal{M}$*
$$\lim\nolimits_{i,\mathcal{U}} m(\alpha(\psi)_i) = \psi(m).$$

PROOF. Let I be the set of tuples (F, G) with F a finite dimensional subspace of $\mathcal{M}^*$ and G a finite dimensional subspace of $\mathcal{M}$. Let $\mathcal{U}$ be an ultrafilter containing all the subsets of I of the form
$$I_{\mathsf{F},\mathsf{G}} = \left\{ (\mathsf{F}', \mathsf{G}') \mid \mathsf{F} \subset \mathsf{F}', \mathsf{G} \subset \mathsf{G}' \right\}.$$
Note that this can be done since these sets have the finite intersection property. For fixed F, G we choose $\omega_{\mathsf{F},\mathsf{G}} : \mathsf{F} \to \mathcal{M}_*$ satisfying the assumptions of Lemma 6.5 for $\delta = (\dim \mathsf{F} \dim \mathsf{G})^{-1}$. Then we define
$$\alpha(\psi) = \left(\omega_{\mathsf{F},\mathsf{G}}(\psi) \right)^\bullet.$$
Note that for $(\mathsf{F}, \mathsf{G}) \in I_{\langle \psi \rangle, \langle 0 \rangle}$ this is well-defined. Hence α is well-defined on $\mathcal{M}^*$. It is easily checked α is linear and completely contractive. By construction we have $\alpha(\mathcal{N}^*) \subset \prod_{i,\mathcal{U}} \mathcal{N}_*$ and $\alpha(\psi)(m) = \lim_{i,\mathcal{U}} m(\alpha(\psi)_i)$ for all $m \in \mathcal{M}$. □

LEMMA 6.7. *There exists normal conditional expectations*
$$\mathsf{E} : \left(\prod\nolimits_{i,\mathcal{U}} \mathcal{M}_* \right)^* \to \left(\prod\nolimits_{i,\mathcal{U}} \mathcal{N}_* \right)^* \quad \text{and} \quad \mathcal{E} : \left(\prod\nolimits_{i,\mathcal{U}} \mathcal{M}_* \right)^* \to \mathcal{M}^{**}.$$
Moreover, they are related to each other by the identity $\mathsf{E}^{**} \circ \mathcal{E} = \mathcal{E} \circ \mathsf{E}$.

PROOF. Let us consider the map
$$\mathsf{E}^{\mathcal{U}} : (x_i)^\bullet \in \prod\nolimits_{i,\mathcal{U}} \mathcal{M} \mapsto (\mathsf{E}(x_i))^\bullet \in \prod\nolimits_{i,\mathcal{U}} \mathcal{N}.$$
By strong density of $\prod_{i,\mathcal{U}} \mathcal{M}$ in $\left(\prod_{i,\mathcal{U}} \mathcal{M}_* \right)^*$, we may define
$$\mathsf{E} : \left(\prod\nolimits_{i,\mathcal{U}} \mathcal{M}_* \right)^* \to \left(\prod\nolimits_{i,\mathcal{U}} \mathcal{N}_* \right)^*$$
with predual $\mathsf{E}^{\mathcal{U}}_*$. On the other hand, the map
$$\gamma : (\varphi_i)^\bullet \in \prod\nolimits_{i,\mathcal{U}} \mathcal{M}_* \mapsto \lim\nolimits_{i,\mathcal{U}} \varphi_i(\cdot) \in \mathcal{M}^*$$
clearly satisfies $\alpha \circ \gamma = 1_{\prod_{i,\mathcal{U}} \mathcal{M}_*}$ with α being the map constructed in Lemma 6.6. Taking adjoints we find that
$$\gamma^* : \mathcal{M}^{**} \to \left(\prod\nolimits_{i,\mathcal{U}} \mathcal{M}_* \right)^*$$

is an injective $*$-homomorphism so that the adjoint $\mathcal{E}$ of α

$$\mathcal{E}: \left(\prod_{i,\mathcal{U}} \mathcal{M}_*\right)^* \to \mathcal{M}^{**}$$

is a normal conditional expectation. We are interested in proving the relation $\mathsf{E}^{**} \circ \mathcal{E} = \mathcal{E} \circ \mathsf{E}$. Note that since $\alpha(\mathcal{N}^*)$ is contained in $\prod_{i,\mathcal{U}} \mathcal{N}_*$, we have

$$\mathcal{E}\left(\prod_{i,\mathcal{U}} \mathcal{N}_*\right)^* \subset \mathcal{N}^{**}.$$

In particular, both maps $\mathsf{E}^{**} \circ \mathcal{E}$ and $\mathcal{E} \circ \mathsf{E}$ end in $\mathcal{N}^{**}$. However, if we *predualize* this identity it turns out that it suffices to see that

$$\alpha \circ \mathsf{E}^* = \mathsf{E}^{\mathcal{U}}_* \circ \alpha|_{\mathcal{N}^*}.$$

This is a reformulation of Lemma 6.6 iii). The proof is complete. $\square$

6.3. Factorization of the space $\mathcal{J}^n_{\infty,1}(\mathcal{M}, \mathsf{E})$

According to [55],

$$L_p(\mathcal{M}_\mathcal{U}) = \prod_{i,\mathcal{U}} L_p(\mathcal{M}) \quad \text{with} \quad \mathcal{M}_\mathcal{U} = \left(\prod_{i,\mathcal{U}} \mathcal{M}_*\right)^* \quad \text{and} \quad 1 \le p < \infty.$$

Then, by Lemma 6.7 the inclusion below is a complete contraction

$$\xi_p : L_p(\mathcal{M}^{**}) \to L_p(\mathcal{M}_\mathcal{U}).$$

LEMMA 6.8. *The map* $\xi_1 : L_1(\mathcal{M}^{**}) \to L_1(\mathcal{M}_\mathcal{U})$ *satisfies*

$$\xi_1\big((ax)(yb)\big) = \xi_1(a^2)^{\frac{1}{2}} xy \, \xi_1(b^2)^{\frac{1}{2}}$$

for all $(x,y) \in L^r_\infty(\mathcal{M}, \mathsf{E}) \times L^c_\infty(\mathcal{M}, \mathsf{E})$ *and all positive elements* $a, b \in L^+_2(\mathcal{N}^{**})$.

PROOF. Accordingly to the terminology used in Lemma 6.7, we may identify $\mathcal{M}^{**}$ with its image $\gamma^*(\mathcal{M}^{**})$ in $\mathcal{M}_\mathcal{U}$. In particular, the map ξ_p can be regarded as an $\mathcal{M}^{**}$ bimodule map satisfying

$$\xi_p(\mathrm{D}_\psi^{1/p}) = \mathrm{D}_{\psi \circ \mathcal{E}}^{1/p}$$

for every functional $\psi \in \mathcal{M}^*$ with associated density $\mathrm{D}_\psi \in L_1(\mathcal{M}^{**})$ so that $\psi \circ \mathcal{E}$ is a functional on $\mathcal{M}_\mathcal{U}$ with associated density $\mathrm{D}_{\psi \circ \mathcal{E}} \in L_1(\mathcal{M}_\mathcal{U})$. Therefore, we have

$$\xi_1(\mathrm{D}_\psi) = \mathrm{D}_{\psi \circ \mathcal{E}}^{1/2} \mathrm{D}_{\psi \circ \mathcal{E}}^{1/2} = \xi_2(\mathrm{D}_\psi^{1/2}) \xi_2(\mathrm{D}_\psi^{1/2}).$$

Now, let us consider $a, b \in L_2(\mathcal{M}^{**})$ and define $\mathrm{D}_\psi = aa^* + b^*b$ with corresponding positive functional ψ. Let $m_1, m_2 \in \mathcal{M}^{**}$ such that $a = \mathrm{D}_\psi^{1/2} m_1$ and $b = m_2 \mathrm{D}_\psi^{1/2}$. Assuming that m_1 and m_2 are ψ-analytic and using the bimodule property of ξ_1 we obtain

$$\begin{aligned}
\xi_1(ab) &= \xi_1\big(\mathrm{D}_\psi^{1/2} m_1 m_2 \mathrm{D}_\psi^{1/2}\big) \\
&= \xi_1\big(\sigma^\psi_{-i/2}(m_1) \mathrm{D}_\psi \sigma^\psi_{i/2}(m_2)\big) \\
&= \sigma^{\psi \circ \mathcal{E}}_{-i/2}(m_1) \mathrm{D}_{\psi \circ \mathcal{E}}^{1/2} \mathrm{D}_{\psi \circ \mathcal{E}}^{1/2} \sigma^{\psi \circ \mathcal{E}}_{i/2}(m_2) \\
&= \mathrm{D}_{\psi \circ \mathcal{E}}^{1/2} m_1 m_2 \mathrm{D}_{\psi \circ \mathcal{E}}^{1/2} \\
&= \xi_2(\mathrm{D}_\psi^{1/2}) m_1 m_2 \xi_2(\mathrm{D}_\psi^{1/2}).
\end{aligned}$$

Thus, by approximation with ψ-analytic elements, we conclude $\xi_1(ab) = \xi_2(a)\xi_2(b)$. Assuming in addition that a, b are positive and $x, y \in \mathcal{M}^{**}$

(6.6) $\quad \xi_1((ax)(yb)) = \xi_2(ax)\xi_2(yb) = \xi_2(a)xy\xi_2(b) = \xi_1(a^2)^{\frac{1}{2}} xy\, \xi_1(b^2)^{\frac{1}{2}}.$

This proves the assertion with $a, b \in L_2^+(\mathcal{M}^{**})$ and $x, y \in \mathcal{M}$. Before considering elements $(x, y) \in L_\infty^r(\mathcal{M}, \mathsf{E}) \times L_\infty^c(\mathcal{M}, \mathsf{E})$, we observe that $\xi_p(L_p(\mathcal{N}^{**}))$ is contained in the space

$$L_p(\mathcal{N}_{\mathcal{U}}) = \prod\nolimits_{i,\mathcal{U}} L_p(\mathcal{N}) \quad \text{with} \quad \mathcal{N}_{\mathcal{U}} = \Big(\prod\nolimits_{i,\mathcal{U}} \mathcal{N}_*\Big)^*.$$

Indeed, according to Lemma 6.7 we know that $\mathsf{E}^{**} \circ \mathcal{E} = \mathcal{E} \circ \mathbf{E}$ so that $\mathcal{E}(L_p(\mathcal{N}_{\mathcal{U}}))$ is contained in $L_p(\mathcal{N}^{**})$. According to (6.6) and the density of $\mathcal{M}$ in $L_\infty^r(\mathcal{M}, \mathsf{E})$ and $L_\infty^c(\mathcal{M}, \mathsf{E})$, to prove the assertion it suffices to see that given $a, b \in L_2^+(\mathcal{N}^{**})$ and $x, y \in \mathcal{M}$ we have

(6.7) $\quad \big\|\xi_1((ax)(yb))\big\|_{L_1(\mathcal{M}_{\mathcal{U}})} \leq \|a\|_{L_2(\mathcal{N}^{**})} \|\mathsf{E}(xx^*)\|_{\mathcal{M}}^{1/2} \|\mathsf{E}(y^*y)\|_{\mathcal{M}}^{1/2} \|b\|_{L_2(\mathcal{N}^{**})}.$

By (6.6) we have

$$\big\|\xi_1((ax)(yb))\big\|_{L_1(\mathcal{M}_{\mathcal{U}})} \leq \operatorname{tr}_{\mathcal{M}_{\mathcal{U}}}\Big(\xi_2(a)xx^*\xi_2(a)\Big)^{1/2} \operatorname{tr}_{\mathcal{M}_{\mathcal{U}}}\Big(\xi_2(b)y^*y\xi_2(b)\Big)^{1/2}.$$

Now, using $\mathsf{E}^{**} \circ \mathcal{E} = \mathcal{E} \circ \mathbf{E}$ and $\xi_2 : L_2(\mathcal{N}^{**}) \hookrightarrow L_2(\mathcal{N}_{\mathcal{U}})$, we have

$$\begin{aligned}
\operatorname{tr}_{\mathcal{M}_{\mathcal{U}}}\Big(\xi_2(a)xx^*\xi_2(a)\Big)^{1/2} &= \operatorname{tr}_{\mathcal{M}_{\mathcal{U}}}\Big(\mathcal{E} \circ \mathbf{E}\big(\xi_2(a)xx^*\xi_2(a)\big)\Big)^{1/2} \\
&= \operatorname{tr}_{\mathcal{M}_{\mathcal{U}}}\Big(\xi_2(a)\mathsf{E}(xx^*)\xi_2(a)\Big)^{1/2} \\
&\leq \|\xi_2(a)\|_{L_2(\mathcal{N}_{\mathcal{U}})} \|\mathsf{E}(xx^*)\|_{\mathcal{M}}^{1/2}.
\end{aligned}$$

Similarly, we have

$$\operatorname{tr}_{\mathcal{M}_{\mathcal{U}}}\Big(\xi_2(b)y^*y\xi_2(b)\Big)^{1/2} \leq \|\mathsf{E}(y^*y)\|_{\mathcal{M}}^{1/2} \|\xi_2(b)\|_{L_2(\mathcal{N}_{\mathcal{U}})}.$$

Therefore, (6.7) follows since ξ_2 is a contraction. This concludes the proof. $\square$

PROPOSITION 6.9. *Any $x \in \mathcal{M}$ satisfies*

$$\|x\|_{L^\infty_{(2,\infty)}(\mathcal{M},\mathsf{E}) \otimes_\mathcal{M} L^\infty_{(\infty,2)}(\mathcal{M},\mathsf{E})} = \|x\|_{L^\infty_{(2,2)}(\mathcal{M},\mathsf{E})}.$$

PROOF. The lower estimate can be proved as in Lemma 6.4. To prove the upper estimate, we consider the following norm on $\mathcal{M}$

$$|||x||| = \inf\left\{ \Big\|\big(\sum\nolimits_k \mathsf{E}(a_k a_k^*)\big)^{1/2}\Big\|_{L_\infty(\mathcal{N})} \Big\|\big(\sum\nolimits_k \mathsf{E}(b_k^* b_k)\big)^{1/2}\Big\|_{L_\infty(\mathcal{N})} \right\}$$

where the infimum runs over all decompositions of x into a finite sum $\sum_k a_k b_k$ with $a_k, b_k \in \mathcal{M}$. Let $x \in \mathcal{M}$ and take a norm one functional $\phi : (\mathcal{M}, ||| \; |||) \to \mathbb{C}$ satisfying $|||x||| = |\phi(x)|$. Then it is clear that it suffices to see that

(6.8) $\quad |\phi(x)| \leq \sup\Big\{\|\alpha x \beta\|_{L_1(\mathcal{M})} \,\big|\, \alpha, \beta \in \mathsf{B}_{L_2(\mathcal{N})}\Big\}.$

Applying the Grothendieck-Pietsch separation argument, we may find states ψ_1 and ψ_2 in $\mathcal{N}^*$ with associated densities D_1 and D_2 in $L_1(\mathcal{N}^{**})$ satisfying the following inequality

$$|\phi(ab)| \leq \psi_1(aa^*)^{\frac{1}{2}} \psi_2(b^*b)^{\frac{1}{2}}.$$

By Kaplansky's density theorem, $D_j^{1/2}\mathcal{M}$ is norm dense in $D_j^{1/2}\mathcal{M}^{**}$ for $j = 1, 2$. Therefore, taking e_j to be the support of D_j for $j = 1, 2$, we may consider the map

$$\Psi : e_1 L_2(\mathcal{M}^{**}) \to e_2 L_2(\mathcal{M}^{**})$$

determined by the relation

$$\phi(ab) = \langle D_2^{\frac{1}{2}} b^*, \Psi(D_1^{\frac{1}{2}} a) \rangle = \mathrm{tr}_{\mathcal{M}^{**}}\left(\Psi(D_1^{\frac{1}{2}} a) b D_2^{\frac{1}{2}}\right).$$

From this it easily follows that $\Psi(D_1^{\frac{1}{2}} am) = \Psi(D_1^{\frac{1}{2}} a)m$ for all $m \in \mathcal{M}$. Hence, by density we deduce that Ψ commutes on $L_2(\mathcal{M}^{**})$ with the right action on $\mathcal{M}^{**}$. In particular, we may find a contraction $m_\Psi \in \mathcal{M}^{**}$ such that

$$\Psi(D_1^{\frac{1}{2}} a) = m_\Psi D_1^{\frac{1}{2}} a.$$

This gives

$$\phi(ab) = \mathrm{tr}_{\mathcal{M}^{**}}\left(m_\Psi D_1^{\frac{1}{2}} ab D_2^{\frac{1}{2}}\right).$$

On the other hand, by Goldstein's theorem there exists a net (m_λ) in the unit ball of $\mathcal{M}$ which converges to m_Ψ in the $\sigma(\mathcal{M}^{**}, \mathcal{M}^*)$ topology. Therefore, since $D_1^{1/2} ab D_2^{1/2} \in L_1(\mathcal{M}^{**})$ we deduce that

$$\phi(ab) = \lim_\lambda \mathrm{tr}_{\mathcal{M}^{**}}\left(m_\lambda D_1^{\frac{1}{2}} ab D_2^{\frac{1}{2}}\right).$$

By Lemma 6.6 we have

$$\mathrm{tr}_{\mathcal{M}^{**}}\left(m_\lambda D_1^{\frac{1}{2}} ab D_2^{\frac{1}{2}}\right) = \psi_{D_1^{\frac{1}{2}} ab D_2^{\frac{1}{2}}}(m_\lambda) = \mathrm{tr}_{\mathcal{M}_\mathcal{U}}\left(m_\lambda \alpha(\psi_{D_1^{\frac{1}{2}} ab D_2^{\frac{1}{2}}})\right).$$

Observing that $\xi_1 = \alpha$ we may apply Lemma 6.8 to obtain

$$\mathrm{tr}_{\mathcal{M}_\mathcal{U}}\left(m_\lambda \alpha(\psi_{D_1^{\frac{1}{2}} ab D_2^{\frac{1}{2}}})\right) = \lim_{i, \mathcal{U}} \mathrm{tr}_{\mathcal{M}}\left(m_\lambda \xi_2(D_1^{\frac{1}{2}})_i\, ab\, \xi_2(D_2^{\frac{1}{2}})_i\right).$$

Therefore we conclude

$$\phi(x) = \lim_\lambda \lim_{i, \mathcal{U}} \mathrm{tr}_\mathcal{M}\left(m_\lambda \xi_2(D_1^{\frac{1}{2}})_i\, x\, \xi_2(D_2^{\frac{1}{2}})_i\right).$$

Moreover, since there exist nets (α_i) and (β_i) in $L_2^+(\mathcal{N})$ such that

$$(\alpha_i)^\bullet = \xi_2(D_1^{\frac{1}{2}}) \quad \text{and} \quad (\beta_i)^\bullet = \xi_2(b^{\frac{1}{2}}),$$

we obtain the following expression for $\phi(x)$

$$\phi(x) = \lim_\lambda \lim_{i, \mathcal{U}} \mathrm{tr}_\mathcal{M}\left(m_\lambda \alpha_i\, x\, \beta_i\right).$$

Recalling that $\lim_{i, \mathcal{U}} \|\alpha_i\|_2 \leq 1$ and $\lim_{i, \mathcal{U}} \|\beta_i\|_2 \leq 1$, we obtain

$$|\phi(x)| \leq \sup\left\{\|\alpha x \beta\|_{L_1(\mathcal{M})} \mid \alpha, \beta \in \mathsf{B}_{L_2(\mathcal{N})}\right\}.$$

This proves (6.8) and implies the assertion. The proof is completed. $\square$

CHAPTER 7

Mixed-norm inequalities

In this chapter we construct an isomorphic embedding
$$\mathcal{J}_{p,q}^n(\mathcal{M}, \mathsf{E}) \hookrightarrow L_p(\mathcal{A}; \ell_q^n),$$
with $\mathcal{A}$ being a sufficiently large von Neumann algebra. After that, we conclude by studying the *right* analogue of inequality (Σ_{pq}) (see the Introduction) in the noncommutative and operator space levels. As we shall see, this appears as a particular case of our embedding of $\mathcal{J}_{p,q}^n(\mathcal{M}, \mathsf{E})$ into $L_p(\mathcal{A}; \ell_q^n)$ when considering the so-called *asymmetric* L_p spaces, a particular case of conditional noncommutative L_p spaces. These results complete the mixed-norm inequalities explored in this paper, which will be instrumental in the last chapter when dealing with operator space L_p embeddings.

7.1. Embedding of $\mathcal{J}_{p,q}^n(\mathcal{M}, \mathsf{E})$ into $L_p(\mathcal{A}; \ell_q^n)$

Now we use the factorization results proved in the previous chapter to identify the spaces $\mathcal{J}_{p,q}^n(\mathcal{M}, \mathsf{E})$ as an interpolation scale in q. This will give rise to an isomorphic embedding of $\mathcal{J}_{p,q}^n(\mathcal{M}, \mathsf{E})$ into the space $L_p(\mathcal{A}; \ell_q^n)$, with $\mathcal{A}$ determined by the map (5.25)
$$u : x \in \mathcal{M} \mapsto \sum_{k=1}^n x_k \otimes \delta_k \in \mathcal{A}_{\oplus n}.$$
Here $x_k = \pi_k(x, -x)$ in the terminology of Chapter 5. That is, $\mathcal{A}_{\oplus n}$ denotes the direct sum $\mathcal{A} \oplus \mathcal{A} \oplus \ldots \oplus \mathcal{A}$ (with n terms) of the $\mathcal{N}$-amalgamated free product $*_{\mathcal{N}} \mathsf{A}_k$ with $\mathsf{A}_k = \mathcal{M} \oplus \mathcal{M}$ for $1 \le k \le n$ and π_k denotes the natural embedding of A_k into $\mathcal{A}$. The following lemma generalizes the main result in [**22**].

LEMMA 7.1. *If $1 \le p \le \infty$, the map*
$$u : x \in \mathcal{J}_{p,1}^n(\mathcal{M}, \mathsf{E}) \mapsto \sum_{k=1}^n x_k \otimes \delta_k \in L_p(\mathcal{A}; \ell_1^n)$$
is an isomorphism with complemented image and constants independent of p, n.

PROOF. Since the result is clear for $p = 1$, we shall assume in what follows that $1 < p \le \infty$. According to Theorem 6.2, we identify the intersection space $\mathcal{J}_{p,1}^n(\mathcal{M}, \mathsf{E})$ with the amalgamated tensor $\mathcal{R}_{2p,1}^n(\mathcal{M}, \mathsf{E}) \otimes_{\mathcal{M}} \mathcal{C}_{2p,1}^n(\mathcal{M}, \mathsf{E})$. Now, using the characterization of $L_p(\mathcal{A}; \ell_1^n)$ given in [**16**], we have

$$\|u(x)\|_{L_p(\mathcal{A};\ell_1^n)} = \inf \left\{ \Big\|\big(\sum_{j,k} a_{kj} a_{kj}^*\big)^{1/2}\Big\|_{L_{2p}(\mathcal{A})} \Big\|\big(\sum_{j,k} b_{kj}^* b_{kj}\big)^{1/2}\Big\|_{L_{2p}(\mathcal{A})} \right\},$$

with the infimum running over all possible decompositions
$$x_k = \sum_j a_{kj} b_{kj}$$
and where sum above is required to converge in the norm topology for $1 \le p < \infty$ and in the weak operator topology otherwise. Then, given any decomposition of x into a finite sum $x = \sum_j \alpha_j \beta_j$ with $\alpha_j \in \mathcal{R}^n_{2p,1}(\mathcal{M},\mathsf{E})$ and $\beta_j \in \mathcal{C}^n_{2p,1}(\mathcal{M},\mathsf{E})$, we have
$$x_k = \sum_j a_{kj} b_{kj} \quad \text{with} \quad a_{kj} = \pi_k(\alpha_j, \alpha_j) \quad \text{and} \quad b_{kj} = \pi_k(\beta_j, -\beta_j).$$
This observation provides the following estimate

(7.1) $$\|u(x)\|_{L_p(\mathcal{A};\ell^n_1)} \le \left\| \sum_k \Lambda_{r,k}(\alpha) \otimes e_{1k} \right\|_{L_{2p}(\mathcal{A};R^n_{2p})}$$
$$\times \left\| \sum_k \Lambda_{c,k}(\beta) \otimes e_{k1} \right\|_{L_{2p}(\mathcal{A};C^n_{2p})}$$

for any possible decomposition $x = \sum_j \alpha_j \beta_j$ and where
$$\Lambda_{r,k}(\alpha) = \pi_k\Big[\big(\sum_j \alpha_j \alpha_j^*\big)^{1/2}, -\big(\sum_j \alpha_j \alpha_j^*\big)^{1/2}\Big],$$
$$\Lambda_{c,k}(\beta) = \pi_k\Big[\big(\sum_j \beta_j^* \beta_j\big)^{1/2}, -\big(\sum_j \beta_j^* \beta_j\big)^{1/2}\Big].$$

We want to reformulate (7.1) in terms of the square functions
$$\mathcal{S}_r(\alpha) = \big(\sum_j \alpha_j \alpha_j^*\big)^{1/2} \quad \text{and} \quad \mathcal{S}_c(\beta) = \big(\sum_j \beta_j^* \beta_j\big)^{1/2}.$$
By the definition of the map u, we find
$$\|u(x)\|_{L_p(\mathcal{A};\ell^n_1)} \le \inf\Big\{ \|u(\mathcal{S}_r(\alpha))\|_{L_{2p}(\mathcal{A};R^n_{2p})} \|u(\mathcal{S}_c(\beta))\|_{L_{2p}(\mathcal{A};C^n_{2p})} \Big\},$$
where the infimum runs over decompositions $x = \sum_k \alpha_j \beta_j$. However, applying Corollary 5.3 as we did in the proof of Lemma 5.14, the spaces $\mathcal{R}^n_{2p,1}(\mathcal{M},\mathsf{E})$ and $\mathcal{C}^n_{2p,1}(\mathcal{M},\mathsf{E})$ are isomorphic to their images via u in $L_{2p}(\mathcal{A};R^n_{2p})$ and $L_{2p}(\mathcal{A};C^n_{2p})$ respectively. In particular, we obtain the following inequality up to a constant independent of p, n
$$\|u(x)\|_{L_p(\mathcal{A};\ell^n_1)} \lesssim \inf\Big\{ \|\mathcal{S}_r(\alpha)\|_{\mathcal{R}^n_{2p,1}(\mathcal{M},\mathsf{E})} \|\mathcal{S}_c(\beta)\|_{\mathcal{C}^n_{2p,1}(\mathcal{M},\mathsf{E})} \Big\}.$$
The right hand side is the norm of x in $\mathcal{R}^n_{2p,1}(\mathcal{M},\mathsf{E}) \otimes_\mathcal{M} \mathcal{C}^n_{2p,1}(\mathcal{M},\mathsf{E})$. Thus, we have proved that u is a bounded map. Now we prove the reverse estimate arguing by duality. First we note that it easily follows from Theorem 4.2 and Remark 4.3 that
$$\mathcal{J}^n_{p,1}(\mathcal{M},\mathsf{E}) = \mathcal{K}^n_{p',\infty}(\mathcal{M},\mathsf{E})^* \quad \text{for} \quad 1 < p \le \infty$$
where, following the notation of [**22**], the latter space is given by
$$\mathcal{K}^n_{p',\infty}(\mathcal{M},\mathsf{E}) = \sum_{u,v \in \{2p',\infty\}} n^{-(\frac{1}{u}+\frac{1}{p}+\frac{1}{v})} L_u(\mathcal{N}) L_{\rho_{uv}}(\mathcal{M}) L_v(\mathcal{N}),$$
with ρ_{uv} determined by $1/u + 1/\rho_{uv} + 1/v = 1/p'$. We claim that

(7.2) $$w : x \in \mathcal{K}^n_{p',\infty}(\mathcal{M},\mathsf{E}) \mapsto \frac{1}{n}\sum_{k=1}^n x_k \otimes \delta_k \in L_{p'}(\mathcal{A};\ell^n_\infty)$$

is a contraction. It is not difficult to see that the result follows from our claim. Indeed, using the anti-linear duality bracket we find that

$$\langle u(x), w(y)\rangle = \frac{1}{n}\sum_{k=1}^{n} \mathrm{tr}_{\mathcal{A}}\bigl(\pi_k(x,-x)^*\pi_k(y,-y)\bigr) = \frac{1}{n}\sum_{k=1}^{n} \mathrm{tr}_{\mathcal{M}}(x^*y) = \langle x, y\rangle.$$

In particular, it turns out that w^*u is the identity map on $\mathcal{J}_{p,1}^n(\mathcal{M},\mathsf{E})$ and uw^* is a bounded projection from $L_p(\mathcal{A};\ell_1^n)$ onto $u(\mathcal{J}_{p,1}^n(\mathcal{M},\mathsf{E}))$. Thus, u is an isomorphism and its image is complemented with constants independent of p, n. Therefore, it remains to prove our claim. Since the initial space of w is a sum of four Banach spaces indexed by the pairs (u,v) with $u,v \in \{2p',\infty\}$, it suffices to see that the restriction of w to each of these spaces is a contraction. Let us start with the case $(u,v)=(\infty,\infty)$. In this case the associated space is simply $n^{-1/p}L_{p'}(\mathcal{M})$ and we have by complex interpolation in $1 \le p' \le \infty$ that

$$\frac{1}{n}\Bigl\|\sum_{k=1}^n x_k \otimes \delta_k\Bigr\|_{L_{p'}(\mathcal{A};\ell_\infty^n)} \le n^{-\frac{1}{p}}\|x\|_{L_{p'}(\mathcal{M})}.$$

Indeed, when $p=1$ we have an identity while for $p=\infty$ the assertion follows by the triangle inequality. Therefore, recalling from [29] that the spaces $L_{p'}(\mathcal{M},\ell_\infty^n)$ form an interpolation scale in p', our claim follows. The space associated to $u = 2p' = v$ is

$$\frac{1}{n}L_{2p'}(\mathcal{N})L_\infty(\mathcal{M})L_{2p'}(\mathcal{N}).$$

Here we use the characterization of the norm of $L_{p'}(\mathcal{A};\ell_\infty^n)$ given in [16]

$$\Bigl\|\sum_{k=1}^n x_k \otimes \delta_k\Bigr\|_{L_{p'}(\mathcal{A};\ell_\infty^n)} = \inf_{x_k=ad_k b}\Bigl\{\|a\|_{L_{2p'}(\mathcal{A})}\Bigl(\sup_{1\le k\le n}\|d_k\|_{L_\infty(\mathcal{A})}\Bigr)\|b\|_{L_{2p'}(\mathcal{A})}\Bigr\}.$$

Then we consider the decompositions of x_k that come from decompositions of x. In other words, any decomposition $x = \alpha y\beta$ with $\alpha,\beta \in L_{2p'}(\mathcal{N})$ and $y \in L_\infty(\mathcal{M})$ gives rise to the decompositions $x_k = \alpha\pi_k(y,-y)\beta$. Hence, since

$$\|\pi_k(y,-y)\|_{L_\infty(\mathcal{A})} = \|y\|_{L_\infty(\mathcal{M})} \quad \text{for} \quad 1 \le k \le n,$$

we obtain the desired estimate

$$\Bigl\|\sum_{k=1}^n x_k \otimes \delta_k\Bigr\|_{L_{p'}(\mathcal{A};\ell_\infty^n)} \le \inf_{x=\alpha y\beta}\Bigl\{\|\alpha\|_{L_{2p'}(\mathcal{N})}\|y\|_{L_\infty(\mathcal{M})}\|\beta\|_{L_{2p'}(\mathcal{N})}\Bigr\} = \|x\|_{2p'\cdot\infty\cdot 2p'}.$$

Finally, it remains to consider the cross terms associated to $(u,v) = (2p',\infty)$ and $(u,v) = (\infty, 2p')$. Since both can be treated using the same arguments, we only consider the case $(u,v) = (2p',\infty)$ whose associated space is

$$\frac{1}{n^{\frac{1}{2}+\frac{1}{2p}}} L_{2p'}(\mathcal{N})L_{2p'}(\mathcal{M})L_\infty(\mathcal{N}).$$

Here we observe that given any $\alpha \in L_{2p'}(\mathcal{N})$, the left multiplication mapping

$$\mathsf{L}_\alpha : \sum_{k=1}^n a_k \otimes \delta_k \in L_{2p'}(\mathcal{A};\ell_{2p'}^n) \mapsto \alpha\sum_{k=1}^n a_k \otimes \delta_k \in L_{p'}(\mathcal{A};\ell_\infty^n)$$

is a bounded map with norm $\le \|\alpha\|_{L_{2p'}(\mathcal{A})}$. Indeed, by complex interpolation we only study the extremal cases. Noting that the result is trivial for $p' = \infty$, it suffices

to see it for $p' = 1$. Let us assume that
$$\sum_{k=1}^{n} \|a_k\|_{L_2(\mathcal{A})}^2 \leq 1.$$
Then we define the invertible operator $\beta = \left(\sum_k a_k^* a_k + \delta \mathrm{D}_\phi\right)^{1/2}$ (D_ϕ being the density associated to the state ϕ of $\mathcal{A}$) so that $a_k = b_k \beta$ for $1 \leq k \leq n$ where the operators $b_1, b_2, \ldots, b_n$ satisfy
$$\sum_{k=1}^{n} b_k^* b_k \leq 1.$$
This provides a factorization $\alpha a_k = \alpha b_k \beta$ from which we deduce
$$\left\|\alpha \sum_{k=1}^{n} a_k \otimes \delta_k\right\|_{L_1(\mathcal{A};\ell_\infty^n)} \leq \|\alpha\|_{L_2(\mathcal{A})} \left(\sup_{1 \leq k \leq n} \|b_k\|_{L_\infty(\mathcal{A})}\right) \|\beta\|_{L_2(\mathcal{A})} \leq (1+\delta) \|\alpha\|_{L_2(\mathcal{A})}.$$
Thus, we conclude letting $\delta \to 0$. Now assume that x is a norm 1 element of
$$\frac{1}{n^{\frac{1}{2} + \frac{1}{2p}}} L_{2p'}(\mathcal{N}) L_{2p'}(\mathcal{M}) L_\infty(\mathcal{N}),$$
so that for any $\delta > 0$ we can find a factorization $x = \alpha y$ such that
$$\|\alpha\|_{L_{2p'}(\mathcal{N})} \leq 1 \quad \text{and} \quad \|y\|_{L_{2p'}(\mathcal{M})} \leq (1+\delta) n^{\frac{1}{2} + \frac{1}{2p}}.$$
Then, since $\alpha \in L_{2p'}(\mathcal{N})$ we have
$$\sum_{k=1}^{n} x_k \otimes \delta_k = \alpha \sum_{k=1}^{n} \pi_k(y, -y) \otimes \delta_k = \mathsf{L}_\alpha\left(\sum_{k=1}^{n} \pi_k(y, -y) \otimes \delta_k\right).$$
In particular,
$$\frac{1}{n}\left\|\sum_{k=1}^{n} x_k \otimes \delta_k\right\|_{L_{p'}(\mathcal{A};\ell_\infty^n)} \leq \frac{1}{n} \|\alpha\|_{L_{2p'}(\mathcal{N})} \left\|\sum_{k=1}^{n} \pi_k(y, -y) \otimes \delta_k\right\|_{L_{2p'}(\mathcal{A};\ell_{2p'}^n)}$$
$$\leq \frac{1}{n}\left(\sum_{k=1}^{n} \|\pi_k(y, -y)\|_{L_{2p'}(\mathcal{A})}^{2p'}\right)^{1/2p'} \leq 1 + \delta.$$
Letting $\delta \to 0$, we conclude that the mapping defined in (7.2) is a contraction. $\square$

THEOREM 7.2. *If* $1 \leq p \leq \infty$, *then*
$$\left[\mathcal{J}_{p,1}^n(\mathcal{M}, \mathsf{E}), \mathcal{J}_{p,p}^n(\mathcal{M}, \mathsf{E})\right]_\theta \simeq \mathcal{J}_{p,q}^n(\mathcal{M}, \mathsf{E})$$
with $1/q = 1 - \theta + \theta/p$ *and with relevant constants independent of* n.

PROOF. By Theorem 4.6 and Observation 4.8, we have
$$\left[\mathcal{J}_{p,1}^n(\mathcal{M}, \mathsf{E}), \mathcal{J}_{p,p}^n(\mathcal{M}, \mathsf{E})\right]_\theta \subset \mathcal{J}_{p,q}^n(\mathcal{M}, \mathsf{E})$$
contractively. To prove the reverse inclusion we consider the map $u : \mathcal{M} \to \mathcal{A}_{\oplus n}$.

STEP 1. Since $\mathcal{J}_{p,p}^n(\mathcal{M}, \mathsf{E}) = n^{\frac{1}{p}} L_p(\mathcal{M})$, we clearly have
$$\|u(x)\|_{L_p(\mathcal{A};\ell_p^n)} = \|x\|_{\mathcal{J}_{p,p}^n(\mathcal{M},\mathsf{E})}.$$
Moreover, $u(\mathcal{J}_{p,p}^n(\mathcal{M}, \mathsf{E}))$ is clearly contractively complemented in $L_p(\mathcal{A};\ell_p^n)$. On the other hand, according to Lemma 7.1, $\mathcal{J}_{p,1}^n(\mathcal{M}, \mathsf{E})$ is isomorphic to its image via u, which is complemented in $L_p(\mathcal{A};\ell_1^n)$ with constants not depending on p, n.

7.1. EMBEDDING OF $\mathcal{J}_{p,q}^n(\mathcal{M},\mathsf{E})$ INTO $L_p(\mathcal{A};\ell_q^n)$

Therefore, the norm of x in the space $[\mathcal{J}_{p,1}^n(\mathcal{M},\mathsf{E}), \mathcal{J}_{p,p}^n(\mathcal{M},\mathsf{E})]_\theta$ turns out to be equivalent to the norm of
$$\sum_{k=1}^n x_k \otimes \delta_k \quad \text{in} \quad L_p(\mathcal{A};\ell_q^n).$$

Thus, it remains to see that

(7.3) $$\Big\|\sum_{k=1}^n x_k \otimes \delta_k\Big\|_{L_p(\mathcal{A};\ell_q^n)} \lesssim \|x\|_{\mathcal{J}_{p,q}^n(\mathcal{M},\mathsf{E})}.$$

STEP 2. We need an auxiliary result. Let us consider the bilinear map
$$\Lambda : (a,b) \in \mathcal{R}_{2p,q}^n(\mathcal{M},\mathsf{E}) \times \mathcal{C}_{2p,q}^n(\mathcal{M},\mathsf{E}) \mapsto \sum_{k=1}^n \pi_k(ab,-ab) \otimes \delta_k \in L_p(\mathcal{A};\ell_q^n).$$

We claim that Λ is bounded by $c(p,q) \sim (p-q)/(pq+q-p)$. By Theorem 5.16, we may use bilinear interpolation and it suffices to show our claim for the extremal values of q. Let us note that, since we are applying Theorem 5.16 for $\mathcal{R}_{2p,q}^n(\mathcal{M},\mathsf{E})$ and $\mathcal{C}_{2p,q}^n(\mathcal{M},\mathsf{E})$, the operator norm of Λ is controlled by the square of the relevant constant in Theorem 5.16, as we have claimed. The boundedness of Λ for $q=1$ follows from Theorem 6.2 and Lemma 7.1. The estimate for $q=p$ is much easier

$$\|\Lambda(a,b)\|_{L_p(\mathcal{A};\ell_p^n)} = \Big(\sum_{k=1}^n \|\pi_k(ab,-ab)\|_p^p\Big)^{\frac{1}{p}} = n^{\frac{1}{p}} \|ab\|_p \le \|a\|_{\mathcal{R}_{2p,p}} \|b\|_{\mathcal{C}_{2p,p}}.$$

STEP 3. Now it is easy to deduce inequality (7.3) from the boundedness of the mapping Λ. Indeed, according to Theorem 6.2 we know that (7.3) is equivalent to

(7.4) $$\Big\|\sum_{k=1}^n x_k \otimes \delta_k\Big\|_{L_p(\mathcal{A};\ell_q^n)} \lesssim \Big\|\Big(\sum_j a_j a_j^*\Big)^{\frac{1}{2}}\Big\|_{\mathcal{R}_{2p,q}^n} \Big\|\Big(\sum_j b_j^* b_j\Big)^{\frac{1}{2}}\Big\|_{\mathcal{C}_{2p,q}^n},$$

for any decomposition of x into a finite sum $\sum_j a_j b_j$. Let us assume that the index j runs from 1 to m for some finite m. Then we consider the matrix amplifications $\hat{\mathcal{N}} = \mathrm{M}_m \otimes \mathcal{N}$ and $\hat{\mathcal{M}} = \mathrm{M}_m \otimes \mathcal{M}$. Similarly, we consider $\hat{\mathcal{A}} = \mathrm{M}_m \otimes \mathcal{A}$. According to (5.3), we have
$$\hat{\mathcal{A}} = \mathrm{M}_m \otimes \mathcal{A}_1 *_{\hat{\mathcal{N}}} \mathrm{M}_m \otimes \mathcal{A}_2 *_{\hat{\mathcal{N}}} \cdots *_{\hat{\mathcal{N}}} \mathrm{M}_m \otimes \mathcal{A}_n = (\hat{\mathcal{M}} \oplus \hat{\mathcal{M}})^{*\hat{\mathcal{N}}}_n.$$

By Step 2, we deduce

(7.5) $$\Big\|\sum_{k=1}^n \mathbf{x}_k \otimes \delta_k\Big\|_{L_p(\hat{\mathcal{A}};\ell_q^n)} \lesssim \|\mathbf{a}\|_{\hat{\mathcal{R}}_{2p,q}^n} \|\mathbf{b}\|_{\hat{\mathcal{C}}_{2p,q}^n}$$

where the elements $\mathbf{x}_k$ are given by $\mathbf{x}_k = \pi_k(\mathbf{ab},-\mathbf{ab})$ and
$$\begin{aligned}\hat{\mathcal{R}}_{2p,q}^n &= \mathcal{R}_{2p,q}^n(\hat{\mathcal{M}}, 1_{\mathrm{M}_m} \otimes \mathsf{E}),\\ \hat{\mathcal{C}}_{2p,q}^n &= \mathcal{C}_{2p,q}^n(\hat{\mathcal{M}}, 1_{\mathrm{M}_m} \otimes \mathsf{E}).\end{aligned}$$

To prove the remaining estimate (7.4) we fix x in $\mathcal{J}_{p,q}^n(\mathcal{M},\mathsf{E})$ and decompose it into a finite sum $\sum_j a_j b_j$ with m terms. Then, we define the row and column matrices
$$\mathbf{a} = \sum_{j=1}^m a_j \otimes e_{1j} \quad \text{and} \quad \mathbf{b} = \sum_{j=1}^m b_j \otimes e_{j1}.$$

According to (7.5), it just remains to show that
$$\Big\|\sum_{k=1}^{n} \mathbf{x}_k \otimes \delta_k\Big\|_{L_p(\hat{\mathcal{A}};\ell_q^n)} = \Big\|\sum_{k=1}^{n} x_k \otimes \delta_k\Big\|_{L_p(\mathcal{A};\ell_q^n)},$$
$$\|\mathbf{a}\|_{\hat{\mathcal{R}}_{2p,q}^n} = \Big\|\Big(\sum_j a_j a_j^*\Big)^{\frac{1}{2}}\Big\|_{\mathcal{R}_{2p,q}^n},$$
$$\|\mathbf{b}\|_{\hat{\mathcal{C}}_{2p,q}^n} = \Big\|\Big(\sum_j b_j^* b_j\Big)^{\frac{1}{2}}\Big\|_{\mathcal{C}_{2p,q}^n}.$$

The first identity follows easily from
$$\begin{aligned}
\mathbf{x}_k &= \pi_k(\mathbf{ab}, -\mathbf{ab}) \\
&= \sum_j \pi_k(a_j b_j \otimes e_{11}, -a_j b_j \otimes e_{11}) \\
&= \pi_k\Big(\sum_j a_j b_j, -\sum_j a_j b_j\Big) \otimes e_{11} \\
&= x_k \otimes e_{11},
\end{aligned}$$
where the third identity holds because π_k is an $\hat{\mathcal{N}}$-bimodule map. The relation $\mathbf{x}_k = x_k \otimes e_{11}$ shows that $\sum_k \mathbf{x}_k \otimes \delta_k$ lives in fact in the $(1,1)$ corner, which by [30] is isometrically isomorphic to $L_p(\mathcal{A};\ell_q^n)$. The identities for $\mathbf{a}$ and $\mathbf{b}$ are proved in the same way, so that we only consider the first of them. We have
$$\|\mathbf{a}\|_{\hat{\mathcal{R}}_{2p,q}^n} = \max\Big\{n^{\frac{1}{2p}}\|\mathbf{a}\|_{L_{2p}(\hat{\mathcal{M}})}, n^{\frac{1}{2q}}\|\mathbf{a}\|_{L_{(\frac{2pq}{p-q},\infty)}^{2p}(\hat{\mathcal{M}},1_{\mathrm{M}_m}\otimes\mathsf{E})}\Big\}.$$
It is clear that
$$\|\mathbf{a}\|_{L_{2p}(\hat{\mathcal{M}})} = \Big\|\Big(\sum_j a_j a_j^*\Big)^{\frac{1}{2}}\Big\|_{L_{2p}(\mathcal{M})}.$$
Therefore, we just need to show that
$$\|\mathbf{a}\|_{L_{(\frac{2pq}{p-q},\infty)}^{2p}(\hat{\mathcal{M}},1_{\mathrm{M}_m}\otimes\mathsf{E})} = \Big\|\Big(\sum_j a_j a_j^*\Big)^{\frac{1}{2}}\Big\|_{L_{(\frac{2pq}{p-q},\infty)}^{2p}(\mathcal{M},\mathsf{E})}.$$
By definition, we have
$$\begin{aligned}
\|\mathbf{a}\|_{L_{(\frac{2pq}{p-q},\infty)}^{2p}(\hat{\mathcal{M}},1_{\mathrm{M}_m}\otimes\mathsf{E})} &= \sup_\alpha \big\|\alpha \mathbf{aa}^*\alpha^*\big\|_q^{\frac{1}{2}} \\
&= \sup_\alpha \Big\|\alpha\Big[\Big(\sum_j a_j a_j^*\Big) \otimes e_{11}\Big]\alpha^*\Big\|_q^{\frac{1}{2}} \\
&= \sup_\alpha \Big\|\alpha\Big[\Big(\sum_j a_j a_j^*\Big)^{\frac{1}{2}} \otimes e_{11}\Big]\Big\|_{2q},
\end{aligned}$$
where the supremum runs over all α such that
$$\|\alpha\|_{L_{\frac{2pq}{p-q}}(\hat{\mathcal{N}})} \leq 1.$$
If (α_{ij}) are the $m \times m$ entries of α, we clearly have
$$\alpha\Big[\Big(\sum_j a_j a_j^*\Big)^{\frac{1}{2}} \otimes e_{11}\Big] = \sum_{s=1}^{m} \alpha_{s1}\Big(\sum_j a_j a_j^*\Big)^{\frac{1}{2}} \otimes e_{s1}.$$
Therefore, we obtain the following estimate
$$\Big\|\alpha\Big[\Big(\sum_j a_j a_j^*\Big)^{\frac{1}{2}} \otimes e_{11}\Big]\Big\|_{L_{2q}(\hat{\mathcal{M}})}$$

$$= \left\| \left(\sum_s \left[\sum_j a_j a_j^* \right]^{\frac{1}{2}} \alpha_{s1}^* \alpha_{s1} \left[\sum_j a_j a_j^* \right]^{\frac{1}{2}} \right)^{\frac{1}{2}} \right\|_{L_{2q}(\mathcal{M})}$$

$$= \left\| \left(\sum_s \alpha_{s1}^* \alpha_{s1} \right)^{\frac{1}{2}} \left(\sum_j a_j a_j^* \right)^{\frac{1}{2}} \right\|_{L_{2q}(\mathcal{M})}$$

$$\leq \left\| \sum_s \alpha_{s1} \otimes e_{s1} \right\|_{L_{\frac{2pq}{p-q}}(\hat{\mathcal{N}})} \left\| \left(\sum_j a_j a_j^* \right)^{\frac{1}{2}} \right\|_{L^{2p}_{(\frac{2pq}{p-q},\infty)}(\mathcal{M},\mathsf{E})}.$$

Hence, the result follows since the column projection is contractive on $L_{\frac{2pq}{p-q}}(\hat{\mathcal{N}})$. $\square$

The following is a major result in this paper.

THEOREM 7.3. *If $1 \leq q \leq p \leq \infty$, the map*

$$u : x \in \mathcal{J}_{p,q}^n(\mathcal{M},\mathsf{E}) \mapsto \sum_{k=1}^n x_k \otimes \delta_k \in L_p(\mathcal{A};\ell_q^n)$$

is an isomorphism with complemented image and constants independent of n.

PROOF. It is an immediate consequence of Lemma 7.1 and Theorem 7.2 $\square$

REMARK 7.4. If $1 \leq q \leq p \leq \infty$, let us define

$$\mathcal{K}_{p',q'}^n(\mathcal{M},\mathsf{E}) = \sum_{u,v \in \{2r,\infty\}} n^{-1+\frac{1}{\rho_{uv}}} L_u(\mathcal{N}) L_{\rho_{uv}}(\mathcal{M}) L_v(\mathcal{N})$$

where $1/r = 1/q - 1/p$ and ρ_{uv} is determined by $1/u + 1/\rho_{uv} + 1/v = 1/p'$. Note that this definition is consistent with the space $\mathcal{K}_{p',\infty}^n(\mathcal{M},\mathsf{E})$ introduced in the proof of Lemma 7.1. Arguing by duality as in [**22**], we easily conclude from Theorem 7.3 that we have an isomorphism with complemented image

$$(7.6) \qquad \xi_{free} : x \in \mathcal{K}_{p',q'}^n(\mathcal{M},\mathsf{E}) \mapsto \frac{1}{n} \sum_{k=1}^n x_k \otimes \delta_k \in L_{p'}(\mathcal{A}_{free};\ell_{q'}^n)$$

where $\mathcal{A}_{free}$ denotes the usual free product algebra $\mathcal{A}$ from Theorem 7.3. Moreover, it is important to note that replacing free products by tensors and freeness by noncommutative independence as in [**22**], Theorem 7.3 and (7.6) hold in the range $1 < p' \leq q' \leq \infty$. In other words, we replace $\mathcal{A}_{free}$ by the tensor product algebra

$$\mathcal{A}_{ind} = \mathcal{M} \otimes \mathcal{M} \otimes \cdots \otimes \mathcal{M}$$

with n terms and x_k is now given by

$$x_k = 1 \otimes \cdots 1 \otimes x \otimes 1 \otimes \cdots \otimes 1$$

where x is placed in the k-th position. The question now is whether or not (7.6) holds in the non-free setting for $p' = 1$. Using recent results from [**18**], we shall see in a forthcoming paper that we have an isomorphism with complemented image

$$\xi_{ind} : x \in \mathcal{K}_{1,2}^n(\mathcal{M},\mathsf{E}) \mapsto \sum_{k=1}^n x_k \otimes \delta_k \in L_{p'}(\mathcal{A}_{ind};\mathrm{OH}_n).$$

Let us note that the same question for $\mathcal{K}_{1,q'}^n(\mathcal{M},\mathsf{E})$ with $q' \neq 2$ is still open.

7.2. Asymmetric L_p spaces and noncommutative (Σ_{pq})

Let $\mathcal{M}$ be a von Neumann algebra equipped with a n.f. state φ. Now let $2 \leq u, v \leq \infty$ be such that $1/p = 1/u + 1/v$ for some $1 \leq p \leq \infty$. Then, we define the *asymmetric L_p space* associated to the pair (u, v) as the $\mathcal{M}$-amalgamated Haagerup tensor product

(7.7) $$L_{(u,v)}(\mathcal{M}) = L_u^r(\mathcal{M}) \otimes_{\mathcal{M},h} L_v^c(\mathcal{M}),$$

where we recall that $L_q^r(\mathcal{M})$ and $L_q^c(\mathcal{M})$ were defined in (1.3) for $2 \leq q \leq \infty$. That is, we consider the quotient of $L_u^r(\mathcal{M}) \otimes_h L_v^c(\mathcal{M})$ by the closed subspace $\mathcal{I}$ generated by the differences $x_1 \gamma \otimes x_2 - x_1 \otimes \gamma x_2$ with $\gamma \in \mathcal{M}$. Recall that we are using the notation $\otimes_{\mathcal{M},h}$ instead of $\otimes_\mathcal{M}$ because, in contrast with the previous chapters, we shall be interested here in the operator space structure rather than in the Banach space one. By a well known factorization argument (see e.g. Lemma 3.5 in [**46**]), the norm of an element x in $L_{(u,v)}(\mathcal{M})$ is given by

$$\|x\|_{(u,v)} = \inf_{x=\alpha\beta} \|\alpha\|_{L_u(\mathcal{M})} \|\beta\|_{L_v(\mathcal{M})}.$$

According to this observation, it turns out that asymmetric L_p spaces arise as a particular case of the amalgamated noncommutative L_p spaces defined in Chapter 2 when we take $q = \infty$ and $\mathcal{N}$ to be $\mathcal{M}$ itself. Asymmetric L_p spaces were introduced in [**22**] for finite matrix algebras. There the amalgamated Haagerup tensor product used in (7.7) was not needed in [**22**] to define the asymmetric Schatten classes. In fact, if $\mathcal{M}$ is the algebra M_m of $m \times m$ matrices and X is an operator space, we can define the *vector-valued asymmetric Schatten class*

$$S_{(u,v)}^m(\mathrm{X}) = C_{u/2}^m \otimes_h \mathrm{X} \otimes_h R_{v/2}^m.$$

Note that this definition is consistent with (7.7). Indeed, recalling that

$$L_u^r(\mathrm{M}_m) = C_{u/2}^m \otimes_h R_m \quad \text{and} \quad L_v^c(\mathrm{M}_m) = C_m \otimes_h R_{v/2}^m,$$

it can be easily checked from the definition of the Haagerup tensor product that we have $L_u^r(\mathrm{M}_m) \otimes_{\mathcal{M},h} L_v^c(\mathrm{M}_m) = S_{(u,v)}^m(\mathbb{C})$ isometrically. Moreover, according to [**22**] any linear map $u: \mathrm{X}_1 \to \mathrm{X}_2$ satisfies

(7.8) $$\|u\|_{cb} = \sup_{n \geq 1} \left\| 1_{\mathrm{M}_n} \otimes u : S_{(u,v)}^n(\mathrm{X}_1) \to S_{(u,v)}^n(\mathrm{X}_2) \right\|.$$

In particular, since it is clear that

$$S_{(u,v)}^n\big(L_u^r(\mathrm{M}_m) \otimes_{\mathcal{M},h} L_v^c(\mathrm{M}_m)\big) = S_{(u,v)}^{mn}(\mathbb{C}) \quad \text{for all} \quad m \geq 1,$$

we have the following completely isometric isomorphism

$$L_{(u,v)}(\mathrm{M}_m) = S_{(u,v)}^m(\mathbb{C}).$$

REMARK 7.5. $L_p(\mathcal{M})$ is completely isometric to $L_{2p}^r(\mathcal{M}) \otimes_{\mathcal{M},h} L_{2p}^c(\mathcal{M})$.

We conclude by generalizing the inequalities (Σ_{pq}) stated in the Introduction to the noncommutative setting. Moreover, we shall seek for a completely isomorphic embedding rather than a Banach space one. As we shall see, this appears as a particular case of Theorem 7.3 module the corresponding identifications. Indeed, let $\mathcal{M}$ be a von Neumann algebra equipped with a n.f. state φ and let us consider the particular situation in which $(\mathcal{M}, \mathcal{N}, \mathsf{E})$ above are replaced by

(7.9) $$(\mathcal{M}_m, \mathcal{N}_m, \mathsf{E}_m) = \Big(\mathrm{M}_m \otimes \mathcal{M}, \mathrm{M}_m, 1_{\mathrm{M}_m} \otimes \varphi \Big).$$

If D_φ is the associated density, we consider the densely defined maps

$$\rho_r : D_\varphi^{1/2}(x_{ij}) \in S_\infty^m(L_2^r(\mathcal{M})) \mapsto (x_{ij}) \in L_\infty^r(\mathcal{M}_m, \mathsf{E}_m),$$
$$\rho_c : (x_{ij})D_\varphi^{1/2} \in S_\infty^m(L_2^c(\mathcal{M})) \mapsto (x_{ij}) \in L_\infty^c(\mathcal{M}_m, \mathsf{E}_m).$$

LEMMA 7.6. *The maps ρ_r and ρ_c extend to isometric isomorphisms.*

PROOF. By [11, p.56] we have

$$\begin{aligned}
\|D_\varphi^{1/2}(z_{ij})\|_{S_\infty^m(L_2^r(\mathcal{M}))} &= \Big\|\Big(\sum_{k=1}^m \mathrm{tr}_\mathcal{M}(D_\varphi^{1/2} z_{ik} z_{jk}^* D_\varphi^{1/2})\Big)^{1/2}\Big\|_{\mathrm{M}_m} \\
&= \Big\|\Big(\sum_{k=1}^m \varphi(z_{ik} z_{jk}^*)\Big)^{1/2}\Big\|_{\mathrm{M}_m} \\
&= \Big\|1_{\mathrm{M}_m} \otimes \varphi\big[(z_{ij})(z_{ij})^*\big]\Big\|_{\mathrm{M}_m}^{1/2} = \|(z_{ij})\|_{L_\infty^r(\mathcal{M}_m,\mathsf{E}_m)}.
\end{aligned}$$

The proof of the second isometric isomorphism follows from [11, p.54] instead. □

Now, assuming that $\mathcal{M}$, $\mathcal{N}$ and E are given as in (7.9), our aim is to identify the intersection space $\mathcal{J}_{p,q}^n(\mathcal{M}_m, \mathsf{E}_m)$ in terms of asymmetric L_p spaces. More concretely, let us define the following intersection of asymmetric spaces

$$\mathcal{J}_{p,q}^n(\mathcal{M}) = \bigcap_{u,v \in \{2p, 2q\}} n^{\frac{1}{u}+\frac{1}{v}} L_{(u,v)}(\mathcal{M}).$$

LEMMA 7.7. *If $m \geq 1$, we have an isometry*

$$S_p^m(\mathcal{J}_{p,q}^n(\mathcal{M})) = \mathcal{J}_{p,q}^n(\mathcal{M}_m, \mathsf{E}_m).$$

PROOF. By definition we have

$$\mathcal{J}_{p,q}^n(\mathcal{M}_m, \mathsf{E}_m) = \bigcap_{u,v \in \{\frac{2pq}{p-q}, \infty\}} n^{\frac{1}{u}+\frac{1}{p}+\frac{1}{v}} L_{(u,v)}^p(\mathcal{M}_m, \mathsf{E}_m).$$

Since the powers of n fit, it clearly suffices to show that

$$\begin{aligned}
S_p^m\big(L_{(2p,2p)}(\mathcal{M})\big) &= L_{(\infty,\infty)}^p(\mathcal{M}_m, \mathsf{E}_m), \\
S_p^m\big(L_{(2p,2q)}(\mathcal{M})\big) &= L_{(\infty,\frac{2pq}{p-q})}^p(\mathcal{M}_m, \mathsf{E}_m), \\
S_p^m\big(L_{(2q,2p)}(\mathcal{M})\big) &= L_{(\frac{2pq}{p-q},\infty)}^p(\mathcal{M}_m, \mathsf{E}_m), \\
S_p^m\big(L_{(2q,2q)}(\mathcal{M})\big) &= L_{(\frac{2pq}{p-q},\frac{2pq}{p-q})}^p(\mathcal{M}_m, \mathsf{E}_m).
\end{aligned}$$

The first isometry follows from Remark 7.5, we have $L_p(\mathcal{M}_m)$ at both sides. The last one follows from Example 4.1 (b), which uses one of Pisier's identities stated in Chapter 1. It remains to see the second and third isometries. By complex interpolation on q, it suffices to assume $q = 1$ since the case $q = p$ has been already considered. In that case, we have to show that

$$\begin{aligned}
S_p^m\big(L_{(2p,2)}(\mathcal{M})\big) &= L_{(\infty,2p')}^p(\mathcal{M}_m, \mathsf{E}_m), \\
S_p^m\big(L_{(2,2p)}(\mathcal{M})\big) &= L_{(2p',\infty)}^p(\mathcal{M}_m, \mathsf{E}_m).
\end{aligned}$$

When $p=1$, the isometry follows again from Remark 7.5. Therefore, by complex interpolation one more time, it suffices to assume that $(p,q) = (\infty, 1)$. In that case, we note that
$$S_\infty^m\big(L_{(\infty,2)}(\mathcal{M})\big) = S_\infty^m\big(L_2^c(\mathcal{M})\big) = L_\infty^c(\mathcal{M}_m, \mathsf{E}_m) = L_{(\infty,2)}^\infty(\mathcal{M}_m, \mathsf{E}_m),$$
$$S_\infty^m\big(L_{(2,\infty)}(\mathcal{M})\big) = S_\infty^m\big(L_2^r(\mathcal{M})\big) = L_\infty^r(\mathcal{M}_m, \mathsf{E}_m) = L_{(2,\infty)}^\infty(\mathcal{M}_m, \mathsf{E}_m).$$
We have used Lemma 7.6 in the second identities. This completes the proof. □

Let $\mathcal{A}_m$ stand for the usual amalgamated free product von Neumann algebra constructed out of $\mathcal{M}_m = \mathrm{M}_m \otimes \mathcal{M}$ with $\mathcal{N}_m = \mathrm{M}_m$. According to (5.3), we have $\mathcal{A}_m = \mathrm{M}_m \otimes \mathcal{A}$ with $\mathcal{A} = (\mathcal{M} \oplus \mathcal{M})^{*n}$. In particular, $L_p(\mathcal{A}_m; \ell_q^n) = S_p^m\big(L_p(\mathcal{A}; \ell_q^n)\big)$ and we deduce the result below from Theorem 7.3 and Lemma 7.7.

THEOREM 7.8. *If $1 \leq q \leq p \leq \infty$, the map*
$$u : x \in \mathcal{J}_{p,q}^n(\mathcal{M}) \mapsto \sum_{k=1}^n x_k \otimes \delta_k \in L_p(\mathcal{A}; \ell_q^n)$$
is a cb-isomorphism with cb-complemented image and constants independent of n.

REMARK 7.9. The o.s.s. of $L_p(\mathcal{A}; \ell_q^n)$ was introduced in Chapter 1.

REMARK 7.10. Let $\mathcal{M}$ be the matrix algebra M_m equipped with its normalized trace τ and let us consider the direct sum $\mathcal{M}_{\oplus n} = \mathcal{M} \oplus \mathcal{M} \oplus \cdots \oplus \mathcal{M}$ with n terms. We equip this algebra with the (non-normalized) trace $\tau_n = \tau \oplus \tau \oplus \cdots \oplus \tau$. In this case, given an operator space X, we could also define the space
$$\mathcal{J}_{p,q}(\mathcal{M}_{\oplus n}; \mathrm{X}) = \bigcap_{u,v \in \{2p, 2q\}} L_u^r(\mathcal{M}_{\oplus n}) \otimes_{\mathcal{M}_{\oplus n}, h} L_\infty(\mathcal{M}_{\oplus n}; \mathrm{X}) \otimes_{\mathcal{M}_{\oplus n}, h} L_v^c(\mathcal{M}_{\oplus n}).$$
The analogue of Theorem 7.8 with
  i) (p,q) being $(p,1)$,
  ii) freeness replaced by noncommutative independence,
  iii) vector-values in some operator space X as explained above,

was the main result in [**22**]. In our situation, a vector-valued analogue of Theorem 7.8 also holds in the context of finite matrix algebras. However, note that the use of free probability requires to define vector-valued L_p spaces for free product von Neumann algebras [**19**], which is beyond the scope of this paper.

REMARK 7.11. The constants in Theorems 7.2, 7.3 and 7.8 are also independent of p, q unless $p \sim \infty$ is large and $q \sim 1$ is small simultaneously. In that case the argument in Theorem 7.2 produces the singularity $(p-q)/(pq+q-p)$ which affects Theorems 7.3 and 7.8. Moreover, it arises as a byproduct of the use of Burkholder inequality in Proposition 5.10, which is used in the proof of Theorem 5.16, a key tool in Theorems 7.2. Of course, this seems to be a removable singularity. It is also worthy of mention that this will not cause any singularity in our main application, the construction of operator space L_p embeddings in the next chapter. Indeed, in that case we will work all the time with $q=2$ and we have
$$(p-q)/(pq+q-p) = (p-2)/(p+2) < 1.$$

CHAPTER 8

Operator space L_p embeddings

Given $1 \leq p < q \leq 2$ and a von Neumann algebra $\mathcal{M}$, we conclude this paper by constructing a completely isomorphic embedding of $L_q(\mathcal{M})$ into $L_p(\mathcal{A})$ for some sufficiently large von Neumann algebra $\mathcal{A}$. In the first section we embed the Schatten class S_q into $L_p(\mathcal{A})$ for some QWEP von Neumann algebra $\mathcal{A}$. Roughly speaking, the proof is almost identical to our argument in [24] after replacing Corollary 1.9 there by Theorem D in the Introduction. Thus, we shall omit some details in our construction. The second paragraph is devoted to the stability of hyperfiniteness and there we will present the transference argument mentioned in the Introduction. Finally, the last paragraph contains our construction for general von Neumann algebras.

8.1. Embedding Schatten classes

We shall prove an L_p version of [24, Theorem D], from which the embedding of the Schatten class S_q into $L_p(\mathcal{A})$ will follow for $1 \leq p < q \leq 2$ and $\mathcal{A}$ a sufficiently large QWEP von Neumann algebra. The main embedding result in Xu's paper [71] claims that any quotient of a subspace of $C_p \oplus_p R_p$ cb-embeds in $L_p(\mathcal{A})$ for some sufficiently large von Neumann algebra $\mathcal{A}$ whenever $1 \leq p < 2$. In particular, if $1 \leq p < q \leq 2$, both R_q and C_q embed completely isomorphically in $L_p(\mathcal{A})$ since both are in $\mathcal{QS}(C_p \oplus_p R_p)$. The last assertion follows as in [24, Lemma 2.1]. More precisely, Xu's construction holds either with $\mathcal{A}$ being the Araki-Woods quasi-free CAR factor and also with Shlyakhtenko's generalization of it in the free setting [59]. In any case, $\mathcal{A}$ can be chosen to be a QWEP type III$_\lambda$ factor, $0 < \lambda \leq 1$.

Our first embedding result generalizes Xu's embedding.

THEOREM 8.1. *If for some* $1 \leq p \leq 2$

$$(X_1, X_2) \in \mathcal{QS}(C_p \oplus_2 \mathrm{OH}) \times \mathcal{QS}(R_p \oplus_2 \mathrm{OH}),$$

there exist a cb-embedding $X_1 \otimes_h X_2 \to L_p(\mathcal{A})$, *for some* QWEP *algebra* $\mathcal{A}$.

SKETCH OF THE PROOF. The argument is very close to that of [24]. By injectivity of the Haagerup tensor product, we may assume that X_1 is a quotient of $C_p \oplus_2 \mathrm{OH}$ and X_2 is a quotient of $R_p \oplus_2 \mathrm{OH}$. Therefore, the cb-isomorphisms below follow from [24, Lemma 2.5] by duality

$$X_1 \simeq_{cb} \mathcal{H}_{11,c_p} \oplus_2 \mathcal{H}_{12,oh} \oplus_2 \Big((\mathcal{K}_{11,c_p} \oplus_2 \mathcal{K}_{12,oh}) / graph(\Lambda_1)^\perp \Big),$$

$$X_2 \simeq_{cb} \mathcal{H}_{21,r_p} \oplus_2 \mathcal{H}_{22,oh} \oplus_2 \Big((\mathcal{K}_{21,r_p} \oplus_2 \mathcal{K}_{22,oh}) / graph(\Lambda_2)^\perp \Big),$$

for certain closed subspaces $\mathcal{H}_{ij}, \mathcal{K}_{ij}$ ($1 \leq i,j \leq 2$) of ℓ_2 and

$$\begin{aligned} \Lambda_1 &: \mathcal{K}_{11,c_{p'}} \to \mathcal{K}_{12,oh}, \\ \Lambda_2 &: \mathcal{K}_{21,r_{p'}} \to \mathcal{K}_{22,oh}, \end{aligned}$$

injective closed densely-defined operators with dense range. Let us set

$$\begin{aligned} \mathcal{Z}_1 &= \left(\mathcal{K}_{11,c_p} \oplus_2 \mathcal{K}_{12,oh}\right)/graph(\Lambda_1)^\perp, \\ \mathcal{Z}_2 &= \left(\mathcal{K}_{21,r_p} \oplus_2 \mathcal{K}_{22,oh}\right)/graph(\Lambda_2)^\perp. \end{aligned}$$

Then, we have the following cb-isometric inclusion

$$\begin{aligned} (8.1) \quad \mathrm{X}_1 \otimes_h \mathrm{X}_2 \subset\ & \mathcal{Z}_1 \otimes_h \mathcal{Z}_2 \\ &\oplus_2\ \mathcal{H}_{11,c_p} \otimes_h \mathrm{X}_2 \\ &\oplus_2\ \mathrm{X}_1 \otimes_h \mathcal{H}_{21,r_p} \\ &\oplus_2\ \mathcal{H}_{12,oh} \otimes_h \mathcal{Z}_2 \\ &\oplus_2\ \mathcal{Z}_1 \otimes_h \mathcal{H}_{22,oh} \\ &\oplus_2\ \mathcal{H}_{12,oh} \otimes_h \mathcal{H}_{22,oh}. \end{aligned}$$

According to [71], we know that OH $\in \mathcal{QS}(C_p \oplus_p R_p)$ and that any element in $\mathcal{QS}(C_p \oplus_p R_p)$ completely embeds in $L_p(\mathcal{A})$ for some QWEP type III factor $\mathcal{A}$. This eliminates the last term in (8.1). The second and third terms embed into $S_p(\mathrm{X}_2)$ and $S_p(\mathrm{X}_1)$ completely isometrically. On the other hand, since OH $\in \mathcal{QS}(C_p \oplus_p R_p)$ and we have by hypothesis

$$\mathrm{X}_1 \in \mathcal{QS}(C_p \oplus_2 \mathrm{OH}) \quad \text{and} \quad \mathrm{X}_2 \in \mathcal{QS}(R_p \oplus_2 \mathrm{OH}),$$

both X_1 and X_2 are cb-isomorphic to an element in $\mathcal{QS}(C_p \oplus_p R_p)$. Applying Xu's theorem [71] one more time, we may eliminate these terms. Finally, for the fourth and fifth terms on the right of (8.1), we apply [24, Lemma 2.6] and the self-duality of OH to rewrite them as particular cases of the first term $\mathcal{Z}_1 \otimes_h \mathcal{Z}_2$. This reduces the proof to construct a complete embedding of $\mathcal{Z}_1 \otimes_h \mathcal{Z}_2$ into $L_p(\mathcal{A})$.

By [24, Lemma 2.8] we may assume that the graphs appearing in the terms $\mathcal{Z}_1$ and $\mathcal{Z}_2$ are graphs of diagonal operators d_{λ_1} and d_{λ_2}. Moreover, exactly as in the proof of [24, Theorem D] we may use polar decomposition, perturbation and complementation and assume that $\mathcal{Z}_1$ and $\mathcal{Z}_2$ are given by

$$\begin{aligned} \mathcal{Z}_1 &= C_p + \ell_2^{oh}(\lambda) = \left(C_p \oplus_2 \mathrm{OH}\right)/\left(C_{p'} \cap \ell_2^{oh}(\lambda)\right)^\perp, \\ \mathcal{Z}_2 &= R_p + \ell_2^{oh}(\lambda) = \left(R_p \oplus_2 \mathrm{OH}\right)/\left(R_{p'} \cap \ell_2^{oh}(\lambda)\right)^\perp, \end{aligned}$$

with $\lambda_1, \lambda_2, \ldots \in \mathbb{R}_+$ strictly positive and

$$\begin{aligned} C_{p'} \cap \ell_2^{oh}(\lambda) &= \mathrm{span}\left\{(\delta_k, \lambda_k \delta_k) \in C_{p'} \oplus_2 \mathrm{OH}\right\}, \\ R_{p'} \cap \ell_2^{oh}(\lambda) &= \mathrm{span}\left\{(\delta_k, \lambda_k \delta_k) \in R_{p'} \oplus_2 \mathrm{OH}\right\}. \end{aligned}$$

Before going on we adapt some terminology from [24]. Given a sequence $\gamma_1, \gamma_2, \ldots$ of strictly positive numbers, the diagonal map $\mathsf{d}_\gamma = \sum_k \gamma_k e_{kk}$ is regarded as the density of a n.s.s.f. weight ψ on $\mathcal{B}(\ell_2)$. We also keep from [24] the same terminology for its restriction ψ_n to the subalgebra $q_n \mathcal{B}(\ell_2) q_n$ (where q_n denotes the projection $\sum_{k=1}^n e_{kk}$) and for the state φ_n determined by the relation $\psi_n = \mathrm{k}_n \varphi_n$ with k_n given by $\sum_{k=1}^n \gamma_k$. If $1 \le p \le 2$, we define

$$\mathcal{J}_{p',2}(\psi_n) = \left\{ \left(d_{\psi_n}^{\frac{1}{2p'}} z d_{\psi_n}^{\frac{1}{2p'}}, d_{\psi_n}^{\frac{1}{2p'}} z d_{\psi_n}^{\frac{1}{4}}, d_{\psi_n}^{\frac{1}{4}} z d_{\psi_n}^{\frac{1}{2p'}}, d_{\psi_n}^{\frac{1}{4}} z d_{\psi_n}^{\frac{1}{4}} \right) \mid z \in q_n \mathcal{B}(\ell_2) q_n \right\}$$

as a subspace of the direct sum

$$\mathcal{L}_{p'}^n = \left(C_{p'}^n \otimes_h R_{p'}^n\right) \oplus_2 \left(C_{p'}^n \otimes_h \mathrm{OH}_n\right) \oplus_2 \left(\mathrm{OH}_n \otimes_h R_{p'}^n\right) \oplus_2 \left(\mathrm{OH}_n \otimes_h \mathrm{OH}_n\right).$$

In other words, we may regard $\mathcal{J}_{p',2}(\psi_n)$ as an intersection of some weighted forms of the asymmetric Schatten classes (see Section 7.2) considered above. On the other hand, by a simple perturbation argument, we may assume as in [24] that $\mathrm{k}_n = \sum_{k=1}^n \gamma_k$ is an integer. Then, if we set $\mathcal{M}_n = q_n \mathcal{B}(\ell_2) q_n$, we easily find the complete isometry

$$\mathcal{J}_{p',2}(\psi_n) = \mathcal{J}_{p',2}^{\mathrm{k}_n}(\mathcal{M}_n),$$

see Section 7.2 again. Hence, according to Theorem 7.8, we obtain a complete embedding of $\mathcal{J}_{p',2}(\psi_n)$ into a cb-complemented subspace of $L_{p'}(\mathcal{A}_n; \mathrm{OH}_{\mathrm{k}_n})$, where $\mathcal{A}_n$ stands for the k_n-fold reduced free product of $q_n \mathcal{B}(\ell_2) q_n \oplus q_n \mathcal{B}(\ell_2) q_n$. Let us now consider the dual space

$$\mathcal{K}_{p,2}(\psi_n) = \mathcal{J}_{p',2}(\psi_n)^*.$$

A simple duality argument provides a cb-embedding

(8.2) $$\omega : x \in \mathcal{K}_{p,2}(\psi_n) \mapsto \frac{1}{\mathrm{k}_n} \sum_{j=1}^{\mathrm{k}_n} x_j \otimes \delta_j \in L_p(\mathcal{A}_n; \mathrm{OH}_{\mathrm{k}_n}),$$

with cb-complemented image and constants independent of n.

Let us now go back to our study of the Haagerup tensor product $\mathcal{Z}_1 \otimes_h \mathcal{Z}_2$. Let us recall that the finite weights ψ_n are restrictions of the n.s.s.f. weight ψ to the subalgebras $q_n \mathcal{B}(\ell_2) q_n$, which are directed by inclusion. In particular, we may consider the direct limit

$$\mathcal{K}_{p,2}(\psi) = \overline{\bigcup_{n \ge 1} \mathcal{K}_{p,2}(\psi_n)}.$$

A fairly simple adaptation of [24, Lemma 2.4] to the L_p case gives

(8.3) $$\mathcal{Z}_1 \otimes_h \mathcal{Z}_2 = \left(C_p + \ell_2^{oh}(\lambda)\right) \otimes_h \left(R_p + \ell_2^{oh}(\lambda)\right) = \mathcal{K}_{p,2}(\psi) \to \prod\nolimits_{n,\mathcal{U}} \mathcal{K}_{p,2}(\psi_n).$$

The last step being the natural embedding of a direct limit into the corresponding ultraproduct. According to [55], this reduces the problem to the finite-dimensional case, which follows from Xu's cb-embedding [71] of OH into $L_p(\mathcal{B})$ since

$$\mathcal{K}_{p,2}(\psi_n) \to L_p(\mathcal{A}_n; \mathrm{OH}_{\mathrm{k}_n}) \to L_p(\mathcal{A}_n \bar\otimes \mathcal{B}) = L_p(\mathcal{A}'_n),$$

see [24] for a rigorous explanation. We have therefore constructed a cb-embedding

$$\mathcal{Z}_1 \otimes_h \mathcal{Z}_2 \to L_p(\mathcal{A}) \quad \text{with} \quad \mathcal{A} = \left(\prod\nolimits_{n,\mathcal{U}} \mathcal{A}'_{n*}\right)^*.$$

The fact that $\mathcal{A}$ is QWEP is justified as in the proof of [24, Theorem D]. $\square$

COROLLARY 8.2. *If $1 \le p < q \le 2$, the Schatten class S_q cb-embeds into $L_p(\mathcal{A})$.*

PROOF. Since $S_q = C_q \otimes_h R_q$, combine [24, Lemma 2.1] and Theorem 8.1. $\square$

8.2. Embedding into the hyperfinite factor

Now we want to show that the cb-embedding $S_q \to L_p(\mathcal{A})$ can be constructed with $\mathcal{A}$ being a hyperfinite type III factor. Moreover, we shall prove some more general results to be used in the next paragraph, where the general cb-embedding will be constructed. We first set a transference argument, based on a Rosenthal type inequality for noncommuting identically distributed random variables in L_1 from [18], which enables us to replace freeness by some sort of independence.

Let $\mathcal{N}$ be a σ-finite von Neumann subalgebra of some algebra $\mathcal{A}$ and let us consider a family $\mathsf{A}_1, \mathsf{A}_2, \ldots$ of von Neumann algebras with $\mathcal{N} \subset \mathsf{A}_k \subset \mathcal{A}$. As usual we require the existence of a n.f. conditional expectation $\mathsf{E}_\mathcal{N} : \mathcal{A} \to \mathcal{N}$. We recall that $(\mathsf{A}_k)_{k\geq 1}$ is a system of *indiscernible independent copies over* $\mathcal{N}$ (*i.i.c.* in short) when

i) If $a \in \langle \mathsf{A}_1, \mathsf{A}_2, \ldots, \mathsf{A}_{k-1}\rangle$ and $b \in \mathsf{A}_k$, we have
$$\mathsf{E}_\mathcal{N}(ab) = \mathsf{E}_\mathcal{N}(a)\mathsf{E}_\mathcal{N}(b).$$

ii) There exist a von Neumann algebra A containing $\mathcal{N}$, a normal faithful conditional expectation $\mathsf{E}_0 : \mathsf{A} \to \mathcal{N}$ and homomorphisms $\pi_k : \mathsf{A} \to \mathsf{A}_k$ such that
$$\mathsf{E}_\mathcal{N} \circ \pi_k = \mathsf{E}_0$$
and the following holds for every strictly increasing function $\alpha : \mathbb{N} \to \mathbb{N}$
$$\mathsf{E}_\mathcal{N}\big(\pi_{j_1}(a_1)\cdots\pi_{j_m}(a_m)\big) = \mathsf{E}_\mathcal{N}\big(\pi_{\alpha(j_1)}(a_1)\cdots\pi_{\alpha(j_m)}(a_m)\big).$$

iii) There exist n.f. conditional expectations $\mathcal{E}_k : \mathcal{A} \to \mathsf{A}_k$ such that
$$\mathsf{E}_\mathcal{N} = \mathsf{E}_0 \pi_k^{-1} \mathcal{E}_k \quad \text{for all} \quad k \geq 1.$$

Further, when the first condition above also holds for a in the algebra generated by $\mathsf{A}_1, \ldots, \mathsf{A}_{k-1}, \mathsf{A}_{k+1}, \ldots$ and the second condition holds for any permutation α of the integers, we shall say that $(\mathsf{A}_k)_{k\geq 1}$ are *symmetrically independent copies* (*s.i.c.* in short) *over* $\mathcal{N}$. In what follows, given a probability space (Ω, μ), we shall write $\varepsilon_1, \varepsilon_2, \ldots$ to denote an independent family of Bernoulli random variables on Ω equidistributed on ± 1. We now present the key inequality in [18].

LEMMA 8.3. *The following inequalities hold for $x \in L_1(\mathsf{A})$*

a) *If $(\mathsf{A}_k)_{k\geq 1}$ are i.i.c. over $\mathcal{N}$, we have*
$$\int_\Omega \Big\|\sum_{k=1}^n \varepsilon_k \pi_k(x)\Big\|_{L_1(\mathcal{A})} d\mu$$
$$\sim \inf_{x=a+b+c} n\|a\|_{L_1(\mathsf{A})} + \sqrt{n}\big\|\mathsf{E}_0(bb^*)^{\frac{1}{2}}\big\|_{L_1(\mathcal{N})} + \sqrt{n}\big\|\mathsf{E}_0(c^*c)^{\frac{1}{2}}\big\|_{L_1(\mathcal{N})}.$$

b) *If moreover, $\mathsf{E}_0(x) = 0$ and $(\mathsf{A}_k)_{k\geq 1}$ are s.i.c. over $\mathcal{N}$, then*
$$\Big\|\sum_{k=1}^n \pi_k(x)\Big\|_{L_1(\mathcal{A})}$$
$$\sim \inf_{x=a+b+c} n\|a\|_{L_1(\mathsf{A})} + \sqrt{n}\big\|\mathsf{E}_0(bb^*)^{\frac{1}{2}}\big\|_{L_1(\mathcal{N})} + \sqrt{n}\big\|\mathsf{E}_0(c^*c)^{\frac{1}{2}}\big\|_{L_1(\mathcal{N})}.$$

PROOF. We claim that
$$\frac{1}{2}\Big\|\sum_{k=1}^n \pi_k(x)\Big\|_{L_1(\mathcal{A})} \leq \Big\|\sum_{k=1}^n \varepsilon_k \pi_k(x)\Big\|_{L_1(\mathcal{A})} \leq 2\Big\|\sum_{k=1}^n \pi_k(x)\Big\|_{L_1(\mathcal{A})}.$$

for any choice of signs $\varepsilon_1, \varepsilon_2, \ldots, \varepsilon_n$ whenever $\mathsf{A}_1, \mathsf{A}_2, \ldots$ are symmetric independent copies of A over $\mathcal{N}$ and $\mathsf{E}_0(x) = 0$. This establishes (a) $\Rightarrow$ (b) and so, since the first assertion is proved in [**18**], it suffices to prove our claim. Such result will follow from the more general statement

$$\Big\| \sum_{k=1}^n \varepsilon_k \pi_k(x_k) \Big\|_{L_1(\mathcal{A})} \leq 2 \Big\| \sum_{k=1}^n \pi_k(x_k) \Big\|_{L_1(\mathcal{A})},$$

for any family $x_1, x_2, \ldots, x_n$ in $L_1(\mathsf{A})$ with $\mathsf{E}_0(x_k) = 0$ for $1 \leq k \leq n$. Indeed, take $x_k = x$ for the upper estimate and $x_k = \varepsilon_k x$ for the lower estimate. Since we assume that $\mathcal{N}$ is σ-finite we may fix a n.f. state φ. We define $\phi = \varphi \circ \mathsf{E}_{\mathcal{N}}$ and $\phi_0 = \varphi \circ \mathsf{E}_0$. According to [**6**] we have

$$\sigma_t^\varphi \circ \mathsf{E}_0 = \mathsf{E}_0 \circ \sigma_t^{\phi_0} \quad \text{and} \quad \sigma_t^\varphi \circ \mathsf{E}_{\mathcal{N}} = \mathsf{E}_{\mathcal{N}} \circ \sigma_t^\phi.$$

Moreover, since $\mathsf{E}_{\mathcal{N}} = \mathsf{E}_{\mathcal{N}} \circ \mathcal{E}_k$ we find $\phi = \phi \circ \mathcal{E}_k$ which implies

$$\sigma_t^\phi \circ \mathcal{E}_k = \mathcal{E}_k \circ \sigma_t^\phi.$$

In particular, $\sigma_t^\phi(\mathsf{A}_k) \subset \mathsf{A}_k$ for $k \geq 1$. Therefore, given any subset S of $\{1, 2, \ldots, n\}$ we find a ϕ-invariant conditional expectation $\mathsf{E}_\mathsf{S} : \mathcal{A} \to \mathcal{A}_\mathsf{S}$ where the von Neumann algebra $\mathcal{A}_\mathsf{S} = \langle \mathsf{A}_k \mid k \in \mathsf{S} \rangle$. We claim that

$$\mathsf{E}_\mathsf{S}(\pi_j(a)) = 0 \quad \text{whenever} \quad \mathsf{E}_0(a) = 0 \quad \text{and} \quad j \notin \mathsf{S}.$$

Indeed, let $b \in \mathcal{A}_\mathsf{S}$ and a as above. Then we deduce from symmetric independence

$$\phi\big(\mathsf{E}_\mathsf{S}(\pi_j(a))b\big) = \phi\big(\pi_j(a)b\big) = \phi\big(\mathsf{E}_{\mathcal{N}}(\pi_j(a)b)\big) = \phi\big(\mathsf{E}_0(a)\mathsf{E}_{\mathcal{N}}(b)\big) = 0.$$

Thus we may apply Doob's trick

$$\Big\| \sum_{k \in \mathsf{S}} \pi_k(x_k) \Big\|_{L_1(\mathcal{A})} = \Big\| \mathsf{E}_\mathsf{S}\Big(\sum_{k=1}^n \pi_k(x_k) \Big) \Big\|_{L_1(\mathcal{A})} \leq \Big\| \sum_{k=1}^n \pi_k(x_k) \Big\|_{L_1(\mathcal{A})}.$$

Then, the claim follows taking $\{1, 2, \ldots, n\} = \mathsf{S}_1 \cup \mathsf{S}_{-1}$ with $\mathsf{S}_\alpha = \{k : \varepsilon_k = \alpha\}$. $\square$

REMARK 8.4. The inequalities above generalize the noncommutative Rosenthal inequality [**30**] to the case $p = 1$ for identically distributed variables and under such notions of noncommutative independence. Of course, in the case $1 < p < 2$ we have much stronger results from [**28, 30**] and there is no need of proving any preliminary result for our aims in this case.

Let us now generalize our previous definition of the space $\mathcal{K}_{p,2}(\psi)$ to general von Neumann algebras. Let $\mathcal{M}$ be a given von Neumann algebra, which we assume σ-finite for the sake of clarity. Let $\mathcal{M}$ be equipped with a n.s.s.f. weight ψ. In other words, ψ is given by an increasing sequence (a net in the general case) of pairs (ψ_n, q_n) such that the q_n's are increasing finite projections in $\mathcal{M}$ with $\lim_n q_n = 1$ in the strong operator topology and $\sigma_t^\psi(q_n) = q_n$. Moreover, the ψ_n's are normal positive functionals on $\mathcal{M}$ with support q_n and satisfying the compatibility condition $\psi_{n+1}(q_n x q_n) = \psi_n(x)$. As above, we shall write k_n for the number $\psi_n(q_n) \in (0, \infty)$ and (again as above) we may and will assume that the k_n's are nondecreasing positive integers. In what follows we shall write d_{ψ_n} for the density on $q_n \mathcal{M} q_n$ associated to the n.f. finite weight ψ_n. If $1 \leq p \leq 2$, we define the space $\mathcal{J}_{p',2}(\psi_n)$ as the closure of

$$\Big\{ \big(d_{\psi_n}^{\frac{1}{2p'}} z d_{\psi_n}^{\frac{1}{2p'}}, d_{\psi_n}^{\frac{1}{2p'}} z d_{\psi_n}^{\frac{1}{4}}, d_{\psi_n}^{\frac{1}{4}} z d_{\psi_n}^{\frac{1}{2p'}}, d_{\psi_n}^{\frac{1}{4}} z d_{\psi_n}^{\frac{1}{4}} \big) \mid z \in q_n \mathcal{M} q_n \Big\}$$

in the direct sum
$$\mathcal{L}^n_{p'} = L_{p'}(q_n\mathcal{M}q_n) \oplus_2 L_{(2p',4)}(q_n\mathcal{M}q_n) \oplus_2 L_{(4,2p')}(q_n\mathcal{M}q_n) \oplus_2 L_2(q_n\mathcal{M}q_n).$$
In other words, after considering the n.f. state φ_n on $q_n\mathcal{M}q_n$ determined by the relation $\psi_n = k_n\varphi_n$ and recalling the definition of the spaces $\mathcal{J}^n_{p,q}(\mathcal{M})$ from Section 7.2, we may regard $\mathcal{J}_{p',2}(\psi_n)$ as the 4-term intersection space
$$\mathcal{J}_{p',2}(\psi_n) = \bigcap_{u,v \in \{2p',4\}} k_n^{\frac{1}{u}+\frac{1}{v}} L_{(u,v)}(q_n\mathcal{M}q_n) = \mathcal{J}^{k_n}_{p',2}(q_n\mathcal{M}q_n).$$
Now we take direct limits and define
$$\mathcal{J}_{p',2}(\psi) = \overline{\bigcup_{n\geq 1} \mathcal{J}_{p',2}(\psi_n)},$$
where the closure is taken with respect to the norm of the space
$$\mathcal{L}_{p'} = L_{p'}(\mathcal{M}) \oplus_2 L_{(2p',4)}(\mathcal{M}) \oplus_2 L_{(4,2p')}(\mathcal{M}) \oplus_2 L_2(\mathcal{M}).$$
To define the space $\mathcal{K}_{p,2}(\psi)$ we also proceed as above and consider
$$\Psi_n : \mathcal{L}^n_p \to L_1(q_n\mathcal{M}q_n)$$
given by
$$\Psi_n(x_1,x_2,x_3,x_4) = d_{\psi_n}^{\frac{1}{2p'}} x_1 d_{\psi_n}^{\frac{1}{2p'}} + d_{\psi_n}^{\frac{1}{2p'}} x_2 d_{\psi_n}^{\frac{1}{4}} + d_{\psi_n}^{\frac{1}{4}} x_3 d_{\psi_n}^{\frac{1}{2p'}} + d_{\psi_n}^{\frac{1}{4}} x_4 d_{\psi_n}^{\frac{1}{4}}.$$
This gives $\ker \Psi_n = \mathcal{J}_{p',2}(\psi_n)^\perp$ and we define
$$\mathcal{K}_{p,2}(\psi_n) = \mathcal{L}^n_p/\ker \Psi_n \quad \text{and} \quad \mathcal{K}_{p,2}(\psi) = \overline{\bigcup_{n\geq 1} \mathcal{K}_{p,2}(\psi_n)},$$
where the latter is understood as a quotient of $\mathcal{L}_p$. In other words, we may regard the space $\mathcal{K}_{p,2}(\psi_n)$ as the sum of the corresponding dual weighted asymmetric L_p spaces considered in the definition of $\mathcal{J}_{p',2}(\psi_n)$
$$(8.4) \qquad \mathcal{K}_{p,2}(\psi_n) = \sum_{u,v \in \{2p,4\}} k_n^{\frac{1}{u}+\frac{1}{v}} L_{(u,v)}(q_n\mathcal{M}q_n).$$
Thus, using $\psi_n = k_n\varphi_n$ backwards and taking direct limits
$$\mathcal{K}_{p,2}(\psi) = L_p(\mathcal{M}) + L_{(2p,4)}(\mathcal{M}) + L_{(4,2p)}(\mathcal{M}) + L_2(\mathcal{M}),$$
where the sum is taken in $L_p(\mathcal{M})$ and the embeddings are given by
$$j_c(x) : x \in L_{(2p,4)}(\mathcal{M}) \mapsto x d_\psi^\beta \in L_p(\mathcal{M}),$$
$$j_r(x) : x \in L_{(4,2p)}(\mathcal{M}) \mapsto d_\psi^\beta x \in L_p(\mathcal{M}),$$
with $\beta = 1/2p - 1/4$, while the embedding of $L_2(\mathcal{M})$ into $L_p(\mathcal{M})$ is given by
$$j_2(x) = d_\psi^\beta x d_\psi^\beta.$$

REMARK 8.5. It will be important below to observe that our definition of $\mathcal{K}_{p,2}(\psi_n)$ is slightly different to the one given in the previous paragraph. Indeed, according to the usual duality bracket $\langle x,y\rangle = \text{tr}(x^*y)$, we should have defined
$$\mathcal{K}_{p,2}(\psi_n) = \sum_{u,v \in \{2p,4\}} k_n^{-\gamma(u,v)} L_{(u,v)}(q_n\mathcal{M}q_n)$$

with
$$\gamma(u,v) = \frac{1}{2(u/2)'} + \frac{1}{2(v/2)'}.$$
This would give $\mathcal{K}_{p,2}(\psi_n) = \mathcal{J}_{p',2}(\psi_n)^*$ and
$$\mathcal{K}_{p,2}(\psi_n) = \frac{1}{\mathsf{k}_n} \sum_{u,v \in \{2p,4\}} \mathsf{k}_n^{\frac{1}{u}+\frac{1}{v}} L_{(u,v)}(q_n \mathcal{M} q_n).$$
However, we prefer to use (8.4) in what follows for notational convenience.

Now we set some notation to distinguish between independent and free random variables. If we fix a positive integer n, the von Neumann algebra $\mathcal{A}^n_{ind}$ will denote the k_n-fold tensor product of $q_n \mathcal{M} q_n$ while $\mathcal{A}^n_{free}$ will be the k_n-fold free product of $q_n \mathcal{M} q_n \oplus q_n \mathcal{M} q_n$. In other words, if we set
$$\widetilde{\mathsf{A}}_{n,j} = q_n \mathcal{M} q_n \quad \text{and} \quad \mathsf{A}_{n,j} = q_n \mathcal{M} q_n \oplus q_n \mathcal{M} q_n$$
for $1 \le j \le \mathsf{k}_n$, we define the following von Neumann algebras
$$\begin{aligned}\mathcal{A}^n_{ind} &= \otimes_j \widetilde{\mathsf{A}}_{n,j}, \\ \mathcal{A}^n_{free} &= *_j \mathsf{A}_{n,j}.\end{aligned}$$
We also consider the natural embeddings
$$\pi^j_{ind} : \widetilde{\mathsf{A}}_{n,j} \to \mathcal{A}^n_{ind} \quad \text{and} \quad \pi^j_{free} : \mathsf{A}_{n,j} \to \mathcal{A}^n_{free}$$
into the j-th components of the algebras $\mathcal{A}^n_{ind}$ and $\mathcal{A}^n_{free}$ respectively.

PROPOSITION 8.6. *If $1 \le p \le 2$, the map*
$$\xi^n_{ind} : x \in \mathcal{K}_{p,2}(\psi_n) \mapsto \sum_{j=1}^{\mathsf{k}_n} \pi^j_{ind}(x) \otimes \delta_j \in L_p(\mathcal{A}^n_{ind}; \mathrm{OH}_{\mathsf{k}_n})$$
is a completely isomorphic embedding with relevant constants independent of n.

Before proceeding with the proof of Proposition 8.6, we need a more in depth discussion on the cb-embedding of OH. Given $1 < p < 2$, Xu constructed in [**71**] a complete embedding of OH into $L_p(\mathcal{A})$ with $\mathcal{A}$ hyperfinite, while for $p=1$ the corresponding cb-embedding was recently constructed in [**18**]. The argument can be sketched with the following chain
$$\mathrm{OH} \hookrightarrow (C_p \oplus_p R_p)/graph(\mathsf{d}_\lambda)^\perp \simeq_{cb} C_p + R_p(\lambda) \hookrightarrow L_p(\mathcal{A}).$$
Indeed, arguing as in [**24**, Lemma 2.1/Remark 2.2] and applying [**24**, Lemma 2.8], we see how to regard OH as a subspace of a quotient of $C_p \oplus_p R_p$ by the annihilator of some diagonal map $\mathsf{d}_\lambda : C_{p'} \to R_{p'}$. By the action of d_λ, the annihilator of its graph is the span of elements of the form $(\delta_k, -\delta_k/\lambda_k)$. This suggest to regard the quotient above as the sum of C_p with a weighted form of R_p. This establishes the cb-isomorphism in the middle. Then, it is natural to guess that the complete embedding into $L_p(\mathcal{A})$ should follow from a *weighted* form of the noncommutative Khintchine inequality. The first inequality of this kind was given by Pisier and Shlyakhtenko in [**50**] for generalized circular variables and further investigated in [**26, 70**]. However, if we want to end up with a hyperfinite von Neumann algebra $\mathcal{A}$, we must replace generalized circulars by their Fermionic analogues. More precisely, given a complex Hilbert space $\mathcal{H}$, we consider its antisymmetric Fock space $\mathcal{F}_{-1}(\mathcal{H})$. Let $c(e)$ and $a(e)$ denote the creation and annihilation operators associated with

a vector $e \in \mathcal{H}$. Given an orthonormal basis $(e_{\pm k})_{k \geq 1}$ of $\mathcal{H}$ and a family $(\mu_k)_{k \geq 1}$ of positive numbers, we set $f_k = c(e_k) + \mu_k \, a(e_{-k})$. The sequence $(f_k)_{k \geq 1}$ satisfies the canonical anticommutation relations and we take $\mathcal{A}$ to be the von Neumann algebra generated by the f_k's. Taking suitable μ_k's depending only on p and the eigenvalues of d_λ, the Khintchine inequality associated to the system of f_k's provides the desired cb-embedding. Namely, let ϕ be the quasi-free state on $\mathcal{A}$ determined by the vacuum and let d_ϕ be the associated density. Then, if $(\delta_k)_{k \geq 1}$ denotes the unit vector basis of OH, the cb-embedding has the form

$$w(\delta_k) = \xi_k \, d_\phi^{\frac{1}{2p}} f_k \, d_\phi^{\frac{1}{2p}} = \xi_k \, f_{p,k}$$

for some scaling factors $(\xi_k)_{k \geq 1}$. The necessary Khintchine type inequalities for $1 < p < 2$ follow from the noncommutative Burkholder inequality [**28**]. In the L_1 case, the key inequalities follow from Lemma 8.3, see [**18**] for details. With this construction, the von Neumann algebra $\mathcal{A}$ turns out to be the Araki-Woods factor arising from the GNS construction applied to the CAR algebra with respect to the quasi-free state ϕ. In fact, using a conditional expectation, we can replace the μ_k's by a sequence $(\mu'_k)_{k \geq 1}$ such that for every rational $0 < \lambda < 1$ there are infinitely many μ'_k's with $\mu'_k = \lambda/(1 + \lambda)$. According to the results in [**1**], we then obtain the hyperfinite type III$_1$ factor $\mathcal{R}$.

On the other hand, there exists a slight modification of this construction which will be used below. Indeed, using the terminology introduced above and following [**18**] there exists a mean-zero $\gamma_p \in L_p(\mathcal{R})$ given by a linear combination of the $f_{p,k}$'s such that

$$w(\delta_j) = \pi_{ind}^j(\gamma_p)$$

defines a completely isomorphic embedding

$$w : \mathrm{OH}_{\mathrm{k}_n} \to L_p(\mathcal{R}_{\otimes \mathrm{k}_n})$$

with constants independent of n. Here $\mathcal{R}_{\otimes \mathrm{k}_n}$ denotes the k_n-fold tensor product of $\mathcal{R}$. Moreover, given any von Neumann algebra $\mathcal{A}$, $id_{L_p(\mathcal{A})} \otimes w$ also defines an isomorphism

$$id_{L_p(\mathcal{A})} \otimes w : L_p(\mathcal{A}; \mathrm{OH}_{\mathrm{k}_n}) \to L_p(\mathcal{A} \bar{\otimes} \mathcal{R}_{\otimes \mathrm{k}_n}).$$

We refer to [**18, 71**] for a more detailed exposition on the cb-embedding of OH.

PROOF OF PROPOSITION 8.6. By Theorem 7.8, this is true for

$$\xi_{free}^n : x \in \mathcal{K}_{p,2}(\psi_n) \mapsto \sum_{j=1}^{\mathrm{k}_n} \pi_{free}^j(x, -x) \otimes \delta_j \in L_p\big(\mathcal{A}_{free}^n; \mathrm{OH}_{\mathrm{k}_n}\big).$$

Indeed, it follows from a simple duality argument (see e.g. Remark 7.4) taking Remark 8.5 into account. According to the preceding discussion on OH, we deduce that

$$w \circ \xi_{free}^n : x \in \mathcal{K}_{p,2}(\psi_n) \mapsto \sum_{j=1}^{n} \pi_{free}^j(x, -x) \otimes \pi_{ind}^j(\gamma_p) \in L_p\big(\mathcal{A}_{free}^n \bar{\otimes} \mathcal{R}_{\otimes \mathrm{k}_n}\big)$$

also provides a cb-isomorphism. Now, we consider

$$\begin{aligned} \widetilde{\mathsf{B}}_{n,j} &= \pi_{ind}^j(\widetilde{\mathsf{A}}_{n,j}) \otimes \pi_{ind}^j(\mathcal{R}), \\ \mathsf{B}_{n,j} &= \pi_{free}^j(\mathsf{A}_{n,j}) \otimes \pi_{ind}^j(\mathcal{R}), \end{aligned}$$

for $1 \leq j \leq k_n$. It is clear from the construction that both families of von Neumann algebras are s.i.c. over the complex field. Therefore, Lemma 8.3/Remark 8.4 apply in both cases (note that the mean-zero condition for the $\widetilde{\mathsf{B}}_{n,j}$'s holds due to the fact that γ_p is mean-zero) and hence

$$\left\| id_{S_p^m} \otimes w \circ \xi_{free}^n(x) \right\|_p \sim_c \left\| id_{S_p^m} \otimes w \circ \xi_{ind}^n(x) \right\|_p$$

holds for every element $x \in S_p^m(\mathcal{K}_{p,2}(\psi_n))$. In particular, we obtain

$$\|x\|_{S_p^m(\mathcal{K}_{p,2}(\psi_n))} \sim \left\| id_{S_p^m} \otimes w \circ \xi_{free}^n(x) \right\|_p \sim \left\| id_{S_p^m} \otimes w \circ \xi_{ind}^n(x) \right\|_p \sim \left\| id_{S_p^m} \otimes \xi_{ind}^n(x) \right\|_p$$

since w and ξ_{free}^n are cb-isomorphic embeddings. This completes the proof. $\square$

REMARK 8.7. Proposition 8.6 extends (8.2) to general von Neumann algebras, where freeness is replaced by noncommutative independence. The only difference between both results is the factor $1/k_n$, which follows from Remark 8.5.

REMARK 8.8. The transference argument applied in the proof of Proposition 8.6 gives a result which might be of independent interest. Given a von Neumann algebra A, let us construct the tensor product $\mathcal{A}_{ind}$ of infinitely many copies of $\mathsf{A} \oplus \mathsf{A}$. Similarly, the free product $\mathcal{A}_{free}$ of infinitely many copies of $\mathsf{A} \oplus \mathsf{A}$ will be considered. Following our terminology, we have maps

$$\pi_{ind}^j : \mathsf{A} \to \mathcal{A}_{ind} \quad \text{and} \quad \pi_{free}^j : \mathsf{A} \to \mathcal{A}_{free}.$$

If $1 < p < q < 2$, we claim that

$$(8.5) \quad \left\| \sum_j \pi_{ind}^j(x, -x) \otimes \delta_j \right\|_{L_p(\mathcal{A}_{ind}; \ell_q)} \sim_{cb} \left\| \sum_j \pi_{free}^j(x, -x) \otimes \delta_j \right\|_{L_p(\mathcal{A}_{free}; \ell_q)},$$

where the symbol $\sim_{cb}$ is used to mean that the equivalence also holds (with absolute constants) when taking the matrix norms arising from the natural operator space structures of the spaces considered. The case $q = 2$ follows by using exactly the same argument as in Proposition 8.6. In fact, the same idea works for general indices. Indeed, we just need to embed ℓ_q into L_p completely isometrically and then use the noncommutative Rosenthal inequality [30]. Recall that the cb-embedding of ℓ_q into L_p is already known at this stage of the paper as a consequence of Corollary 8.2. At the time of this writing, it is still open whether or not (8.5) is still valid for other values of (p, q).

Our main goal in this paragraph is to generalize the complete embedding in Proposition 8.6 to the direct limit $\mathcal{K}_{p,2}(\psi)$. Of course, this is possible using an ultraproduct procedure. However, this would not preserve hyperfiniteness. We will now explain how the proof of Proposition 8.6 allows to factorize the cb-embedding $\mathcal{K}_{p,2}(\psi_n) \to L_p(\mathcal{A}_{ind}^n \bar{\otimes} \mathcal{R}_{\otimes k_n})$ via a three term K-functional. We will combine this with the concept of noncommutative Poisson random measure from [20] to produce a complete embedding which preserves the direct limit mentioned above.

Let us consider the operator space

$$\mathcal{K}_{rc_p}^p(\psi_n) = k_n^{\frac{1}{p}} L_p(q_n \mathcal{M} q_n) + k_n^{\frac{1}{2}} L_2^{r_p}(q_n \mathcal{M} q_n) + k_n^{\frac{1}{2}} L_2^{c_p}(q_n \mathcal{M} q_n),$$

where the norms in the L_p spaces considered above are calculated with respect to the state φ_n arising from the relation $\psi_n = k_n \varphi_n$. More precisely, the operator space structure is determined by

$$\|x\|_{S_p^m(\mathcal{K}_{rc_p}^p(\psi_n))} = \inf \left\{ k_n^{\frac{1}{p}} \|x_1\|_{S_p^m(L_p)} + k_n^{\frac{1}{2}} \|x_2\|_{S_p^m(L_2^{r_p})} + k_n^{\frac{1}{2}} \|x_3\|_{S_p^m(L_2^{c_p})} \right\},$$

where the infimum runs over all possible decompositions
$$x = x_1 + d_{\varphi_n}^\alpha x_2 + x_3 d_{\varphi_n}^\alpha,$$
with d_{φ_n} standing for the density associated to φ_n and $\alpha = 1/p - 1/2$. Note that $\mathcal{K}_{rc_p}^p(\psi_n)$ coincides algebraically with $L_p(q_n \mathcal{M} q_n)$. There exists a close relation between $\mathcal{K}_{rc_p}^p(\psi_n)$ and conditional L_p spaces. Indeed, let us consider the conditional expectation $\mathsf{E}_{\varphi_n} : \mathrm{M}_m(q_n \mathcal{M} q_n) \to \mathrm{M}_m$ given by
$$\mathsf{E}_{\varphi_n}\big((x_{ij})\big) = \big(\varphi_n(x_{ij})\big) = \Big(\frac{\psi_n(x_{ij})}{\mathrm{k}_n}\Big).$$

LEMMA 8.9. *We have isometries*
$$\begin{aligned}
S_p^m\big(L_2^{r_p}(q_n \mathcal{M} q_n)\big) &= m^{\frac{1}{p}} L_p^r\big(\mathrm{M}_m(q_n \mathcal{M} q_n), \mathsf{E}_{\varphi_n}\big), \\
S_p^m\big(L_2^{c_p}(q_n \mathcal{M} q_n)\big) &= m^{\frac{1}{p}} L_p^c\big(\mathrm{M}_m(q_n \mathcal{M} q_n), \mathsf{E}_{\varphi_n}\big).
\end{aligned}$$
Moreover, these isometries are densely determined by
$$\begin{aligned}
\big\|d_{\varphi_n}^{\frac{1}{2}-\frac{1}{2p}} a d_{\varphi_n}^{\frac{1}{2p}}\big\|_{S_p^m(L_2^{r_p}(q_n \mathcal{M} q_n))} &= m^{\frac{1}{p}} \big\|d_{\varphi_n}^{\frac{1}{2p}} a d_{\varphi_n}^{\frac{1}{2p}}\big\|_{L_p^r(\mathrm{M}_m(q_n \mathcal{M} q_n), \mathsf{E}_{\varphi_n})}, \\
\big\|d_{\varphi_n}^{\frac{1}{2p}} a d_{\varphi_n}^{\frac{1}{2}-\frac{1}{2p}}\big\|_{S_p^m(L_2^{c_p}(q_n \mathcal{M} q_n))} &= m^{\frac{1}{p}} \big\|d_{\varphi_n}^{\frac{1}{2p}} a d_{\varphi_n}^{\frac{1}{2p}}\big\|_{L_p^c(\mathrm{M}_m(q_n \mathcal{M} q_n), \mathsf{E}_{\varphi_n})}.
\end{aligned}$$
In particular, using the relation $d_{\varphi_n}^{\frac{1}{p}} = d_{\varphi_n}^{\frac{1}{2}} d_{\varphi_n}^\alpha$, we conclude
$$\|x\|_{S_p^m(\mathcal{K}_{rc_p}^p(\psi_n))}$$
$$= \inf_{x = x_p + x_r + x_c} \Big\{ \mathrm{k}_n^{\frac{1}{p}} \|x_p\|_p + \mathrm{k}_n^{\frac{1}{2}} \big\|\mathsf{E}_{\varphi_n}(x_r x_r^*)^{\frac{1}{2}}\big\|_{S_p^m} + \mathrm{k}_n^{\frac{1}{2}} \big\|\mathsf{E}_{\varphi_n}(x_c^* x_c)^{\frac{1}{2}}\big\|_{S_p^m} \Big\}.$$

PROOF. We have
$$\begin{aligned}
\big\|d_{\varphi_n}^{\frac{1}{2}-\frac{1}{2p}} a d_{\varphi_n}^{\frac{1}{2p}}\big\|_{S_p^m(L_2^{r_p}(q_n \mathcal{M} q_n))} &= \Big\|\mathrm{tr}_{\mathcal{M}}\big(d_{\varphi_n}^{\frac{1}{2}-\frac{1}{2p}} a d_{\varphi_n}^{\frac{1}{p}} a^* d_{\varphi_n}^{\frac{1}{2}-\frac{1}{2p}}\big)^{\frac{1}{2}}\Big\|_{S_p^m} \\
&= \Big\|(id_{\mathrm{M}_m} \otimes \varphi_n)\big(d_{\varphi_n}^{-\frac{1}{2p}} a d_{\varphi_n}^{\frac{1}{p}} a^* d_{\varphi_n}^{-\frac{1}{2p}}\big)^{\frac{1}{2}}\Big\|_{S_p^m}.
\end{aligned}$$
Then, normalizing the trace on M_m and recalling that
$$\Big\|(id_{\mathrm{M}_m} \otimes \varphi_n)\big(d_{\varphi_n}^{-\frac{1}{2p}} a d_{\varphi_n}^{\frac{1}{p}} a^* d_{\varphi_n}^{-\frac{1}{2p}}\big)^{\frac{1}{2}}\Big\|_{S_p^m} = m^{\frac{1}{p}} \Big\|\mathsf{E}_{\varphi_n}\big(d_{\varphi_n}^{\frac{1}{2p}} a d_{\varphi_n}^{\frac{1}{p}} a^* d_{\varphi_n}^{\frac{1}{2p}}\big)^{\frac{1}{2}}\Big\|_{L_p(\mathrm{M}_m(q_n \mathcal{M} q_n))}$$
when regarding the conditional expectation as a mapping
$$\mathsf{E}_{\varphi_n} : L_p\big(\mathrm{M}_m(q_n \mathcal{M} q_n)\big) \to L_p(\mathrm{M}_m),$$
we deduce the assertion. The column case is proved in the same way. □

PROPOSITION 8.10. *Let $\mathcal{R}$ be the hyperfinite III_1 factor and ϕ the quasi-free state on $\mathcal{R}$ considered above. Let us consider the space $\mathcal{K}_{rc_p}^p(\phi \otimes \psi_n)$, defined as we did above. Then, there exists a completely isomorphic embedding*
$$\rho_n : \mathcal{K}_{p,2}(\psi_n) \to \mathcal{K}_{rc_p}^p(\phi \otimes \psi_n).$$
Moreover, the relevant constants in ρ_n are independent of n.

PROOF. We have
$$\mathcal{K}^p_{rc_p}(\phi \otimes \psi_n) = \mathrm{k}_n^{\frac{1}{p}} L_p(\mathcal{R}\bar{\otimes} q_n \mathcal{M} q_n) + \mathrm{k}_n^{\frac{1}{2}} L_2^{r_p}(\mathcal{R}\bar{\otimes} q_n \mathcal{M} q_n) + \mathrm{k}_n^{\frac{1}{2}} L_2^{c_p}(\mathcal{R}\bar{\otimes} q_n \mathcal{M} q_n).$$

The embedding is given by $\rho_n(x) = \gamma_p \otimes x$, with γ_p the element of $L_p(\mathcal{R})$ introduced after Proposition 8.6. Indeed, taking $\mathsf{E}_{\phi \otimes \varphi_n} : \mathrm{M}_m(\mathcal{R}\bar{\otimes} q_n \mathcal{M} q_n) \to \mathrm{M}_m$ and letting $x \in S_p^m(\mathcal{K}_{p,2}(\psi_n))$, we may apply Lemma 8.9 and obtain

$\|\gamma_p \otimes x\|_{S_p^m(\mathcal{K}^p_{rc_p}(\phi \otimes \psi_n))}$
$$= \inf_{\gamma_p \otimes x = x_p + x_r + x_c} \Big\{ \mathrm{k}_n^{\frac{1}{p}} \|x_p\|_p + \mathrm{k}_n^{\frac{1}{2}} \big\| \mathsf{E}_{\phi \otimes \varphi_n}(x_r x_r^*)^{\frac{1}{2}} \big\|_{S_p^m} + \mathrm{k}_n^{\frac{1}{2}} \big\| \mathsf{E}_{\phi \otimes \varphi_n}(x_c^* x_c)^{\frac{1}{2}} \big\|_{S_p^m} \Big\}.$$

Therefore, Lemma 8.3 and Remark 8.4 give

$$\|\gamma_p \otimes x\|_{S_p^m(\mathcal{K}^p_{rc_p}(\phi \otimes \psi_n))} \sim \Big\| \sum_{k=1}^{\mathrm{k}_n} \pi^j_{ind}(\gamma_p \otimes x) \Big\|_p \sim \Big\| \sum_{j=1}^{\mathrm{k}_n} \pi^j_{ind}(\gamma_p) \otimes \pi^j_{free}(x, -x) \Big\|_p.$$

Hence, the assertion follows as in Proposition 8.6. The proof is complete. □

Let $\mathcal{M}$ be a von Neumann algebra equipped with a n.s.s.f. weight ψ and let us write $(\psi_n, q_n)_{n \geq 1}$ for the associated sequence of q_n-supported weights. Then we define the following direct limit

$$\mathcal{K}^p_{rc_p}(\psi) = \overline{\bigcup_{n \geq 1} \mathcal{K}^p_{rc_p}(\psi_n)}.$$

We are interested in a cb-embedding $\mathcal{K}^p_{rc_p}(\psi) \to L_p(\mathcal{A})$ preserving hyperfiniteness. In the construction, we shall use a noncommutative Poisson random measure. Let us briefly review the main properties of this notion from [20] before stating our result. Let $\mathcal{M}^f_{sa}$ stand for the subspace of self-adjoint elements in $\mathcal{M}$ which are ψ-finitely supported. Let $\mathcal{M}_\pi$ denote the projection lattice of $\mathcal{M}$. We write $e \perp f$ for orthogonal projections. A *noncommutative Poisson random measure* is a map $\lambda : (\mathcal{M}, \psi) \to L_1(\mathcal{A}, \Phi_\psi)$, where $(\mathcal{A}, \Phi_\psi)$ is a noncommutative probability space and the following conditions hold

i) $\lambda : \mathcal{M}^f_{sa} \to L_1(\mathcal{A})$ is linear.
ii) $\Phi_\psi(e^{i\lambda(x)}) = \exp(\psi(e^{ix} - 1))$ for $x \in \mathcal{M}^f_{sa}$.
iii) If $e, f \in \mathcal{M}_\pi$ and $e \perp f$, $\lambda(e\mathcal{M}e)''$ and $\lambda(f\mathcal{M}f)''$ are strongly independent.

These properties are not yet enough to characterize λ, see below. Let us recall that two von Neumann subalgebras $\mathcal{A}_1, \mathcal{A}_2$ of $\mathcal{A}$ are called strongly independent if $a_1 a_2 = a_2 a_1$ and $\Phi_\psi(a_1 a_2) = \Phi_\psi(a_1)\Phi_\psi(a_2)$ for any pair (a_1, a_2) in $\mathcal{A}_1 \times \mathcal{A}_2$. The construction of λ follows by a direct limit argument. Indeed, let us show how to produce $\lambda : (q_n \mathcal{M} q_n, \psi_n) \to L_1(\mathcal{A}_n, \Phi_{\psi_n})$. We define

$$\mathcal{A}_n = M_{sym}(q_n \mathcal{M} q_n) = \prod_{k=0}^{\infty} \otimes^k_{sym} q_n \mathcal{M} q_n,$$

where $\otimes^k_{sym} q_n \mathcal{M} q_n$ denotes the subspace of symmetric tensors in the k-fold tensor product $(q_n \mathcal{M} q_n)_{\otimes k}$. In other words, if $\mathcal{S}_k$ is the symmetric group of permutations of k elements, the space $\otimes^k_{sym} q_n \mathcal{M} q_n$ is the range of the conditional expectation

$$\mathcal{E}_k(x_1 \otimes x_2 \otimes \cdots \otimes x_k) = \frac{1}{k!} \sum_{\pi \in \mathcal{S}_k} x_{\pi(1)} \otimes x_{\pi(2)} \otimes \cdots \otimes x_{\pi(k)}.$$

Then we set

$$\lambda(x) = (\lambda_k(x))_{k \geq 0} \in M_{sym}(q_n \mathcal{M} q_n) \quad \text{with} \quad \lambda_k(x) = \sum_{j=1}^{k} \pi_{ind}^j(x),$$

and properties (i), (ii) and (iii) hold when working with the state

$$\Phi_{\psi_n}\big((z_k)_{k \geq 1}\big) = \sum_{k=0}^{\infty} \frac{\exp(-\psi_n(1))}{k!} \underbrace{\psi_n \otimes \cdots \otimes \psi_n}_{k \text{ times}}(z_k).$$

In the following, it will be important to know the moments with respect to this state. Given $m \geq 1$, $\Pi(m)$ will be the set of partitions of $\{1, 2, \ldots, m\}$. On the other hand, given an ordered family $(x_\alpha)_{\alpha \in \Lambda}$ in $\mathcal{M}$, we shall write

$$\overrightarrow{\prod_{\alpha \in \Lambda}} x_\alpha$$

for the directed product of the x_α's. Then, the moments are given by the formula

iv) $\Phi_{\psi_n}\big(\lambda(x_1) \lambda(x_2) \cdots \lambda(x_m)\big) = \displaystyle\sum_{\substack{\sigma \in \Pi(m) \\ \sigma = \{\sigma_1, \ldots, \sigma_r\}}} \prod_{k=1}^{r} \psi_n\Big(\overrightarrow{\prod_{j \in \sigma_k}} x_j \Big).$

Now we can say that properties (i)-(iv) determine the Poisson random measure λ for any given *n.s.s.f.* weight ψ in $\mathcal{M}$. According to a uniqueness result from [18] which provides a noncommutative form of the Hamburger moment problem, it turns out that there exists a state preserving embedding

$$M_{sym}(q_{n_1} \mathcal{M} q_{n_1}) \to M_{sym}(q_{n_2} \mathcal{M} q_{n_2})$$

for $n_1 \leq n_2$ and such that the map $\lambda = \lambda_{n_1}$ constructed for q_{n_1} may be obtained as a restriction of λ_{n_2}. This allows to take direct limits. More precisely, let us define the algebra $M_{sym}(\mathcal{M})$ as the ultra-weak closure of the direct limit of the $M_{sym}(q_n \mathcal{M} q_n)$'s. Then, there exists a *n.f.* state Φ_ψ on $M_{sym}(\mathcal{M})$ and a map λ which assigns to every self-adjoint operator x (with $\mathrm{supp}(x) \leq e$ for some ψ-finite projection e in $\mathcal{M}$) a self-adjoint unbounded operator $\lambda(x)$ affiliated to $M_{sym}(\mathcal{M})$ and such that

$$\Phi_\psi(e^{i\lambda(x)}) = \exp(\psi(e^{ix} - 1)).$$

THEOREM 8.11. *Let $1 \leq p \leq 2$ and ψ be a n.s.s.f. weight on $\mathcal{M}$. Then there exists a von Neumann algebra $\mathcal{A}$, which is hyperfinite when $\mathcal{M}$ is hyperfinite, and a completely isomorphic embedding*

$$\mathcal{K}_{rc_p}^p(\psi) \to L_p(\mathcal{A}).$$

PROOF. Let us set the s-fold tensor product

$$\mathcal{B}_{n,s} = \Big(L_\infty[0,1] \bar{\otimes} [q_n \mathcal{M} q_n \oplus q_n \mathcal{M} q_n]\Big)_{\otimes s}.$$

Given $s \geq \mathrm{k}_n$, we define the mapping $\Lambda_{n,s} : \mathcal{K}_{rc_p}^p(\psi_n) \to L_p(\mathcal{B}_{n,s})$ by

$$\begin{aligned}\Lambda_{n,s}\big(d_{\varphi_n}^{\frac{1}{2p}} x d_{\varphi_n}^{\frac{1}{2p}}\big) &= \sum_{j=1}^{s} \pi_{ind}^j \Big(1_{[0, \mathrm{k}_n/s]} \otimes d_{\varphi_n}^{\frac{1}{2p}}(x, -x) d_{\varphi_n}^{\frac{1}{2p}}\Big) \\ &= \sum_{j=1}^{s} d_{n,s}^{\frac{1}{2p}} \pi_{ind}^j \big(1_{[0, \mathrm{k}_n/s]} \otimes (x, -x)\big) d_{n,s}^{\frac{1}{2p}},\end{aligned}$$

where $d_{n,s}$ is the density associated to the s-fold tensor product state

$$\phi_{n,s} = \Big[\int_0^1 \cdot\, dt \otimes \frac{1}{2}(\varphi_n \oplus \varphi_n)\Big]_{\otimes s} = \underbrace{\phi_n \otimes \phi_n \otimes \cdots \otimes \phi_n}_{s \text{ times}}.$$

If we tensor $\Lambda_{n,s}$ with the identity map on M_m, the resulting mapping gives a sum of symmetrically independent mean-zero random variables over M_m. Therefore, taking $x \in S_p^m(\mathcal{K}_{rc_p}^p(\psi_n))$ and applying Lemma 8.3/Remark 8.4

$$\big\|\Lambda_{n,s}\big(d_{\varphi_n}^{\frac{1}{2p}} x d_{\varphi_n}^{\frac{1}{2p}}\big)\big\|_p \sim \inf\Big\{s^{\frac{1}{p}}\|a\|_p + s^{\frac{1}{2}}\big\|\mathsf{E}_{\phi_n}(bb^*)^{\frac{1}{2}}\big\|_{S_p^m} + s^{\frac{1}{2}}\big\|\mathsf{E}_{\phi_n}(c^*c)^{\frac{1}{2}}\big\|_{S_p^m}\Big\},$$

where E_{ϕ_n} denotes the conditional expectation

$$\mathsf{E}_{\phi_n} : \mathrm{M}_m\Big(L_\infty[0,1]\bar\otimes[q_n\mathcal{M}q_n \oplus q_n\mathcal{M}q_n]\Big) \to \mathrm{M}_m$$

and the infimum runs over all possible decompositions

(8.6) $$1_{[0,\mathrm{k}_n/s]} \otimes d_{\varphi_n}^{\frac{1}{2p}}(x,-x) d_{\varphi_n}^{\frac{1}{2p}} = a + b + c.$$

Multiplying at both sizes of (8.6) by $1_{[0,\mathrm{k}_n/s]} \otimes 1$, we obtain a new decomposition which vanishes over $(\mathrm{k}_n/s, 1]$. Thus, since this clearly improves the infimum above, we may assume this property in all decompositions considered. Moreover, we claim that we can also restrict the infimum above to those decompositions $a+b+c$ which are constant on $[0, \mathrm{k}_n/s]$. Indeed, given any decomposition of the form (8.6) we take averages at both sizes and produce another decomposition $a_0 + b_0 + c_0$ given by the relations

$$(a_0, b_0, c_0) = 1_{[0,\mathrm{k}_n/s]} \otimes \frac{s}{\mathrm{k}_n}\int_0^{\frac{\mathrm{k}_n}{s}} \big(a(t), b(t), c(t)\big)\, dt.$$

Then, our claim is a consequence of the inequalities

$$\|a_0\|_p \leq \|a\|_p,$$
$$\big\|\mathsf{E}_{\phi_n}(b_0 b_0^*)^{\frac{1}{2}}\big\|_{S_p^m} \leq \big\|\mathsf{E}_{\phi_n}(bb^*)^{\frac{1}{2}}\big\|_{S_p^m},$$
$$\big\|\mathsf{E}_{\phi_n}(c_0^* c_0)^{\frac{1}{2}}\big\|_{S_p^m} \leq \big\|\mathsf{E}_{\phi_n}(c^*c)^{\frac{1}{2}}\big\|_{S_p^m}.$$

The first one is justified by means of the inequality

$$\lambda^{\frac{1}{p}}\Big\|\frac{1}{\lambda}\int_0^\lambda a(t)\, dt\Big\|_{L_p(\mathcal{M})} \leq \|a\|_{L_p([0,\lambda]\bar\otimes\mathcal{M})},$$

which follows easily by complex interpolation. The two other inequalities arise as a consequence of Kadison's inequality $\mathsf{E}(x)\mathsf{E}(x^*) \leq \mathsf{E}(xx^*)$ applied to the conditional expectation

$$\mathsf{E} = \frac{1}{\lambda}\int_0^\lambda \cdot\, dt.$$

Our considerations allow us to assume

$$(a,b,c) = 1_{[0,\mathrm{k}_n/s]} \otimes (x_p, x_r, x_c)$$

for some $x_p, x_r, x_c \in S_p^m(L_p(q_n\mathcal{M}q_n))$. This gives rise to

$$\big\|\Lambda_{n,s}\big(d_{\varphi_n}^{\frac{1}{2p}} x d_{\varphi_n}^{\frac{1}{2p}}\big)\big\|_p \sim \inf\Big\{\mathrm{k}_n^{\frac{1}{p}}\|x_p\|_p + \mathrm{k}_n^{\frac{1}{2}}\big\|\mathsf{E}_{\varphi_n}(x_r x_r^*)^{\frac{1}{2}}\big\|_{S_p^m} + \mathrm{k}_n^{\frac{1}{2}}\big\|\mathsf{E}_{\varphi_n}(x_c^* x_c)^{\frac{1}{2}}\big\|_{S_p^m}\Big\},$$

where the infimum runs over all possible decompositions
$$d_{\varphi_n}^{\frac{1}{2p}} x d_{\varphi_n}^{\frac{1}{2p}} = x_p + x_r + x_c.$$
This shows that $\Lambda_{n,s}: \mathcal{K}_{rc_p}^p(\psi_n) \to L_p(\mathcal{B}_{n,s})$ is a completely isomorphic embedding with constants independent of n or s. We are not ready yet to take direct limits. Before that, we use the algebraic central limit theorem to identify the moments in the limit as $s \to \infty$. To calculate the joint moments we set
$$\zeta_n = \frac{1}{2}(\varphi_n \oplus \varphi_n) \quad \text{and} \quad \zeta_{n,s} = \underbrace{\zeta_n \otimes \zeta_n \otimes \cdots \otimes \zeta_n}_{s \text{ times}}$$
and recall that the map $\Lambda_{n,s}$ corresponds to
$$u_{n,s}(x) = \sum_{j=1}^{s} \pi_{ind}^j \Big(1_{[0, k_n/s]} \otimes (x, -x) \Big).$$
Then, the joint moments are given by

$$\phi_{n,s}\big(u_{n,s}(x_1)\cdots u_{n,s}(x_m)\big)$$
$$= \sum_{j_1,j_2,\ldots,j_m=1}^{s} \int_{[0,1]^s} \prod_{i=1}^{m} \pi_{ind}^{j_i}(1_{[0,k_n/s]})(t)\, dt\, \zeta_{n,s}\Big(\prod_{1 \le i \le m}^{\rightarrow} \pi_{ind}^{j_i}(x_i, -x_i)\Big)$$
$$= \sum_{\substack{\sigma \in \Pi(m) \\ \sigma=\{\sigma_1,\ldots,\sigma_r\}}} \sum_{(j_1,\ldots,j_m)\sim\sigma} \Big(\frac{k_n}{s}\Big)^r \prod_{k=1}^{r} \zeta_n\Big[\Big(\prod_{i \in \sigma_k}^{\rightarrow} x_i, (-1)^{|\sigma_k|} \prod_{i \in \sigma_k}^{\rightarrow} x_i\Big)\Big],$$

where $|\sigma_k|$ denotes the cardinality of σ_k and we write $(j_1,\ldots,j_m) \sim \sigma$ when $j_a = j_b$ if and only if there exits $1 \le k \le r$ such that $a, b \in \sigma_k$. Therefore, recalling that $\zeta_n = \frac{1}{2}(\varphi_n \oplus \varphi_n)$, the only partitions which contribute to the sum above are the even partitions satisfying $|\sigma_k| \in 2\mathbb{N}$ for $1 \le k \le r$. Let us write $\Pi_e(m)$ for the set of even partitions. Then, using $\psi_n = k_n \varphi_n$ we deduce

$$\phi_{n,s}\big(u_{n,s}(x_1)\cdots u_{n,s}(x_m)\big) = \sum_{\substack{\sigma \in \Pi_e(m) \\ \sigma=\{\sigma_1,\ldots,\sigma_r\}}} \frac{|\{(j_1,\ldots,j_m) \sim \sigma\}|}{s^r} \prod_{k=1}^{r} \psi_n\Big(\prod_{i \in \sigma_k}^{\rightarrow} x_i\Big)$$
$$= \sum_{\substack{\sigma \in \Pi_e(m) \\ \sigma=\{\sigma_1,\ldots,\sigma_r\}}} \frac{s!}{s^r(s-r)!} \prod_{k=1}^{r} \psi_n\Big(\prod_{i \in \sigma_k}^{\rightarrow} x_i\Big).$$

Therefore, taking limits
$$\lim_{s \to \infty} \phi_{n,s}\big(u_{n,s}(x_1)\cdots u_{n,s}(x_m)\big) = \sum_{\substack{\sigma \in \Pi_e(m) \\ \sigma=\{\sigma_1,\ldots,\sigma_r\}}} \prod_{k=1}^{r} \psi_n\Big(\prod_{i \in \sigma_k}^{\rightarrow} x_i\Big).$$

These moments coincide with the moments of the Poisson random process
$$\lambda: (q_n \mathcal{M} q_n, \psi_n) \to L_1(\mathcal{M}_{sym}(q_n \mathcal{M} q_n), \Phi_{\psi_n}).$$

Hence, the noncommutative version of the Hamburger moment problem from [18] provides a state preserving homomorphism between the von Neumann algebra which generate the operators
$$\Big\{ e^{iu_{n,s}(x)} \mid x \in q_n \mathcal{M} q_n, s \ge 1 \Big\}$$

and the von Neumann subalgebra of $M_{sym}(q_n\mathcal{M}q_n)$ generated by

$$\left\{e^{i\lambda(x)} \mid x \in (q_n\mathcal{M}q_n)_{sa}^f\right\}.$$

In particular, taking $\mathcal{A}_n = M_{sym}(q_n\mathcal{M}q_n)$

$$\|d_{\psi_n}^{\frac{1}{2p}} x d_{\psi_n}^{\frac{1}{2p}}\|_{S_p^m(\mathcal{K}_{rc_p}^p(\psi_n))}$$
$$\sim \lim_{s\to\infty} \|\Lambda_{n,s}(d_{\psi_n}^{\frac{1}{2p}} x d_{\psi_n}^{\frac{1}{2p}})\|_{S_p^m(L_p(\mathcal{B}_{n,s}))} = \|d_{\Phi_{\psi_n}}^{\frac{1}{2p}} \lambda(x) d_{\Phi_{\psi_n}}^{\frac{1}{2p}}\|_{S_p^m(L_p(\mathcal{A}_n))}.$$

Now, we use from [20] that

$$\big(M_{sym}(\mathcal{M}), \Phi_\psi\big) = \overline{\bigcup_{n\geq 1} \big(M_{sym}(q_n\mathcal{M}q_n), \Phi_{\psi_n}\big)}$$

exists. Therefore, the map

$$\Lambda\big(d_\psi^{\frac{1}{2p}} x d_\psi^{\frac{1}{2p}}\big) = d_{\Phi_\psi}^{\frac{1}{2p}} \lambda(x) d_{\Phi_\psi}^{\frac{1}{2p}}$$

extends to a complete embedding

$$\mathcal{K}_{rc_p}^p(\psi) = \lim_n \mathcal{K}_{rc_p}^p(\psi_n) \to L_p(M_{sym}(\mathcal{M})).$$

Moreover, if $\mathcal{M}$ is hyperfinite so is $M_{sym}(q_n\mathcal{M}q_n)$ and hence $M_{sym}(\mathcal{M})$. □

COROLLARY 8.12. *Let $1 \leq p \leq 2$ and ψ be a n.s.s.f. weight on $\mathcal{M}$. Then there exists a von Neumann algebra $\mathcal{A}$, which is hyperfinite when $\mathcal{M}$ is hyperfinite, and a completely isomorphic embedding*

$$\mathcal{K}_{p,2}(\psi) \to L_p(\mathcal{A}).$$

PROOF. Let us set

$$\mathcal{RB}_{n,s} = \Big(L_\infty[0,1]\bar\otimes\big[(\mathcal{R}\bar\otimes q_n\mathcal{M}q_n) \oplus (\mathcal{R}\bar\otimes q_n\mathcal{M}q_n)\big]\Big)_{\otimes s}.$$

By Proposition 8.10 and Theorem 8.11, the map

$$\Lambda_{n,s} \circ \rho_n : \mathcal{K}_{p,2}(\psi_n) \to \mathcal{K}_{rc_p}^p(\phi \otimes \psi_n) \to L_p(\mathcal{RB}_{n,s})$$

provides a complete isomorphism with constant independent of n and s. Using the algebraic central limit theorem to take limits in s and the noncommutative version of the Hamburger moment problem one more time, we obtain a complete embedding

$$\Lambda_n \circ \rho_n : \mathcal{K}_{p,2}(\psi_n) \to \mathcal{K}_{rc_p}^p(\phi \otimes \psi_n) \to L_p\big(M_{sym}(\mathcal{R}\bar\otimes q_n\mathcal{M}q_n)\big).$$

Taking direct limits we obtain a cb-embedding which preserves hyperfiniteness. □

COROLLARY 8.13. *If $1 \leq p < q \leq 2$, S_q cb-embeds in $L_p(\mathcal{A})$ with $\mathcal{A}$ hyperfinite.*

PROOF. By the complete isometry $S_q = C_q \otimes_h R_q$ and [24, Remark 2.2], it suffices to embed the first term $\mathcal{Z}_1 \otimes \mathcal{Z}_2$ on the right of (8.1) into $L_p(\mathcal{A})$ for some hyperfinite von Neumann algebra $\mathcal{A}$. However, following the proof of Theorem 8.1, we know that $\mathcal{Z}_1 \otimes \mathcal{Z}_2$ embeds completely isomorphically into $\mathcal{K}_{p,2}(\psi)$, where ψ denotes some n.s.s.f. weight on $\mathcal{B}(\ell_2)$. Therefore, the assertion follows from Corollary 8.12. □

REMARK 8.14. Theorem 8.1 easily generalizes to the context of Corollary 8.13. More precisely, given operator spaces $X_1 \in \mathcal{QS}(C_p \oplus_2 \mathrm{OH})$ and $X_2 \in \mathcal{QS}(R_p \oplus_2 \mathrm{OH})$ and combining the techniques applied so far, it is rather easy to find a hyperfinite type III factor $\mathcal{A}$ and a completely isomorphic embedding

$$X_1 \otimes_h X_2 \to L_p(\mathcal{A}).$$

REMARK 8.15. In contrast with Corollary 8.2, where free products are used, the complete embedding of S_q into L_p given in Corollary 8.13 provides estimates on the dimension of $\mathcal{A}$ in the cb-embedding

$$S_q^m \to L_p(\mathcal{A}).$$

Indeed, a quick look at our construction shows that

$$S_q^m = C_q^m \otimes_h R_q^m \to \mathcal{K}_{p,2}(\psi_n) \to L_p(\mathcal{A}_{ind}^n; \mathrm{OH}_{k_n}) = L_p(\mathrm{M}_n^{\otimes k_n}; \mathrm{OH}_{k_n}),$$

with $n \sim m \log m$, see [18] for this last assertion. This chain essentially follows from [24, Remark 2.2], the complete isometry (8.3) and Proposition 8.6. On the other hand, given any parameter $\gamma > 1/2k_n$ and according once more to [18], we know that OH_{k_n} embeds completely isomorphically into $S_p^{w_n}$ for $w_n = k_n^{\gamma k_n}$ with constants depending only on γ and that $k_n \sim n^{\alpha_p}$. Combining the embeddings mentioned so far, we have found a complete embedding $S_q^m \to S_p^M$ with

$$M \sim m^{\beta_p m^{\alpha_p}}.$$

8.3. Embedding for general von Neumann algebras

Let $1 \le p < q \le 2$ as in the statement of our main result in the Introduction. We will encode complex interpolation in a suitable graph. This follows Pisier's approach [49] to the main result in [17]. Indeed, given $0 < \theta < 1$, let μ_θ be the harmonic measure of the point $z = \theta$. This is a probability measure on the boundary $\partial \mathcal{S}$ (with density given by the Poisson kernel in the strip) that can be written as $\mu_\theta = (1-\theta)\mu_0 + \theta\mu_1$, with μ_j being probability measures supported by ∂_j and such that

$$f(\theta) = \int_{\partial \mathcal{S}} f \, d\mu_\theta$$

for any bounded analytic f extended non-tangentially to $\partial \mathcal{S}$. Let

$$\mathcal{H}_2 = \Big\{ (f|_{\partial_0}, f|_{\partial_1}) \mid f : \mathcal{S} \to \mathbb{C} \text{ analytic} \Big\} \subset L_2(\partial_0) \oplus L_2(\partial_1).$$

We need operator-valued versions of this space given by subspaces

$$\mathcal{H}_{2p',2}^r(\mathcal{M}, \theta) \subset \Big(L_2^{c_{p'}}(\partial_0) \otimes_h L_{2p'}^r(\mathcal{M}) \Big) \oplus \Big(L_2^{oh}(\partial_1) \otimes_h L_4^r(\mathcal{M}) \Big) = \mathcal{O}_{p',0}^r \oplus \mathcal{O}_{p',1}^r,$$

$$\mathcal{H}_{2p',2}^c(\mathcal{M}, \theta) \subset \Big(L_{2p'}^c(\mathcal{M}) \otimes_h L_2^{r_{p'}}(\partial_0) \Big) \oplus \Big(L_4^c(\mathcal{M}) \otimes_h L_2^{oh}(\partial_1) \Big) = \mathcal{O}_{p',0}^c \oplus \mathcal{O}_{p',1}^c.$$

More precisely, if $\mathcal{M}$ comes equipped with a n.s.s.f. weight ψ and d_ψ denotes the associated density, $\mathcal{H}_{2p',2}^r(\mathcal{M}, \theta)$ is the subspace of all pairs (f_0, f_1) of functions in $\mathcal{O}_{p',0}^r \oplus \mathcal{O}_{p',1}^r$ such that for every scalar-valued analytic function $g : \mathcal{S} \to \mathbb{C}$ (extended non-tangentially to the boundary) with $g(\theta) = 0$, we have

$$(1-\theta)\int_{\partial_0} g(z) \, d_\psi^{\frac{1}{4} - \frac{1}{2p'}} f_0(z) \, d\mu_0(z) + \theta \int_{\partial_1} g(z) f_1(z) \, d\mu_1(z) = 0.$$

8.3. EMBEDDING FOR GENERAL VON NEUMANN ALGEBRAS

Similarly, the condition on $\mathcal{H}^c_{2p',2}(\mathcal{M},\theta)$ is

$$(1-\theta)\int_{\partial_0} g(z)f_0(z)\,d_\psi^{\frac{1}{4}-\frac{1}{2p'}}\,d\mu_0(z) + \theta\int_{\partial_1} g(z)f_1(z)\,d\mu_1(z) = 0.$$

We shall also need to consider the subspaces

$$\mathcal{H}_{r,0} = \left\{(f_0,f_1)\in\mathcal{H}^r_{2p',2}(\mathcal{M},\theta)\,\Big|\,(1-\theta)\int_{\partial_0} d_\psi^{\frac{1}{4}-\frac{1}{2p'}} f_0 d\mu_0 + \theta\int_{\partial_1} f_1 d\mu_1 = 0\right\},$$

$$\mathcal{H}_{c,0} = \left\{(f_0,f_1)\in\mathcal{H}^c_{2p',2}(\mathcal{M},\theta)\,\Big|\,(1-\theta)\int_{\partial_0} f_0 d_\psi^{\frac{1}{4}-\frac{1}{2p'}} d\mu_0 + \theta\int_{\partial_1} f_1 d\mu_1 = 0\right\}.$$

REMARK 8.16. In order to make all the forthcoming duality arguments work, we need to introduce a slight modification of these spaces for $p=1$. Indeed, in that case the spaces defined above must be regarded as subspaces of

$$\mathcal{H}^r_{\infty,2}(\mathcal{M},\theta) \subset \left(L_2^c(\partial_0)\bar\otimes\mathcal{M}\right) \oplus \left(L_2^{oh}(\partial_1)\otimes_h L_4^r(\mathcal{M})\right),$$

$$\mathcal{H}^c_{\infty,2}(\mathcal{M},\theta) \subset \left(\mathcal{M}\bar\otimes L_2^r(\partial_0)\right) \oplus \left(L_4^c(\mathcal{M})\otimes_h L_2^{oh}(\partial_1)\right).$$

The von Neumann algebra tensor product used above is the weak closure of the minimal tensor product, which in this particular case coincides with the Haagerup tensor product since we have either a column space on the left or a row space on the right. In particular, the only difference is that we are taking the closure in the weak operator topology.

LEMMA 8.17. *Let $\mathcal{M}$ be a finite von Neumann algebra equipped with a n.f. state φ and let d_φ be the associated density. If $2\leq q' < p'$ and $\frac{1}{2q'} = \frac{1-\theta}{2p'} + \frac{\theta}{4}$, we have complete contractions*

$$u_r: d_\varphi^{\frac{1}{2q'}} x\in L_{2q'}^r(\mathcal{M}) \mapsto \left(1\otimes d_\varphi^{\frac{1}{2p'}} x, 1\otimes d_\varphi^{\frac{1}{4}} x\right) + \mathcal{H}_{r,0} \in \mathcal{H}^r_{2p',2}(\mathcal{M},\theta)/\mathcal{H}_{r,0},$$

$$u_c: xd_\varphi^{\frac{1}{2q'}} \in L_{2q'}^c(\mathcal{M}) \mapsto \left(xd_\varphi^{\frac{1}{2p'}}\otimes 1, xd_\varphi^{\frac{1}{4}}\otimes 1\right) + \mathcal{H}_{c,0} \in \mathcal{H}^c_{2p',2}(\mathcal{M},\theta)/\mathcal{H}_{c,0}.$$

PROOF. By symmetry, it suffices to consider the column case. Let x be an element in $\mathrm{M}_m(L_{2q'}^c(\mathcal{M}))$ of norm less than 1. According to our choice of $0 < \theta < 1$, we find that

$$\mathrm{M}_m(L_{2q'}^c(\mathcal{M})) = \left[\mathrm{M}_m(L_{2p'}^c(\mathcal{M})), \mathrm{M}_m(L_4^c(\mathcal{M}))\right]_\theta.$$

Thus, there exists $f:\mathcal{S}\to\mathrm{M}_m(\mathcal{M})$ analytic such that $f(\theta)=x$ and

$$\max\left\{\sup_{z\in\partial_0}\|f(z)d_\varphi^{\frac{1}{2p'}}\|_{\mathrm{M}_m(L_{2p'}^c(\mathcal{M}))}, \sup_{z\in\partial_1}\|f(z)d_\varphi^{\frac{1}{4}}\|_{\mathrm{M}_m(L_4^c(\mathcal{M}))}\right\} \leq 1.$$

If $1\leq s\leq\infty$ and $j\in\{0,1\}$, we claim that

$$(8.7) \qquad \|f_{|\partial_j} d_\varphi^{\frac{1}{2s}}\|_{\mathrm{M}_m(L_{2s}^c(\mathcal{M})\otimes_h L_2^{rs}(\partial_j))} \leq \sup_{z\in\partial_j} \|f(z)d_\varphi^{\frac{1}{2s}}\|_{\mathrm{M}_m(L_{2s}^c(\mathcal{M}))}.$$

Before proving our claim, let us finish the proof. Taking $f_j = f_{|\partial_j}$, we have

$$\left(f_0 d_\varphi^{\frac{1}{2p'}}, f_1 d_\varphi^{\frac{1}{4}}\right) - \left(xd_\varphi^{\frac{1}{2p'}}\otimes 1, xd_\varphi^{\frac{1}{4}}\otimes 1\right) \in \mathcal{H}_{c,0}.$$

Indeed by analyticity, we have

$$(1-\theta)\int_{\partial_0} f_0\, d_\varphi^{\frac{1}{2p'}} d_\varphi^{\frac{1}{4}-\frac{1}{2p'}} d\mu_0 + \theta\int_{\partial_1} f_1\, d_\varphi^{\frac{1}{4}} d\mu_1 = \int_{\partial\mathcal{S}} f d_\varphi^{\frac{1}{4}} d\mu_\theta = f(\theta) d_\varphi^{\frac{1}{4}} = xd_\varphi^{\frac{1}{4}}.$$

This implies from (8.7) applied to $(s,j) = (p',0)$ and $(s,j) = (2,1)$ that
$$\left\| u_c(xd_\varphi^{\frac{1}{2q'}}) \right\|_{M_m(\mathcal{H}_{2p',2}^c/\mathcal{H}_{c,0})} \leq \left\| (f_0 d_\varphi^{\frac{1}{2p'}}, f_1 d_\varphi^{\frac{1}{4}}) \right\|_{M_m(\mathcal{H}_{2p',2}^c)} \leq 1.$$

Hence, it remains to prove our claim (8.7). We must show that the identity map $L_\infty(\partial_j; L_{2s}^c(\mathcal{M})) \to L_{2s}^c(\mathcal{M}) \otimes_h L_2^{r_s}(\partial_j)$ is a complete contraction. By complex interpolation, we have

$$\begin{aligned}
L_\infty(\partial_j; L_{2s}^c(\mathcal{M})) &= \left[L_\infty(\partial_j; \mathcal{M}), L_\infty(\partial_j; L_2^c(\mathcal{M})) \right]_{\frac{1}{s}}, \\
L_{2s}^c(\mathcal{M}) \otimes_h L_2^{r_s}(\partial_j) &= \left[\mathcal{M} \otimes_h L_2^r(\partial_j), L_2^c(\mathcal{M}) \otimes_h L_2^c(\partial_j) \right]_{\frac{1}{s}} \\
&= \left[\mathcal{M} \otimes_{\min} L_2^r(\partial_j), L_2^c(\mathcal{M}) \otimes_{\min} L_2^c(\partial_j) \right]_{\frac{1}{s}}.
\end{aligned}$$

In other words, we must study the identity mappings

$$\begin{aligned}
\mathcal{M} \otimes_{\min} L_\infty(\partial_j) &\to \mathcal{M} \otimes_{\min} L_2^r(\partial_j), \\
L_2^c(\mathcal{M}) \otimes_{\min} L_\infty(\partial_j) &\to L_2^c(\mathcal{M}) \otimes_{\min} L_2^c(\partial_j).
\end{aligned}$$

However, this automatically reduces to see that we have complete contractions

$$\begin{aligned}
L_\infty(\partial_j) &\to L_2^r(\partial_j), \\
L_\infty(\partial_j) &\to L_2^c(\partial_j).
\end{aligned}$$

Therefore it suffices to observe that

$$\|f\|^2_{M_m(L_2^r(\partial_j))} = \left\| \int_{\partial_j} ff^* d\mu_j \right\|_{M_m} \leq \mu_j(\partial_j) \sup_{z \in \partial_j} \|f(z)\|^2_{M_m} = \|f\|^2_{M_m(L_\infty(\partial_j))},$$

$$\|f\|^2_{M_m(L_2^c(\partial_j))} = \left\| \int_{\partial_j} f^* f d\mu_j \right\|_{M_m} \leq \mu_j(\partial_j) \sup_{z \in \partial_j} \|f(z)\|^2_{M_m} = \|f\|^2_{M_m(L_\infty(\partial_j))}.$$

This completes the proof for $1 < p \leq 2$. In the case $p = 1$, we have overlooked the fact that the definition of $\mathcal{H}_{\infty,2}^c(\mathcal{M}, \theta)$ (see Remark 8.16 above) is slightly different. The only consequence of this point is that we also need the inequality

$$\left\| f_{|\partial_0} \right\|_{M_m(\mathcal{M} \bar{\otimes} L_2^r(\partial_0))} \leq \sup_{z \in \partial_0} \|f(z)\|_{M_m(\mathcal{M})}.$$

However, this is proved once more as above. The proof is complete. $\square$

To continue with the argument we need to introduce some predual spaces. This requires to extend our definition (1.3) to the case $1 \leq q \leq 2$. This is easily done as follows
$$\begin{aligned}
L_q^r(\mathcal{M}) &= \left[L_1(\mathcal{M}), L_2^r(\mathcal{M}) \right]_{\frac{2}{q'}}, \\
L_q^c(\mathcal{M}) &= \left[L_1(\mathcal{M}), L_2^c(\mathcal{M}) \right]_{\frac{2}{q'}}.
\end{aligned}$$

Lemma 8.17 is closely related to [24, Lemma 2.1] and a similar result holds on the preduals. More precisely, we begin by defining the operator-valued Hardy spaces which arise as subspaces

$$\mathcal{H}_{2p,2}^c(\mathcal{M}, \theta) \subset \left(L_2^{c_p}(\partial_0) \otimes_h L_{\frac{2p}{p+1}}^c(\mathcal{M}) \right) \oplus \left(L_2^{oh}(\partial_1) \otimes_h L_{\frac{4}{3}}^c(\mathcal{M}) \right),$$

$$\mathcal{H}_{2p,2}^r(\mathcal{M}, \theta) \subset \left(L_{\frac{2p}{p+1}}^r(\mathcal{M}) \otimes_h L_2^{r_p}(\partial_0) \right) \oplus \left(L_{\frac{4}{3}}^r(\mathcal{M}) \otimes_h L_2^{oh}(\partial_1) \right),$$

formed by pairs (f_0, f_1) respectively satisfying

$$(1 - \theta) \int_{\partial_0} g(z) f_0(z) \, d\mu_0(z) + \theta \int_{\partial_1} g(z) d_\varphi^{\frac{p+1}{2p} - \frac{3}{4}} f_1(z) \, d\mu_1(z) = 0,$$

8.3. EMBEDDING FOR GENERAL VON NEUMANN ALGEBRAS

$$(1-\theta)\int_{\partial_0} g(z)f_0(z)\,d\mu_0(z) + \theta \int_{\partial_1} g(z)f_1(z)d_\varphi^{\frac{p+1}{2p}-\frac{3}{4}}\,d\mu_1(z) \;=\; 0,$$

for all scalar-valued analytic function $g : \mathcal{S} \to \mathbb{C}$ (extended non-tangentially to the boundary) with $g(\theta) = 0$. The subspaces $\mathcal{H}'_{r,0}$ and $\mathcal{H}'_{c,0}$ are defined accordingly. In other words, we have

$$\mathcal{H}'_{c,0} \;=\; \Big\{ (f_0,f_1) \in \mathcal{H}^c_{2p,2}(\mathcal{M},\theta) \,\big|\, (1-\theta)\int_{\partial_0} f_0\,d\mu_0 + \theta \int_{\partial_1} d_\varphi^{\frac{p+1}{2p}-\frac{3}{4}} f_1\,d\mu_1 = 0 \Big\},$$

$$\mathcal{H}'_{r,0} \;=\; \Big\{ (f_0,f_1) \in \mathcal{H}^r_{2p,2}(\mathcal{M},\theta) \,\big|\, (1-\theta)\int_{\partial_0} f_0\,d\mu_0 + \theta \int_{\partial_1} f_1 d_\varphi^{\frac{p+1}{2p}-\frac{3}{4}}\,d\mu_1 = 0 \Big\}.$$

LEMMA 8.18. *Let $\mathcal{M}$ be a finite von Neumann algebra equipped with a n.f. state φ and let d_φ be the associated density. Taking the same values for p, q and θ as above, we have complete contractions*

$$w_r : d_\varphi^{\frac{q+1}{2q}} x \in L^c_{\frac{2q}{q+1}}(\mathcal{M}) \mapsto \Big(1 \otimes d_\varphi^{\frac{p+1}{2p}} x, 1 \otimes d_\varphi^{\frac{3}{4}} x\Big) + \mathcal{H}'_{c,0} \in \mathcal{H}^c_{2p,2}(\mathcal{M},\theta)/\mathcal{H}'_{c,0},$$

$$w_c : x d_\varphi^{\frac{q+1}{2q}} \in L^r_{\frac{2q}{q+1}}(\mathcal{M}) \mapsto \Big(x d_\varphi^{\frac{p+1}{2p}} \otimes 1, x d_\varphi^{\frac{3}{4}} \otimes 1\Big) + \mathcal{H}'_{r,0} \in \mathcal{H}^r_{2p,2}(\mathcal{M},\theta)/\mathcal{H}'_{r,0}.$$

PROOF. If $1 \le s \le \infty$ and $\mathcal{M}_m = \mathrm{M}_m(\mathcal{M})$, we have

$$\begin{aligned}
(8.8) \qquad S_1^m\big(L^r_{\frac{2s}{s+1}}(\mathcal{M})\big) &= S_1^m\Big(\big[L_1(\mathcal{M}), L_2^r(\mathcal{M})\big]_{\frac{1}{s'}}\Big) \\
&= \Big[S_1^m(L_1(\mathcal{M})), S_2^m(L_2(\mathcal{M}))S_2^m\Big]_{\frac{1}{s'}} \\
&= S_{\frac{2s}{s+1}}^m\big(L_{\frac{2s}{s+1}}(\mathcal{M})\big)S_{2s'}^m = L_{\frac{2s}{s+1}}(\mathcal{M}_m)S_{2s'}^m.
\end{aligned}$$

Indeed, the second isometry follows by dualizing the second isometry below

$$\begin{aligned}
(8.9) \qquad \big\| d_\varphi^{\frac{1}{2}}(x_{ij}) \big\|_{\mathrm{M}_m(L_2^r(\mathcal{M}))} &= \sup_{\|\alpha\|_{S_2^m} \le 1} \Big\| d_\varphi^{\frac{1}{2}}\Big(\sum_{k=1}^m \alpha_{ik} x_{kj}\Big) \Big\|_{L_2(\mathcal{M}_m)}, \\
\big\| (x_{ij}) d_\varphi^{\frac{1}{2}} \big\|_{\mathrm{M}_m(L_2^c(\mathcal{M}))} &= \sup_{\|\beta\|_{S_2^m} \le 1} \Big\| \Big(\sum_{k=1}^m x_{ik}\beta_{kj}\Big) d_\varphi^{\frac{1}{2}} \Big\|_{L_2(\mathcal{M}_m)}.
\end{aligned}$$

The first one is needed for the analysis of the mapping w_r. The isometries (8.9) are well-known in operator space theory, see e.g. p.56 in [**11**]. Alternatively, one may argue directly as we did in Lemma 8.9. The third isometry in (8.8) follows from Theorem 3.2. Now, let us consider an element of norm less than 1

$$x d_\varphi^{\frac{q+1}{2q}} \in S_1^m\big(L^r_{\frac{2q}{q+1}}(\mathcal{M})\big).$$

The isometry above provides a factorization

$$x d_\varphi^{\frac{q+1}{2q}} = \alpha\beta\gamma\delta \in L_{\frac{4}{3}}(\mathcal{M}_m) L_{\rho_1}(\mathcal{M}_m) S_{2p'}^m S_{\rho_2}^m$$

with $\alpha, \beta, \gamma, \delta$ in the unit balls of their respective spaces with

$$\frac{1}{\rho_1} = \frac{q+1}{2q} - \frac{3}{4} \quad \text{and} \quad \frac{1}{\rho_2} = \frac{1}{2q'} - \frac{1}{2p'}.$$

Moreover, by polar decomposition and approximation, we may assume that β and δ are strictly positive elements. In particular, motivated by the complex interpolation isometry

$$L_{\frac{2q}{q+1}}(\mathcal{M}_m) S_{2q'}^m = \Big[L_{\frac{2p}{p+1}}(\mathcal{M}_m) S_{2p'}^m, L_{\frac{4}{3}}(\mathcal{M}_m) S_4^m\Big]_\theta = [\mathrm{X}_0, \mathrm{X}_1]_\theta,$$

we take $(\beta_\theta, \delta_\theta) = (\beta^{1/(1-\theta)}, \delta^{1/\theta})$ and define
$$f : z \in \mathcal{S} \mapsto \alpha \beta_\theta^{1-z} \gamma \delta_\theta^z \in X_0 + X_1.$$
Since f is analytic and $f(\theta) = xd_\varphi^{\frac{q+1}{2q}}$, we conclude
$$(f_{|\partial_0}, f_{|\partial_1}) \in \left(xd_\varphi^{\frac{p+1}{2p}} \otimes 1, xd_\varphi^{\frac{3}{4}} \otimes 1\right) + \mathcal{H}'_{r,0}.$$
Therefore, taking $f_{|\partial_j} = f_j$ we obtain the following estimate
$$\left\|w_c\big(xd_\varphi^{\frac{q+1}{2q}}\big)\right\|_{S_1^m(\mathcal{H}_{2p,2}^r/\mathcal{H}'_{r,0})}$$
$$\leq \max\left\{\|f_0\|_{S_1^m(L^r_{\frac{2p}{p+1}}(\mathcal{M})\otimes_h L_2^{rp}(\partial_0))}, \|f_1\|_{S_1^m(L^r_{\frac{4}{3}}(\mathcal{M})\otimes_h L_2^{qh}(\partial_1))}\right\}$$
$$= \max\left\{\|\alpha\beta_\theta\gamma\|_{S_1^m(L^r_{\frac{2p}{p+1}}(\mathcal{M}))}, \|\alpha\gamma\delta_\theta\|_{S_1^m(L^r_{\frac{4}{3}}(\mathcal{M}))}\right\} \leq 1,$$

where the last inequality follows from (8.8) and the fact that $(\alpha\beta_\theta, \gamma\delta_\theta)$ are in the unit balls of $L_{2p/p+1}(\mathcal{M}_m)$ and S_4^m respectively. The assertion for the mapping w_r is proved similarly. The proof is complete. $\square$

PROPOSITION 8.19. *Let $\mathcal{M}$ be a finite von Neumann algebra equipped with a n.f. state φ and let d_φ be the associated density. If $2 \leq q' < p'$ and $\frac{1}{2q'} = \frac{1-\theta}{2p'} + \frac{\theta}{4}$, we have complete isomorphisms*

$$u_r : d_\varphi^{\frac{1}{2q'}} x \in L^r_{2q'}(\mathcal{M}) \mapsto \left(1 \otimes d_\varphi^{\frac{1}{2p'}} x, 1 \otimes d_\varphi^{\frac{1}{4}} x\right) + \mathcal{H}_{r,0} \in \mathcal{H}^r_{2p',2}(\mathcal{M},\theta)/\mathcal{H}_{r,0},$$

$$u_c : xd_\varphi^{\frac{1}{2q'}} \in L^c_{2q'}(\mathcal{M}) \mapsto \left(xd_\varphi^{\frac{1}{2p'}} \otimes 1, xd_\varphi^{\frac{1}{4}} \otimes 1\right) + \mathcal{H}_{c,0} \in \mathcal{H}^c_{2p',2}(\mathcal{M},\theta)/\mathcal{H}_{c,0}.$$

PROOF. This follows easily from the identity
$$\mathrm{tr}_\mathcal{M}(x^*y) = \int_{\partial\mathcal{S}} \mathrm{tr}_\mathcal{M}\big(f(\bar{z})^* g(z)\big) d\mu_\theta(z),$$
valid for any pair of analytic functions f and g such that $(f(\theta), g(\theta)) = (x, y)$. Indeed, according to the definition of the mappings u_r, u_c, w_r, w_c, this means that we have
$$\langle u_r(x_1), w_r(y_1)\rangle = \langle x_1, y_1\rangle$$
$$\langle u_c(x_2), w_c(y_2)\rangle = \langle x_2, y_2\rangle$$
for any
$$(x_1, x_2, y_1, y_2) \in L^r_{2q'}(\mathcal{M}) \times L^c_{2q'}(\mathcal{M}) \times L^c_{\frac{2q}{q+1}}(\mathcal{M}) \times L^r_{\frac{2q}{q+1}}(\mathcal{M}).$$
In particular, we deduce
$$w_r^* u_r = id_{L^r_{2q'}(\mathcal{M})} \quad \text{and} \quad w_c^* u_c = id_{L^c_{2q'}(\mathcal{M})}.$$
Therefore, the assertion follows combining Lemma 8.17 and Lemma 8.18. $\square$

Let $\mathcal{M}$ be a von Neumann algebra equipped with a *n.f.* state φ and let $\mathcal{M}_m$ be the tensor product $\mathrm{M}_m \otimes \mathcal{M}$. Then, if $\mathsf{E}_m = id_{\mathrm{M}_m} \otimes \varphi : \mathcal{M}_m \to \mathrm{M}_m$ denotes the associated conditional expectation, the following generalizes Lemma 7.7.

8.3. EMBEDDING FOR GENERAL VON NEUMANN ALGEBRAS

LEMMA 8.20. *If $1/r = 1/2 - 1/p'$, we have isometries*
$$L_{(2r,\infty)}^{2p'}(\mathcal{M}_m, \mathsf{E}_m) = C_{p'}^m \otimes_h L_4^r(\mathcal{M}) \otimes_h R_m,$$
$$L_{(\infty,2r)}^{2p'}(\mathcal{M}_m, \mathsf{E}_m) = C_m \otimes_h L_4^c(\mathcal{M}) \otimes_h R_{p'}^m.$$

PROOF. By Kouba's theorem, it is clear that the spaces on the right form complex interpolation families with respect to $2 \leq p' \leq \infty$. Let us see that the same happens for the conditional L_p spaces on the left. Indeed, according to Theorem 4.6 we have isometries
$$L_{(2r,\infty)}^{2p'}(\mathcal{M}_m, \mathsf{E}_m) = \left[L_{(4,\infty)}^{\infty}(\mathcal{M}_m, \mathsf{E}_m), L_{(\infty,\infty)}^{4}(\mathcal{M}_m, \mathsf{E}_m)\right]_{2/p'},$$
$$L_{(\infty,2r)}^{2p'}(\mathcal{M}_m, \mathsf{E}_m) = \left[L_{(\infty,4)}^{\infty}(\mathcal{M}_m, \mathsf{E}_m), L_{(\infty,\infty)}^{4}(\mathcal{M}_m, \mathsf{E}_m)\right]_{2/p'}.$$

This reduces the problem to the cases $p' = 2$ and $p' = \infty$. If $p' = 2$, we have $r = \infty$ and both spaces on the left coincide with $L_{(\infty,\infty)}^{4}(\mathcal{M}_m, \mathsf{E}_m) = L_4(\mathcal{M}_m)$. On the other hand, the spaces on the right are respectively given by
$$C_2^m \otimes_h L_4^r(\mathcal{M}) \otimes_h R_m = \left[C_m \otimes_h \mathcal{M} \otimes_h R_m, R_m \otimes_h L_2^r(\mathcal{M}) \otimes_h R_m\right]_\theta,$$
$$C_m \otimes_h L_4^c(\mathcal{M}) \otimes_h R_{p'}^m = \left[C_m \otimes_h \mathcal{M} \otimes_h R_m, C_m \otimes_h L_2^c(\mathcal{M}) \otimes_h C_m\right]_\theta.$$

In other words, we find the spaces $L_4^r(\mathcal{M}_m)$ and $L_4^c(\mathcal{M}_m)$ which coincide with $L_4(\mathcal{M}_m)$ at the Banach space level. It remains to check the validity of the assertion for $p' = \infty$, but this is exactly the content of [**24**, Lemma 1.8]. □

Let $\mathcal{M}$ be a von Neumann algebra equipped with a n.f. state φ and let $\mathcal{N}$ be a von Neumann subalgebra of $\mathcal{M}$. Let $\mathsf{E} : \mathcal{M} \to \mathcal{N}$ denote the corresponding conditional expectation. In Chapter 5, we defined the spaces
$$\mathcal{R}_{2p',2}^n(\mathcal{M}, \mathsf{E}) = n^{\frac{1}{2p'}} L_{2p'}^r(\mathcal{M}) \cap n^{\frac{1}{4}} L_{(2r,\infty)}^{2p'}(\mathcal{M}, \mathsf{E}),$$
$$\mathcal{C}_{2p',2}^n(\mathcal{M}, \mathsf{E}) = n^{\frac{1}{2p'}} L_{2p'}^c(\mathcal{M}) \cap n^{\frac{1}{4}} L_{(\infty,2r)}^{2p'}(\mathcal{M}, \mathsf{E}),$$
with $1/r = 1/2 - 1/p'$ and in Chapter 6 we proved the isomorphism
$$(8.10) \qquad \mathcal{J}_{p',2}^n(\mathcal{M}, \mathsf{E}) \simeq \mathcal{R}_{2p',2}^n(\mathcal{M}, \mathsf{E}) \otimes_{\mathcal{M}} \mathcal{C}_{2p',2}^n(\mathcal{M}, \mathsf{E}).$$
In fact, to be completely fair we should say that we have slightly modified the definition of $\mathcal{R}_{2p',2}^n(\mathcal{M}, \mathsf{E})$ and $\mathcal{C}_{2p',2}^n(\mathcal{M}, \mathsf{E})$ by considering the row/column o.s.s. of $L_{2p'}(\mathcal{M})$. However, the new definition coincides with the former one at the Banach space level. Hence, since we do not even have an operator space structure for these spaces, our modification is only motivated by notational convenience below. Namely, inspired by Lemma 8.20, we introduce the operator spaces
$$\mathcal{R}_{2p',2}^n(\mathcal{M}) = n^{\frac{1}{2p'}} L_{2p'}^r(\mathcal{M}) \cap n^{\frac{1}{4}} L_4^r(\mathcal{M}),$$
$$\mathcal{C}_{2p',2}^n(\mathcal{M}) = n^{\frac{1}{2p'}} L_{2p'}^c(\mathcal{M}) \cap n^{\frac{1}{4}} L_4^c(\mathcal{M}).$$
These spaces give rise to the complete isomorphism
$$(8.11) \qquad \mathcal{J}_{p',2}^n(\mathcal{M}) \simeq_{cb} \mathcal{R}_{2p',2}^n(\mathcal{M}) \otimes_{\mathcal{M},h} \mathcal{C}_{2p',2}^n(\mathcal{M}).$$
Indeed, taking $(\mathcal{M}, \mathcal{N}, \mathsf{E}) = (\mathcal{M}_m, \mathsf{M}_m, \mathsf{E}_m)$ in (8.10) we have
$$S_{p'}^m(\mathcal{J}_{p',2}^n(\mathcal{M})) = \mathcal{J}_{p',2}^n(\mathcal{M}_m, \mathsf{E}_m)$$
$$\simeq \mathcal{R}_{2p',2}^n(\mathcal{M}_m, \mathsf{E}_m) \otimes_{\mathcal{M}_m} \mathcal{C}_{2p',2}^n(\mathcal{M}_m, \mathsf{E}_m)$$

$$\begin{aligned}
&= S^m_{(2p',\infty)}(\mathcal{R}^n_{2p',2}(\mathcal{M})) \otimes_{\mathcal{M}_m} S^m_{(\infty,2p')}(\mathcal{C}^n_{2p',2}(\mathcal{M})) \\
&= C^m_{p'} \otimes_h \left(\mathcal{R}^n_{2p',2}(\mathcal{M}) \otimes_{\mathcal{M},h} \mathcal{C}^n_{2p',2}(\mathcal{M}) \right) \otimes_h R^m_{p'} \\
&= S^m_{p'}(\mathcal{R}^n_{2p',2}(\mathcal{M}) \otimes_{\mathcal{M},h} \mathcal{C}^n_{2p',2}(\mathcal{M})),
\end{aligned}$$

where the third identity follows from Lemma 8.20 after taking in consideration our *new* definition of $\mathcal{R}^n_{2p',2}(\mathcal{M},\mathsf{E})$ and $\mathcal{C}^n_{2p',2}(\mathcal{M},\mathsf{E})$. In other words, (8.10) and (8.11) are the amalgamated and operator space versions of the same factorization isomorphism. Now assume that $\mathcal{M}$ is equipped with a *n.s.s.f.* weight ψ, given by the increasing sequence $(\psi_n, q_n)_{n \geq 1}$. Then, we may generalize the factorization result above in the usual way. Namely, assuming by approximation that $\mathrm{k}_n = \psi_n(1)$ are positive integers, we define

$$\begin{aligned}
\mathcal{R}_{2p',2}(\psi_n) &= \mathrm{k}_n^{\frac{1}{2p'}} L^r_{2p'}(q_n \mathcal{M} q_n) \cap \mathrm{k}_n^{\frac{1}{4}} L^r_4(q_n \mathcal{M} q_n), \\
\mathcal{C}_{2p',2}(\psi_n) &= \mathrm{k}_n^{\frac{1}{2p'}} L^c_{2p'}(q_n \mathcal{M} q_n) \cap \mathrm{k}_n^{\frac{1}{4}} L^c_4(q_n \mathcal{M} q_n).
\end{aligned}$$

This gives the complete isomorphism

$$\mathcal{J}_{p',2}(\psi_n) \simeq_{cb} \mathcal{R}_{2p',2}(\psi_n) \otimes_{\mathcal{M},h} \mathcal{C}_{2p',2}(\psi_n) = \bigcap_{u,v \in \{2p',4\}} \mathrm{k}_n^{\frac{1}{u} + \frac{1}{v}} L_{(u,v)}(q_n \mathcal{M} q_n).$$

Then, taking direct limits we obtain the space

$$\mathcal{J}_{p',2}(\psi) = \mathcal{R}_{2p',2}(\psi) \otimes_{\mathcal{M},h} \mathcal{C}_{2p',2}(\psi) = \bigcap_{u,v \in \{2p',4\}} L_{(u,v)}(\mathcal{M}).$$

LEMMA 8.21. *Let $\mathcal{M}$ be a von Neumann algebra equipped with a n.s.s.f. weight ψ. Then, there exists a n.s.s.f. weight ξ on $\mathcal{B}(\ell_2)$ such that the following complete isomorphisms hold*

$$\begin{aligned}
\mathcal{H}^r_{2p',2}(\mathcal{M},\theta) \otimes_h R &\simeq_{cb} \mathcal{R}_{2p',2}(\psi \otimes \xi), \\
C \otimes_h \mathcal{H}^c_{2p',2}(\mathcal{M},\theta) &\simeq_{cb} \mathcal{C}_{2p',2}(\psi \otimes \xi).
\end{aligned}$$

PROOF. By symmetry, we only consider the column case. Let us first observe that $\mathcal{H}_2$ is indeed the graph of an injective closed densely-defined (unbounded) operator with dense range. This is quite similar to [**24**, Remark 2.2]. It follows from the three lines lemma that for $z = a + ib$

$$|f(z)| \leq \|f_{|\partial_0}\|^{1-a}_{L_2(\partial_0,\mu_a)} \|f_{|\partial_1}\|^a_{L_2(\partial_1,\mu_a)}.$$

Since μ_a and μ_θ have the same null sets, we deduce that

$$\pi_j(f) = f_{|\partial_j} \in L_2(\partial_j, \mu_\theta)$$

are injective for $j = 0, 1$ when restricted to analytic functions. Thus, the mapping $\Lambda(\pi_0(f)) = \pi_1(f)$ is an injective closed densely-defined operator with dense range and $\mathcal{H}_2$ is its graph. Let $\Lambda = u|\Lambda|$ be the polar decomposition. Since $M_u = 1 \otimes u^*$ defines a complete isometry (recall that Λ has dense range)

$$L^c_4(\mathcal{M}) \otimes_h L^{oh}_2(\partial_1) \to L^c_4(\mathcal{M}) \otimes_h L^{oh}_2(\partial_0),$$

we may replace Λ by $|\Lambda|$ in the definition of $\mathcal{H}^c_{2p',2}(\mathcal{M},\theta)$. Using [**24**, Lemma 2.8], we may also replace $L_2(\partial_0)$ by ℓ_2 and the operator $|\Lambda|$ by a diagonal operator d_λ. These considerations provide a cb-isomorphism

$$C \otimes_h \mathcal{H}^c_{2p',2}(\mathcal{M},\theta) \simeq_{cb} \left(C \otimes_h L^c_{2p'}(\mathcal{M}) \otimes_h R_{p'} \right) \cap \left(C \otimes_h L^c_4(\mathcal{M}) \otimes_h \ell^{oh}_2(\lambda) \right),$$

8.3. EMBEDDING FOR GENERAL VON NEUMANN ALGEBRAS   151

where $\ell_2^{oh}(\lambda)$ is the weighted form of OH which arises from the action of d_λ. The assertion follows by a direct limit argument. Indeed, the n.s.s.f. weight ψ on $\mathcal{M}$ is given by the sequence $(\psi_n, q_n)_{n \geq 1}$. On the other hand, we may consider the n.s.s.f. weight ξ on $\mathcal{B}(\ell_2)$ determined by the sequence $(\xi_n, \pi_n)_{n \geq 1}$, where π_n is the projection onto the first n coordinates and ξ_n is the finite weight on $\pi_n \mathcal{B}(\ell_2) \pi_n$ given by

$$\xi_n\Big(\pi_n\big(\sum_{ij} x_{ij} e_{ij}\big) \pi_n\Big) = \sum_{k=1}^n \gamma_k x_{kk} \quad \text{with} \quad \gamma_k^{\frac{1}{4} - \frac{1}{2p'}} = \lambda_k.$$

Let us define the parameters $k_n' = \xi_n(1)$ and $w_n = k_n k_n'$. Then, arguing as we did in [**24**, Lemma 2.4], it turns out that the intersection space above is the direct limit of the following sequence of spaces

$$w_n^{\frac{1}{2p'}} \Big(L_{2p'}^c\big(q_n \mathcal{M} q_n \bar{\otimes} \pi_n \mathcal{B}(\ell_2) \pi_n\big)\Big) \cap w_n^{\frac{1}{4}} \Big(L_4^c\big(q_n \mathcal{M} q_n \bar{\otimes} \pi_n \mathcal{B}(\ell_2) \pi_n\big)\Big).$$

However, the latter space is $\mathcal{C}_{2p',2}(\psi_n \otimes \xi_n)$. This completes the proof. $\square$

PROPOSITION 8.22. *The predual space of*

$$\mathcal{H}_{2p',2}^r(\mathcal{M}, \theta) \otimes_{\mathcal{M},h} \mathcal{H}_{2p',2}^c(\mathcal{M}, \theta)$$

embeds completely isomorphically into $\mathcal{K}_{rc_p}^p(\phi \otimes \psi \otimes \xi)$ *for some n.s.s.f. weight* ξ *on* $\mathcal{B}(\ell_2)$ *and where* ϕ *denotes the quasi-free state over the hyperfinite* III_1 *factor* $\mathcal{R}$ *considered in Proposition 8.10.*

PROOF. According to Lemma 8.21

$$\mathcal{H}_{2p',2}^r(\mathcal{M}, \theta) \otimes_{\mathcal{M},h} \mathcal{H}_{2p',2}^c(\mathcal{M}, \theta)$$
$$= \Big(\mathcal{H}_{2p',2}^r(\mathcal{M}, \theta) \otimes_h R\Big) \otimes_{\mathcal{M} \bar{\otimes} \mathcal{B}(\ell_2), h} \Big(C \otimes_h \mathcal{H}_{2p',2}^c(\mathcal{M}, \theta)\Big)$$
$$\simeq_{cb} \mathcal{R}_{2p',2}(\psi \otimes \xi) \otimes_{\mathcal{M} \bar{\otimes} \mathcal{B}(\ell_2), h} \mathcal{C}_{2p',2}(\psi \otimes \xi) \simeq_{cb} \mathcal{J}_{p',2}(\psi \otimes \xi).$$

However, $\mathcal{J}_{p',2}(\psi \otimes \xi)$ is a direct limit of spaces

$$\mathcal{J}_{p',2}(\psi_n \otimes \xi_n) = \mathcal{J}_{p',2}^{w_n}\Big(q_n \mathcal{M} q_n \bar{\otimes} \pi_n \mathcal{B}(\ell_2) \pi_n\Big).$$

According to Proposition 8.10, the direct limit

$$\mathcal{K}_{p,2}(\psi \otimes \xi) = \lim_n \mathcal{K}_{p,2}^n(\psi_n \otimes \xi_n)$$

of the corresponding predual spaces cb-embeds into $\mathcal{K}_{rc_p}^p(\phi \otimes \psi \otimes \xi)$. $\square$

Now we are ready to prove our main result. In the proof we shall need to work with certain quotient of $\mathcal{H}_{2p',2}^r(\mathcal{M}, \theta) \otimes_{\mathcal{M},h} \mathcal{H}_{2p',2}^c(\mathcal{M}, \theta)$. Namely, recalling the subspaces $\mathcal{H}_{r,0}$ and $\mathcal{H}_{c,0}$, we set

$$\mathcal{Q}_{2p',2}(\mathcal{M}, \theta) = \big(\mathcal{H}_{2p',2}^r(\mathcal{M}, \theta)/\mathcal{H}_{r,0}\big) \otimes_{\mathcal{M},h} \big(\mathcal{H}_{2p',2}^c(\mathcal{M}, \theta)/\mathcal{H}_{c,0}\big).$$

We claim that $\mathcal{Q}_{2p',2}(\mathcal{M}, \theta)$ is a quotient of $\mathcal{H}_{2p',2}^r(\mathcal{M}, \theta) \otimes_{\mathcal{M},h} \mathcal{H}_{2p',2}^c(\mathcal{M}, \theta)$. Indeed, according to the definition of the $\mathcal{M}$-amalgamated Haagerup tensor product of two operator spaces (see Chapter 6), we may write $\mathcal{Q}_{2p',2}(\mathcal{M}, \theta)$ as a quotient of the Haagerup tensor product

$$\Lambda_{2p',2}(\mathcal{M}, \theta) = \big(\mathcal{H}_{2p',2}^r(\mathcal{M}, \theta)/\mathcal{H}_{r,0}\big) \otimes_h \big(\mathcal{H}_{2p',2}^c(\mathcal{M}, \theta)/\mathcal{H}_{c,0}\big)$$

by the closed subspace spanned by the differences $x_1 \gamma \otimes x_2 - x_1 \otimes \gamma x_2$, with $\gamma \in \mathcal{M}$. Therefore, it suffices to see that the space $\Lambda_{2p',2}(\mathcal{M}, \theta)$ is a quotient of $\mathcal{H}^r_{2p',2}(\mathcal{M},\theta) \otimes_h \mathcal{H}^c_{2p',2}(\mathcal{M},\theta)$. However, this follows from the projectivity of the Haagerup tensor product and our claim follows.

Let us prove our main embedding result.

THEOREM 8.23. *Let $1 \le p < q \le 2$ and let $\mathcal{M}$ be a von Neumann algebra. Then, there exists a sufficiently large von Neumann algebra $\mathcal{A}$ and a completely isomorphic embedding of $L_q(\mathcal{M})$ into $L_p(\mathcal{A})$, where both spaces are equipped with their respective natural operator space structures. Moreover, we have*

  i) *If $\mathcal{M}$ is QWEP, we can choose $\mathcal{A}$ to be QWEP.*
  ii) *If $\mathcal{M}$ is hyperfinite, we can choose $\mathcal{A}$ to be hyperfinite.*

PROOF. Let us first assume that $\mathcal{M}$ is finite. According to Theorem 8.11 and Proposition 8.22, it suffices to prove that the operator space $L_{q'}(\mathcal{M})$ is completely isomorphic to a quotient of
$$\mathcal{H}^r_{2p',2}(\mathcal{M},\theta) \otimes_\mathcal{M} \mathcal{H}^c_{2p',2}(\mathcal{M},\theta).$$
This follows from Proposition 8.19 since
$$L_{q'}(\mathcal{M}) \simeq_{cb} L^r_{2q'}(\mathcal{M}) \otimes_{\mathcal{M},h} L^c_{2q'}(\mathcal{M}) \simeq_{cb} \mathcal{Q}_{2p',2}(\mathcal{M},\theta).$$
The construction of the cb-embedding for a general von Neumann algebra $\mathcal{M}$ can be obtained by using Haagerup's approximation theorem [**12**] and the fact that direct limits are stable in our construction. Indeed, Haagerup theorem shows that for every σ-finite von Neumann algebra $\mathcal{M}$, the space $L_q(\mathcal{M})$ is complemented in a direct limit of L_q spaces over finite von Neumann algebras. Finally, if $\mathcal{M}$ is any von Neumann algebra, we observe that $L_q(\mathcal{M})$ can always be written as a direct limit of L_q spaces associated to σ-finite von Neumann algebras. On the other hand, the stability of hyperfiniteness follows directly from our construction. Indeed, our construction goes as follows
$$L_q(\mathcal{M}) \to \Big(\mathcal{H}^r_{2p',2}(\mathcal{M},\theta) \otimes_{\mathcal{M},h} \mathcal{H}^c_{2p',2}(\mathcal{M},\theta)\Big)_* \to \mathcal{K}^p_{rc_p}(\phi \otimes \psi \otimes \xi) \to L_p(\mathcal{A})$$
where the first embedding follows as above, the second from Proposition 8.22 and the last one from Theorem 8.11. In particular, it turns out that the von Neumann algebra $\mathcal{A}$ is of the form
$$\mathcal{A} = M_{sym}\big(\mathcal{R} \bar\otimes \mathcal{M} \bar\otimes \mathcal{B}(\ell_2)\big),$$
which is hyperfinite when $\mathcal{M}$ is hyperfinite and is a factor when $\mathcal{M}$ is a factor. Finally, it remains to justify that the QWEP is preserved. If $\mathcal{M}$ is QWEP, there exists a completely isometric embedding of $L_q(\mathcal{M})$ into $L_q(\mathcal{M}_\mathcal{U})$ with $\mathcal{M}_\mathcal{U}$ of the form
$$\mathcal{M}_\mathcal{U} = \Big(\prod\nolimits_{n,\mathcal{U}} S^n_1\Big)^*.$$
Since we know from Corollary 8.2 that the Schatten class S^n_q embeds completely isomorphically into $L_p(\mathcal{A}_n)$ with relevant constants independent of n and $\mathcal{A}_n$ being QWEP, we find a completely isomorphic embedding
$$L_q(\mathcal{M}) \to L_p(\mathcal{A}_\mathcal{U}) \quad \text{with} \quad \mathcal{A}_\mathcal{U} = \Big(\prod\nolimits_{n,\mathcal{U}} \mathcal{A}_{n*}\Big)^*.$$
This proves the assertion since $\mathcal{A}_\mathcal{U}$ is QWEP. The proof is complete. □

Bibliography

1. H. Araki and E.J. Woods, *A classification of factors.* Publ. Res. Inst. Math. Sci. Ser. A **4** (1968/69), 51-130. MR0244773 (39:6087)
2. J. Bergh and J. Löfström, Interpolation Spaces. Springer-Verlag, Berlin, 1976. MR0482275 (58:2349)
3. J. Bretagnolle, D. Dacunha-Castelle and J.L. Krivine, *Lois stables et espaces L^p.* Ann. Inst. H. Poincaré **2** (1966), 231-259. MR0203757 (34:3605)
4. D.L. Burkholder, *Distribution function inequalities for martingales.* Ann. Probab. **1** (1973), 19-42. MR0365692 (51:1944)
5. D.L. Burkholder and R.F. Gundy, *Extrapolation and interpolation of quasi-linear operators on martingales.* Acta Math. **124** (1970), 249-304. MR0440695 (55:13567)
6. A. Connes, *Une classification des facteurs de type III.* Ann. Sci. École Norm. Sup. **6** (1973), 133-252. MR0341115 (49:5865)
7. A. Defant and M. Junge, *Maximal theorems of Menchoff-Rademacher type in non-commutative L_q-spaces.* J. Funct. Anal. **206** (2004), 322-355. MR2021850 (2005k:46169)
8. A. Devinatz, *The factorization of operator valued analytic functions.* Ann. of Math. **73** (1961), 458-495. MR0126702 (23:A3997)
9. K. Dykema, *Exactness of reduced amalgamated free product C^*-algebras.* Forum Math **16** (2004), 121-189. MR2039095 (2004m:46142)
10. E.G. Effros, M. Junge and Z.J. Ruan, *Integral mappings and the principle of local reflexivity for noncommutative L_1 spaces.* Ann. of Math. **151** (2000), 59-92. MR1745018 (2000m:46120)
11. E.G. Effros and Z.J. Ruan, Operator Spaces. London Math. Soc. Monogr. **23**, Oxford University Press, 2000. MR1793753 (2002a:46082)
12. U. Haagerup, *Non-commutative integration theory.* Unpublished manuscript (1978). See also Haagerup's Lecture given at the Symposium in Pure Mathematics of the Amer. Math. Soc. Queens University, Kingston, Ontario, 1980.
13. U. Haagerup, *L_p spaces associated with an arbitrary von Neumann algebra.* Algèbres d'opérateurs et leurs applications en physique mathématique, CNRS (1979), 175-184. MR560633 (81e:46050)
14. U. Haagerup, H.P. Rosenthal and F.A. Sukochev, Banach Embedding Properties of Noncommutative L_p-Spaces. Mem. Amer. Math. Soc. **163**, 2003. MR1963854 (2004f:46076)
15. M. Junge, *Embeddings of non-commutative L_p-spaces into non-commutative L_1-spaces, $1 < p < 2$.* Geom. Funct. Anal. **10** (2000), 389-406. MR1771425 (2001f:46098)
16. M. Junge, *Doob's inequality for non-commutative martingales.* J. reine angew. Math. **549** (2002), 149-190. MR1916654 (2003k:46097)
17. M. Junge, *Embedding of the operator space OH and the logarithmic 'little Grothendieck inequality'.* Invent. Math. **161** (2005), 225-286. MR2180450 (2006i:47130)
18. M. Junge, *Operator spaces and Araki-Woods factors: A quantum probabilistic approach.* Int. Math. Res. Pap. 2006. MR2268491
19. M. Junge, Vector-valued L_p spaces over QWEP von Neumann algebras. In progress.
20. M. Junge, Noncommutative Poisson random measure. In progress.
21. M. Junge and M. Musat, *A noncommutative version of the John-Nirenberg theorem.* Trans. Amer. Math. Soc. **359** (2007), 115-142. MR2247885 (2007m:46103)
22. M. Junge and J. Parcet, *The norm of sums of independent noncommutative random variables in $L_p(\ell_1)$.* J. Funct. Anal. **221** (2005), 366-406. MR2124869 (2005k:46151)
23. M. Junge and J. Parcet, *Rosenthal's theorem for subspaces of noncommutatuve L_p.* Duke Math. J. **141** (2008), 75-122. MR2372148

24. M. Junge and J. Parcet, Operator space embedding of Schatten p-classes into von Neumann algebra preduals. Geom. Funct. Anal. **18** (2008), 522-551. MR2421547
25. M. Junge and J. Parcet, A transference method in quantum probability. Preprint 2008.
26. M. Junge, J. Parcet and Q. Xu, *Rosenthal type inequalities for free chaos*. Ann. Probab. **35** (2007), 1374-1437. MR2330976
27. M. Junge and Z.J. Ruan, *Decomposable maps on non-commutative L_p spaces*. Contemp. Math. **365** (2004), 355-381. MR2106828 (2005k:46150)
28. M. Junge and Q. Xu, *Noncommutative Burkholder/Rosenthal inequalities*. Ann. Probab. **31** (2003), 948-995. MR1964955 (2004f:46078)
29. M. Junge and Q. Xu, *Noncommutative maximal ergodic theorems*. J. Amer. Math. Soc. **20** (2007), 385-439. MR2276775 (2007k:46109)
30. M. Junge and Q. Xu, *Noncommutative Burkholder/Rosenthal inequalities II: Applications*. Israel J. Math. **167** (2008), 227-282. MR2448025
31. R.V. Kadison and J.R. Ringrose, Fundamentals of the Theory of Operator Algebras I and II. Grad. Stud. Math., **15** & **16**, American Mathematical Society, 1997. MR1468229 (98f:46001a); MR1468230 (98f:46001b)
32. H. Kosaki, *Applications of the complex interpolation method to a von Neumann algebra*. J. Funct. Anal. **56** (1984), 29-78. MR735704 (86a:46085)
33. S. Kwapień, *Isomorphic characterizations of inner product spaces by orthogonal series with vector valued coefficients*. Studia Math. **44** (1972), 583-595. MR0341039 (49:5789)
34. E.C. Lance, Hilbert C*-modules. Cambridge University Press, 1995. MR1325694 (96k:46100)
35. J. Lindenstrauss and L. Tzafriri, Classical Banach Spaces I and II. Springer-Verlag, 1996. MR0500056 (58:17766)
36. F. Lust-Piquard, *Inégalités de Khintchine dans C_p ($1 < p < \infty$)*. C.R. Acad. Sci. Paris **303** (1986), 289-292. MR859804 (87j:47032)
37. F. Lust-Piquard and G. Pisier, *Non-commutative Khintchine and Paley inequalities*. Ark. Mat. **29** (1991), 241-260. MR1150376 (94b:46011)
38. M. Musat, *Interpolation between non-commutative BMO and non-commutative L_p-spaces*. J. Funct. Anal. **202** (2003), 195- 225. MR1994770 (2004g:46081)
39. J. Parcet, *B-convex operator spaces*. Proc. Edinburgh Math. Soc. **46** (2003), 649-668. MR2013959 (2004k:46095)
40. J. Parcet and G. Pisier, *Non-commutative Khintchine type inequalities associated with free groups*. Indiana Univ. Math. J. **54** (2005), 531-556. MR2136820 (2006a:46071)
41. J. Parcet and N. Randrianantoanina, *Gundy's decomposition for non-commutative martingales and applications*. Proc. London Math. Soc. **93** (2006), 227-252. MR2235948 (2007b:46115)
42. G. Pedersen and M. Takesaki, *The Radon-Nikodym theorem for von Neumann algebras*. Acta Math. **130** (1973), 53-87. MR0412827 (54:948)
43. G. Pisier, Factorization of Linear Operators and Geometry of Banach Spaces. CBMS (Regional conferences of the A.M.S.) **60**, 1987. MR829919 (88a:47020)
44. G. Pisier, *Projections from a von Neumann algebra onto a subalgebra*. Bull. Soc. Math. France **123** (1995), 139-153. MR1330791 (96f:46111)
45. G. Pisier, The Operator Hilbert Space OH, Complex Interpolation and Tensor Norms. Mem. Amer. Math. Soc. **122** (1996). MR1342022 (97a:46024)
46. G. Pisier, Non-Commutative Vector Valued L_p-Spaces and Completely p-Summing Maps. Astérisque **247** (1998).
47. G. Pisier, Introduction to Operator Space Theory. Cambridge University Press, 2003. MR2006539 (2004k:46097)
48. G. Pisier, *The operator Hilbert space OH and type III von Neumann algebras*. Bull. London Math. Soc. **36** (2004), 455-459. MR2069007 (2005c:46082)
49. G. Pisier, *Completely bounded maps into certain Hilbertian operator spaces*. Internat. Math. Res. Notices **74** (2004), 3983-4018. MR2103799 (2005g:46114)
50. G. Pisier and D. Shlyakhtenko, *Grothendieck's theorem for operator spaces*. Invent. Math. **150** (2002), 185-217. MR1930886 (2004k:46096)
51. G. Pisier and Q. Xu, *Non-commutative martingale inequalities*. Comm. Math. Phys. **189** (1997), 667-698. MR1482934 (98m:46079)

52. G. Pisier and Q. Xu, *Non-commutative L_p-spaces*. Handbook of the Geometry of Banach Spaces II (Ed. W.B. Johnson and J. Lindenstrauss) North-Holland (2003), 1459-1517. MR1999201 (2004i:46095)
53. N. Randrianantoanina, *Non-commutative martingale transforms*. J. Funct. Anal. **194** (2002), 181-212. MR1929141 (2003m:46098)
54. N. Randrianantoanina, *Conditioned square functions for noncommutative martingales*. Ann. Probab. **35** (2007), 1039-1070. MR2319715 (2009d:46112)
55. Y. Raynaud, *On ultrapowers of non-commutative L_p spaces*. J. Operator Theory **48** (2002), 41-68. MR1926043 (2003i:46069)
56. H.P. Rosenthal, *On the subspaces of L^p ($p > 2$) spanned by sequences of independent random variables*. Israel J. Math. **8** (1970), 273-303. MR0271721 (42:6602)
57. H.P. Rosenthal, *On subspaces of L_p*. Ann. of Math. **97** (1973), 344-373. MR0312222 (47:784)
58. Z.J. Ruan, *Subspaces of C^*-algebras*. J. Funct. Anal. **76** (1988), 217-230. MR923053 (89h:46082)
59. D. Shlyakhtenko, *Free quasi-free states*. Pacific J. Math. **177** (1997), 329-368. MR1444786 (98b:46086)
60. M. Takesaki, Tomita's theory of Modular Hilbert Algebras and Its Applications. Lecture Notes in Mathematics **128**, Springer, 1970. MR0270168 (42:5061)
61. M. Takesaki, *Conditional expectations in von Neumann algebras*. J. Func. Anal. **9** (1972), 306-321. MR0303307 (46:2445)
62. M. Takesaki, Theory of operator algebras I. Springer-Verlag, New York, 1979. MR548728 (81e:46038)
63. M. Terp, L_p spaces associated with von Neumann algebras. Math. Institute Copenhagen University, 1981.
64. M. Terp, *Interpolation spaces between a von Neumann algebra and its predual*. J. Operator Theory **8** (1982), 327-360. MR677418 (85b:46075)
65. D.V. Voiculescu, *Symmetries of some reduced free product C^*-algebras*. Operator Algebras and Their Connections with Topology and Ergodic Theory, Springer-Verlag **1132** (1985), 556-588. MR799593 (87d:46075)
66. D.V. Voiculescu, *A strengthened asymptotic freeness result for random matrices with applications to free entropy*. Internat. Math. Res. Notices **1** (1998), 41-63. MR1601878 (2000d:46080)
67. D.V. Voiculescu, K. Dykema and A. Nica, Free random variables. CRM Monograph Series **1**, American Mathematical Society, 1992. MR1217253 (94c:46133)
68. Q. Xu, *Recent devepolment on non-commutative martingale inequalities*. Functional Space Theory and its Applications. Proceedings of International Conference & 13th Academic Symposium in China. Ed. Research Information Ltd UK. Wuhan 2003, 283-314.
69. Q. Xu, *A description of $(C_p[L_p(M)], R_p[L_p(M)])_\theta$*. Proc. Roy. Soc. Edinburgh Sect. A **135** (2005), 1073-1083. MR2187225 (2006i:47131)
70. Q. Xu, *Operator-space Grothendieck inequalities for noncommutative L_p-spaces*. Duke Math. J. **131** (2006), 525-574. MR2219250 (2007b:46101)
71. Q. Xu, *Embedding of C_q and R_q into noncommutative L_p-spaces, $1 \leq p < q \leq 2$*. Math. Ann. **335** (2006), 109-131. MR2217686 (2007m:46095)

Editorial Information

To be published in the *Memoirs*, a paper must be correct, new, nontrivial, and significant. Further, it must be well written and of interest to a substantial number of mathematicians. Piecemeal results, such as an inconclusive step toward an unproved major theorem or a minor variation on a known result, are in general not acceptable for publication.

Papers appearing in *Memoirs* are generally at least 80 and not more than 200 published pages in length. Papers less than 80 or more than 200 published pages require the approval of the Managing Editor of the Transactions/Memoirs Editorial Board. Published pages are the same size as those generated in the style files provided for $\mathcal{AMS}$-LaTeX or $\mathcal{AMS}$-TeX.

Information on the backlog for this journal can be found on the AMS website starting from http://www.ams.org/memo.

In an effort to make articles available as quickly as possible, *Memoir* articles are posted individually to the AMS website before publication.

A Consent to Publish and Copyright Agreement is required before a paper will be published in the *Memoirs*. After a paper is accepted for publication, the Providence office will send a Consent to Publish and Copyright Agreement to all authors of the paper. By submitting a paper to the *Memoirs*, authors certify that the results have not been submitted to nor are they under consideration for publication by another journal, conference proceedings, or similar publication.

Information for Authors

Memoirs is an author-prepared publication. Once formatted for print and on-line publication, articles will be published as is with the addition of AMS-prepared frontmatter and backmatter. Articles are not copyedited; however, confirmation copy will be sent to the authors.

Initial submission. The AMS uses Centralized Manuscript Processing for initial submissions. Authors should submit a PDF file using the Initial Manuscript Submission form found at www.ams.org/peer-review-submission, or send one copy of the manuscript to the following address: Centralized Manuscript Processing, MEMOIRS OF THE AMS, 201 Charles Street, Providence, RI 02904-2294 USA. If a paper copy is being forwarded to the AMS, indicate that it is for *Memoirs* and include the name of the corresponding author, contact information such as email address or mailing address, and the name of an appropriate Editor to review the paper (see the list of Editors below).

The paper must contain a *descriptive title* and an *abstract* that summarizes the article in language suitable for workers in the general field (algebra, analysis, etc.). The *descriptive title* should be short, but informative; useless or vague phrases such as "some remarks about" or "concerning" should be avoided. The *abstract* should be at least one complete sentence, and at most 300 words. Included with the footnotes to the paper should be the 2010 *Mathematics Subject Classification* representing the primary and secondary subjects of the article. The classifications are accessible from www.ams.org/msc/. The Mathematics Subject Classification footnote may be followed by a list of *key words and phrases* describing the subject matter of the article and taken from it. Journal abbreviations used in bibliographies are listed in the latest *Mathematical Reviews* annual index. The series abbreviations are also accessible from www.ams.org/msnhtml/serials.pdf. To help in preparing and verifying references, the AMS offers MR Lookup, a Reference Tool for Linking, at www.ams.org/mrlookup/.

Electronically prepared manuscripts. The AMS encourages electronically prepared manuscripts, with a strong preference for $\mathcal{AMS}$-LaTeX. To this end, the Society has prepared $\mathcal{AMS}$-LaTeX author packages for each AMS publication. Author packages include instructions for preparing electronic manuscripts, samples, and a style file that generates the particular design specifications of that publication series. Though $\mathcal{AMS}$-LaTeX is the highly preferred format of TeX, author packages are also available in $\mathcal{AMS}$-TeX.

Authors may retrieve an author package for *Memoirs of the AMS* from `www.ams.org/journals/memo/memoauthorpac.html` or via FTP to `ftp.ams.org` (login as `anonymous`, enter your complete email address as password, and type `cd pub/author-info`). The *AMS Author Handbook* and the *Instruction Manual* are available in PDF format from the author package link. The author package can also be obtained free of charge by sending email to `tech-support@ams.org` (Internet) or from the Publication Division, American Mathematical Society, 201 Charles St., Providence, RI 02904-2294, USA. When requesting an author package, please specify $\mathcal{AMS}$-LaTeX or $\mathcal{AMS}$-TeX and the publication in which your paper will appear. Please be sure to include your complete mailing address.

After acceptance. The source files for the final version of the electronic manuscript should be sent to the Providence office immediately after the paper has been accepted for publication. The author should also submit a PDF of the final version of the paper to the editor, who will forward a copy to the Providence office.

Accepted electronically prepared files can be submitted via the web at `www.ams.org/submit-book-journal/`, sent via FTP, or sent on CD-Rom or diskette to the Electronic Prepress Department, American Mathematical Society, 201 Charles Street, Providence, RI 02904-2294 USA. TeX source files and graphic files can be transferred over the Internet by FTP to the Internet node `ftp.ams.org` (130.44.1.100). When sending a manuscript electronically via CD-Rom or diskette, please be sure to include a message indicating that the paper is for the *Memoirs*.

Electronic graphics. Comprehensive instructions on preparing graphics are available at `www.ams.org/authors/journals.html`. A few of the major requirements are given here.

Submit files for graphics as EPS (Encapsulated PostScript) files. This includes graphics originated via a graphics application as well as scanned photographs or other computer-generated images. If this is not possible, TIFF files are acceptable as long as they can be opened in Adobe Photoshop or Illustrator.

Authors using graphics packages for the creation of electronic art should also avoid the use of any lines thinner than 0.5 points in width. Many graphics packages allow the user to specify a "hairline" for a very thin line. Hairlines often look acceptable when proofed on a typical laser printer. However, when produced on a high-resolution laser imagesetter, hairlines become nearly invisible and will be lost entirely in the final printing process.

Screens should be set to values between 15% and 85%. Screens which fall outside of this range are too light or too dark to print correctly. Variations of screens within a graphic should be no less than 10%.

Inquiries. Any inquiries concerning a paper that has been accepted for publication should be sent to `memo-query@ams.org` or directly to the Electronic Prepress Department, American Mathematical Society, 201 Charles St., Providence, RI 02904-2294 USA.

Editors

This journal is designed particularly for long research papers, normally at least 80 pages in length, and groups of cognate papers in pure and applied mathematics. Papers intended for publication in the *Memoirs* should be addressed to one of the following editors. The AMS uses Centralized Manuscript Processing for initial submissions to AMS journals. Authors should follow instructions listed on the Initial Submission page found at www.ams.org/memo/memosubmit.html.

Algebra, to ALEXANDER KLESHCHEV, Department of Mathematics, University of Oregon, Eugene, OR 97403-1222; e-mail: ams@noether.uoregon.edu

Algebraic geometry, to DAN ABRAMOVICH, Department of Mathematics, Brown University, Box 1917, Providence, RI 02912; e-mail: amsedit@math.brown.edu

Algebraic geometry and its applications, to MINA TEICHER, Emmy Noether Research Institute for Mathematics, Bar-Ilan University, Ramat-Gan 52900, Israel; e-mail: teicher@macs.biu.ac.il

Algebraic topology, to ALEJANDRO ADEM, Department of Mathematics, University of British Columbia, Room 121, 1984 Mathematics Road, Vancouver, British Columbia, Canada V6T 1Z2; e-mail: adem@math.ubc.ca

Combinatorics, to JOHN R. STEMBRIDGE, Department of Mathematics, University of Michigan, Ann Arbor, Michigan 48109-1109; e-mail: JRS@umich.edu

Commutative and homological algebra, to LUCHEZAR L. AVRAMOV, Department of Mathematics, University of Nebraska, Lincoln, NE 68588-0130; e-mail: avramov@math.unl.edu

Complex analysis and harmonic analysis, to ALEXANDER NAGEL, Department of Mathematics, University of Wisconsin, 480 Lincoln Drive, Madison, WI 53706-1313; e-mail: nagel@math.wisc.edu

Differential geometry and global analysis, to CHRIS WOODWARD, Department of Mathematics, Rutgers University, 110 Frelinghuysen Road, Piscataway, NJ 08854; e-mail: ctw@math.rutgers.edu

Dynamical systems and ergodic theory and complex analysis, to YUNPING JIANG, Department of Mathematics, CUNY Queens College and Graduate Center, 65-30 Kissena Blvd., Flushing, NY 11367; e-mail: Yunping.Jiang@qc.cuny.edu

Functional analysis and operator algebras, to DIMITRI SHLYAKHTENKO, Department of Mathematics, University of California, Los Angeles, CA 90095; e-mail: shlyakht@math.ucla.edu

Geometric analysis, to WILLIAM P. MINICOZZI II, Department of Mathematics, Johns Hopkins University, 3400 N. Charles St., Baltimore, MD 21218; e-mail: trans@math.jhu.edu

Geometric topology, to MARK FEIGHN, Math Department, Rutgers University, Newark, NJ 07102; e-mail: feighn@andromeda.rutgers.edu

Harmonic analysis, representation theory, and Lie theory, to ROBERT J. STANTON, Department of Mathematics, The Ohio State University, 231 West 18th Avenue, Columbus, OH 43210-1174; e-mail: stanton@math.ohio-state.edu

Logic, to STEFFEN LEMPP, Department of Mathematics, University of Wisconsin, 480 Lincoln Drive, Madison, Wisconsin 53706-1388; e-mail: lempp@math.wisc.edu

Number theory, to JONATHAN ROGAWSKI, Department of Mathematics, University of California, Los Angeles, CA 90095; e-mail: jonr@math.ucla.edu

Number theory, to SHANKAR SEN, Department of Mathematics, 505 Malott Hall, Cornell University, Ithaca, NY 14853; e-mail: ss70@cornell.edu

Partial differential equations, to GUSTAVO PONCE, Department of Mathematics, South Hall, Room 6607, University of California, Santa Barbara, CA 93106; e-mail: ponce@math.ucsb.edu

Partial differential equations and dynamical systems, to PETER POLACIK, School of Mathematics, University of Minnesota, Minneapolis, MN 55455; e-mail: polacik@math.umn.edu

Probability and statistics, to RICHARD BASS, Department of Mathematics, University of Connecticut, Storrs, CT 06269-3009; e-mail: bass@math.uconn.edu

Real analysis and partial differential equations, to DANIEL TATARU, Department of Mathematics, University of California, Berkeley, Berkeley, CA 94720; e-mail: tataru@math.berkeley.edu

All other communications to the editors, should be addressed to the Managing Editor, ROBERT GURALNICK, Department of Mathematics, University of Southern California, Los Angeles, CA 90089-1113; e-mail: guralnic@math.usc.edu.

Titles in This Series

956 **Richard Montgomery and Michail Zhitomirskii,** Points and curves in the Monster tower, 2010

955 **Martin R. Bridson and Daniel Groves,** The quadratic isoperimetric inequality for mapping tori of free group automorphisms, 2010

954 **Volker Mayer and Mariusz Urbański,** Thermodynamical formalism and multifractal analysis for meromorphic functions of finite order, 2010

953 **Marius Junge and Javier Parcet,** Mixed-norm inequalities and operator space L_p embedding theory, 2010

952 **Martin W. Liebeck, Cheryl E. Praeger, and Jan Saxl,** Regular subgroups of primitive permutation groups, 2010

951 **Pierre Magal and Shigui Ruan,** Center manifolds for semilinear equations with non-dense domain and applications to Hopf bifurcation in age structured models, 2009

950 **Cédric Villani,** Hypocoercivity, 2009

949 **Drew Armstrong,** Generalized noncrossing partitions and combinatorics of Coxeter groups, 2009

948 **Nan-Kuo Ho and Chiu-Chu Melissa Liu,** Yang-Mills connections on orientable and nonorientable surfaces, 2009

947 **W. Turner,** Rock blocks, 2009

946 **Jay Jorgenson and Serge Lang,** Heat Eisenstein series on $SL_n(C)$, 2009

945 **Tobias H. Jäger,** The creation of strange non-chaotic attractors in non-smooth saddle-node bifurcations, 2009

944 **Yuri Kifer,** Large deviations and adiabatic transitions for dynamical systems and Markov processes in fully coupled averaging, 2009

943 **István Berkes and Michel Weber,** On the convergence of $\sum c_k f(n_k x)$, 2009

942 **Dirk Kussin,** Noncommutative curves of genus zero: Related to finite dimensional algebras, 2009

941 **Gelu Popescu,** Unitary invariants in multivariable operator theory, 2009

940 **Gérard Iooss and Pavel I. Plotnikov,** Small divisor problem in the theory of three-dimensional water gravity waves, 2009

939 **I. D. Suprunenko,** The minimal polynomials of unipotent elements in irreducible representations of the classical groups in odd characteristic, 2009

938 **Antonino Morassi and Edi Rosset,** Uniqueness and stability in determining a rigid inclusion in an elastic body, 2009

937 **Skip Garibaldi,** Cohomological invariants: Exceptional groups and spin groups, 2009

936 **André Martinez and Vania Sordoni,** Twisted pseudodifferential calculus and application to the quantum evolution of molecules, 2009

935 **Mihai Ciucu,** The scaling limit of the correlation of holes on the triangular lattice with periodic boundary conditions, 2009

934 **Arjen Doelman, Björn Sandstede, Arnd Scheel, and Guido Schneider,** The dynamics of modulated wave trains, 2009

933 **Luchezar Stoyanov,** Scattering resonances for several small convex bodies and the Lax-Phillips conjecture, 2009

932 **Jun Kigami,** Volume doubling measures and heat kernel estimates of self-similar sets, 2009

931 **Robert C. Dalang and Marta Sanz-Solé,** Hölder-Sobolv regularity of the solution to the stochastic wave equation in dimension three, 2009

930 **Volkmar Liebscher,** Random sets and invariants for (type II) continuous tensor product systems of Hilbert spaces, 2009

929 **Richard F. Bass, Xia Chen, and Jay Rosen,** Moderate deviations for the range of planar random walks, 2009

928 **Ulrich Bunke,** Index theory, eta forms, and Deligne cohomology, 2009

927 **N. Chernov and D. Dolgopyat,** Brownian Brownian motion-I, 2009

TITLES IN THIS SERIES

- 926 **Riccardo Benedetti and Francesco Bonsante,** Canonical wick rotations in 3-dimensional gravity, 2009
- 925 **Sergey Zelik and Alexander Mielke,** Multi-pulse evolution and space-time chaos in dissipative systems, 2009
- 924 **Pierre-Emmanuel Caprace,** "Abstract" homomorphisms of split Kac-Moody groups, 2009
- 923 **Michael Jöllenbeck and Volkmar Welker,** Minimal resolutions via algebraic discrete Morse theory, 2009
- 922 **Ph. Barbe and W. P. McCormick,** Asymptotic expansions for infinite weighted convolutions of heavy tail distributions and applications, 2009
- 921 **Thomas Lehmkuhl,** Compactification of the Drinfeld modular surfaces, 2009
- 920 **Georgia Benkart, Thomas Gregory, and Alexander Premet,** The recognition theorem for graded Lie algebras in prime characteristic, 2009
- 919 **Roelof W. Bruggeman and Roberto J. Miatello,** Sum formula for SL_2 over a totally real number field, 2009
- 918 **Jonathan Brundan and Alexander Kleshchev,** Representations of shifted Yangians and finite W-algebras, 2008
- 917 **Salah-Eldin A. Mohammed, Tusheng Zhang, and Huaizhong Zhao,** The stable manifold theorem for semilinear stochastic evolution equations and stochastic partial differential equations, 2008
- 916 **Yoshikata Kida,** The mapping class group from the viewpoint of measure equivalence theory, 2008
- 915 **Sergiu Aizicovici, Nikolaos S. Papageorgiou, and Vasile Staicu,** Degree theory for operators of monotone type and nonlinear elliptic equations with inequality constraints, 2008
- 914 **E. Shargorodsky and J. F. Toland,** Bernoulli free-boundary problems, 2008
- 913 **Ethan Akin, Joseph Auslander, and Eli Glasner,** The topological dynamics of Ellis actions, 2008
- 912 **Igor Chueshov and Irena Lasiecka,** Long-time behavior of second order evolution equations with nonlinear damping, 2008
- 911 **John Locker,** Eigenvalues and completeness for regular and simply irregular two-point differential operators, 2008
- 910 **Joel Friedman,** A proof of Alon's second eigenvalue conjecture and related problems, 2008
- 909 **Cameron McA. Gordon and Ying-Qing Wu,** Toroidal Dehn fillings on hyperbolic 3-manifolds, 2008
- 908 **J.-L. Waldspurger,** L'endoscopie tordue n'est pas si tordue, 2008
- 907 **Yuanhua Wang and Fei Xu,** Spinor genera in characteristic 2, 2008
- 906 **Raphaël S. Ponge,** Heisenberg calculus and spectral theory of hypoelliptic operators on Heisenberg manifolds, 2008
- 905 **Dominic Verity,** Complicial sets characterising the simplicial nerves of strict ω-categories, 2008
- 904 **William M. Goldman and Eugene Z. Xia,** Rank one Higgs bundles and representations of fundamental groups of Riemann surfaces, 2008
- 903 **Gail Letzter,** Invariant differential operators for quantum symmetric spaces, 2008
- 902 **Bertrand Toën and Gabriele Vezzosi,** Homotopical algebraic geometry II: Geometric stacks and applications, 2008
- 901 **Ron Donagi and Tony Pantev (with an appendix by Dmitry Arinkin),** Torus fibrations, gerbes, and duality, 2008

For a complete list of titles in this series, visit the
AMS Bookstore at **www.ams.org/bookstore/**.